Aquaculture Marketing Handbook

D1196124

Aquaculture Marketing Handbook

Carole R. Engle
Kwamena Quagrainie

Carole R. Engle, Ph.D., is Professor, Director of the Aquaculture/Fisheries Center and Chairperson of the Department of Aquaculture and Fisheries at the University of Arkansas at Pine Bluff. She has more than 25 years of experience in the analysis of economics and marketing issues related to aquaculture. She has worked in 20 different countries in all major world regions, but much of her work has focused on the economics and marketing issues of U.S. and Latin American aquaculture businesses. She continues to teach an undergraduate and a graduate course in Aquaculture Economics and Marketing in addition to research and Extension responsibilities in the same area. She is the current President of the International Association of Aquaculture Economics and Management, past-President of the U.S. Aquaculture Society, a Chapter of the World Aquaculture Society, and was the recipient of the Joseph P. McCraren Award of the National Aquaculture Association and Researcher of the Year Award from the Catfish Farmers of America.

Kwamena Quagrainie, Ph.D., is Director of Aquaculture Marketing at Purdue University. He has more than 10 years of industry and research experience in economics and marketing issues related to both agriculture and aquaculture. He has worked in Canada and some African countries on economics and marketing issues of agriculture and aquaculture businesses. He teaches quantitative methods in Aquaculture Economics/Marketing and has research responsibilities in agribusiness/aquaculture marketing.

©2006 Blackwell Publishing
All rights reserved

Blackwell Publishing Professional
2121 State Avenue, Ames, Iowa 50014, USA

Orders: 1-800-862-6657
Office: 1-515-292-0140
Fax: 1-515-292-3348
Web site: www.blackwellprofessional.com

Blackwell Publishing Ltd
9600 Garsington Road, Oxford OX4 2DQ, UK
Tel.: +44 (0)1865 776868

Blackwell Publishing Asia
550 Swanston Street, Carlton, Victoria 3053, Australia
Tel.: +61 (0)3 8359 1011

Authorization to photocopy items for internal or personal use, or the internal or personal use of specific clients, is granted by Blackwell Publishing, provided that the base fee of $.10 per copy is paid directly to the Copyright Clearance Center, 222 Rosewood Drive, Danvers, MA 01923. For those organizations that have been granted a photocopy license by CCC, a separate system of payments has been arranged. The fee codes for users of the Transactional Reporting Service are ISBN-13: 978-0-8138-1604-3; ISBN-10: 0-8138-1604-1/2006 $.10.

First edition, 2006

Library of Congress Cataloging-in-Publication Data

Engle, Carole Ruth, 1952-
Aquaculture marketing handbook / Carole R. Engle, Kwamena Quagrainie.— 1st ed.
p. cm.
Includes bibliographical references and index.
ISBN-13: 978-0-8138-1604-3 (alk. paper)
ISBN-10: 0-8138-1604-1 (alk. paper)
1. Aquaculture industry. 2. Seafood industry. 3. Aquaculture--Marketing. 4. Seafood—Marketing. 5. Aquaculture industry—United States. 6. Seafood industry—United States. I. Quagrainie, Kwamena. II. Title.

HD9450.5.E54 2006
664′.94′0688—dc22

2005015283

The last digit is the print number: 9 8 7 6 5 4 3 2 1

This book is dedicated to our familes:
Nathan, Reina, Eric, and Cody, and Mildred and Glenn Wambold and
Gifty, Sabina, Joshua, Damaris, and Joseph

Contents

Foreword

The growth and development of aquaculture industries around the world has transformed and in some cases dominated local, national, and international seafood markets. Yet many aquaculture growers have little formal training in marketing and lack the tools to adequately evaluate and adapt to changing market conditions. Moreover, the stunning diversity of cultured aquatic species, combined with strong local preferences for seafood, present a confusing arena within which to design successful marketing strategies. The dynamic nature of seafood markets adds additional uncertainty to marketing decisions.

As aquaculture businesses have emerged from subsistence-level, cottage industries and hobbies into commercial businesses, interest in marketing has grown. Small-scale businesses continue to seek to capture niche markets, while larger industry segments have sought to create demand for their products outside traditional consumption areas. Vertically integrated aquaculture companies have targeted export markets to compete with products grown locally or to introduce their products into larger, more lucrative markets. The international trade in aquaculture products such as salmon, shrimp, and tilapia has grown to have a substantial effect on prices and consumption in other markets. Conflicts have developed as this trade has grown over time.

Markets, marketing, and trade have become ever more important to growing aquaculture industries worldwide. Yet the diversity and idiosyncrasies of seafood markets require information that is specific to aquaculture and seafood markets. The *Aquaculture Marketing Handbook* has been written as a step toward filling a void in understanding both basic marketing concepts as well as markets for specific aquaculture products. It is intended to serve as a guide, textbook, and reference for critical sources of marketing data, information, and research on seafood and aquaculture markets. Most of all, we hope that this handbook will spark new ideas and creative new marketing solutions for those segments of aquaculture searching for new markets and strategies.

Preface

Many aquaculture practitioners recognize the need to understand product marketing. Nevertheless, many aquaculture growers tend to equate marketing with sales and do not have an adequate level of understanding of fundamental marketing principles. Thus, they lack the tools to adequately evaluate and adapt to changing market conditions.

This book presents fundamental principles of marketing from a practical how-to perspective for those who are already in the business, those who might be in the aquaculture business someday, and students, scientists, and Extension personnel seeking to understand aquaculture markets and marketing. Aquaculture market synopses are used throughout the book to add the complexities of the real world into each chapter. This book can be used as a text for academic classes, as the basis for marketing workshops, as a self-study guide for aquaculture entrepreneurs, and as a reference book for commercial aquaculture businesspersons.

The book contains both an annotated bibliography and a Webliography. These describe the marketing information and sources of data available at the time of writing of this book. The bibliography and Webliography include descriptions of key references or a site's contents and use. Data sources on prices, demographics, trends, and other critical data are described and evaluated in the Webliography.

Chapter 1 presents an overview of seafood and aquaculture markets and marketing. This chapter establishes the uniqueness, scope, and diversity of seafood and aquaculture markets. It establishes the global nature of seafood markets and provides an overview of characteristics and trends. The synopsis on tilapia provides an encouraging example of a market that was developed in a relatively short period of time and that continues to grow at a rapid pace.

Chapter 2 defines the fundamental economic concepts of demand, supply, and price discovery mechanisms of the market. Determinants of demand and supply are reviewed. Inclusion of this chapter allows readers unfamiliar with economics to understand the use of these terms in later chapters. Readers who are already familiar with these concepts can forego Chapter 2. The synopsis on shrimp presents an overview of a complex but highly valued aquaculture product.

Chapter 3 reviews fundamental marketing concepts. This chapter lays the groundwork for defining both technical terms and conceptual understandings for the discussion that follows in subsequent chapters. The synopsis on salmon presents an example of an aquaculture product that has come to dominate the overall salmon market.

Chapter 4 provides an overview of aquaculture supply considerations. Aquaculture products present some unusual characteristics and challenges that have implications for successful marketing strategies. The synopsis on baitfish markets provides an excellent example of the need to constantly align farm production to meet the changing demand of different market segments.

Chapter 5 reviews the structure of the processing sector for aquaculture products. Processing innovation, branding, and associated challenges are discussed. The synopsis on catfish draws upon the history of the development of the catfish processing sector and supply coordination to present examples of these challenges.

Chapter 6 discusses the dynamics of channel organization, ownership, and control in aquaculture marketing. Contrasts are made with trends in agribusiness marketing. The trout synopsis highlights a mature industry that has moved into value-added products to develop new markets.

Chapter 7 covers marketing by farmer groups from cooperatives to bargaining groups to generic advertising. Aquaculture marketing initiatives are contrasted with those of other farm commodity groups. The growing markets for farmed seabass and seabream are discussed in the synopsis.

Chapter 8 discusses the wholesale market for aquaculture products. The respective roles of food service distributors and brokers are defined, described, and

contrasted. The synopsis on carp adds a different dimension to the discussion on wholesaling and distribution of aquaculture products.

Chapter 9 summarizes recent trends in marketing. The role of imports in seafood markets, increased wholesale-retail integration, food traceability, and technological innovations in distribution are changing the nature of seafood marketing. The giant clam market synopsis introduces the complex interactions among aquaculture, the environment, and markets. As a threatened species, giant clams are protected from trade. These same restrictions may limit development of markets for giant clam aquaculture.

Chapter 10 provides a brief overview of the theoretical underpinnings of international trade. Trade disputes related to aquaculture products are chronicled and discussed. The crawfish market synopsis traces the entry of Chinese crawfish into these markets, the impact on the U.S. crawfish industry, and the subsequent anti-dumping lawsuit and its effect in the United States.

Chapter 11 outlines policies and regulations that govern aquaculture marketing. The synopsis on mussels provides an example of product differentiation and branding in the French shellfish market.

Chapter 12 develops a framework and methods to develop market plans based on well-conceived market strategies. The synopsis on hybrid striped bass provides an example of an aquaculture product filling a market niche that became available with the demise of a popular fishery.

Chapter 13 reviews various research methodologies used in marketing research. The intent of this chapter is to make the reader generally familiar with the terms involved, to provide some idea of which methodologies are best suited to which market questions, and to offer some understanding of interpretations. The synopsis on ornamental fish markets further demonstrates the complexity and diversity of aquatic products, their markets, and the challenges faced.

We sincerely appreciate the following individuals who reviewed this book and provided helpful and thoughtful suggestions: Robert Pomeroy, Jo Sadler, Aloyce Kaliba, and Nathaniel Wiese.

Carole Engle and Kwamena Quagrainie
Pine Bluff, Arkansas

Aquaculture Marketing Handbook

1
Seafood and Aquaculture Markets

Successful industries must be successful in marketing their products. A market can be defined in a number of ways. It can be a location, as in the Fulton Fish Market in New York City or the Tsukiji Market in Tokyo, Japan; a product, as in the jumbo shrimp market; a time, as in the Lenten season market in the United States or the European Christmas market; or a level, such as the retail or wholesale market.

This handbook combines marketing concepts and theory with practical examples from aquaculture. This introductory chapter provides an overview of seafood and aquaculture markets worldwide, the global supply of major seafood and aquaculture species, the location of major markets, and international trade volumes and partners. The chapter continues with a discussion of characteristics of aquaculture products and the market competition between wild-caught and farmed fish. The chapter concludes by summarizing trends in consumption of seafood and aquaculture products.

GLOBAL TRENDS IN SEAFOOD AND AQUACULTURE MARKETS

The earliest recorded evidence of aquaculture dates back to 900 B.C. (Bardach et al. 1972). Although this date places aquaculture as an ancient technology, it is still quite young when compared to terrestrial agriculture. Diamond (1999) shows that domesticated species of both crops and animals were being cultivated by 8500 B.C. (Table 1.1). Southwest Asia and China served as the birthplace of much agriculture and aquaculture. Diamond theorized that the relative availability of game was one of the factors affecting the development of human history. Areas with abundant game provided little incentive for developing alternative food species, but areas with sparse game would provide greater returns to the effort in developing farming techniques. For most species of fish, scarcities due to overfishing have become evident only in the latter part of the 1900s. Thus, the incentive to explore and invest in the domesticated production of aquatic plants and animals has been of recent origin. The overall level of scientific and technological progress in the development of aquaculture technologies in the 1900s has allowed aquaculture industries to blossom.

Demand for fisheries products has grown as the world's population has continued to grow. However, capture fisheries worldwide appear to have leveled off over the last 15 years while the quantity of aquaculture production supplied to the world market has continued to increase (Fig. 1.1). World capture fisheries increased most rapidly during the late 1950s through the end of the 1960s. From that point, capture fisheries continued to increase, but at a slower rate until reaching slightly more than 90 million metric tons in 1989. Since then, world capture fisheries have fluctuated from 86 million to 97 million metric tons, averaging about 92 million metric tons. It is clear that most of the increase in the world supply of fish and seafood is due to the expansion of aquaculture production, not to capture fisheries.

Global aquaculture production has increased each year since 1984 at an average annual rate of 10% (FAO 2004). By comparison, animal livestock production has increased by only 3%/yr, and capture fisheries production by only 1.6%/year. Over the past five years alone, the annual growth rate of cultured finfish and shellfish production increased from an annual rate of 5–6% in 1990–1991 to 14% in 1994–1995. Data for capture fisheries over this same time period show either zero or negative growth.

All aquatic farming combined represented a 3% share of the world harvest of fish, shellfish, and seaweeds in 1950 (FAO 2004). By 2001, this share had increased to 34% and consisted of a record 48.4 million metric tons (MT) of total farmed aquatic production. Of this, the greatest increase was for finfish and shellfish production that constituted 37.9 million MT in addition to seaweed production of 11.9 million MT in 2001. The total value of aquaculture production worldwide increased to $61.5 billion in 2001.

Table 1.1. Dates of Domestication of Various Plant and Animal Crops Important in the Cultural Development of Humans.

Area	Domesticated		Earliest attested date of domestication
	Plants	Animals	
Independent origins of domestication			
Southwest Asia	wheat, pea, olive	sheep, goat	8500 B.C.
China	rice, millet	pig, silkworm	By 7500 B.C.
Mesoamerica	corn, beans, squash	turkey	By 3500 B.C.
Andes & Amazonia	potato, manioc	llama, guinea pig	By 3500 B.C.
Eastern U.S.	sunflower, goosefoot	none	2500 B.C.
Sahel	sorghum, African rice	Guinea fowl	By 5000 B.C.
Tropical West Africa	African yams, oil palm	none	By 3000 B.C.
Ethiopia	coffee, tea	none	unknown
New Guinea	sugar cane, banana	none	7000 B.C.
Local demonstration following arrival of founder crops from elsewhere			
Western Europe	poppy, oat	none	6000-3500 B.C.
Indus Valley	sesame, eggplant	humped cattle	7000 B.C.
Egypt	sycamore fig, chufa	donkey, cat	6000 B.C.

SOURCE: Diamond 1999.

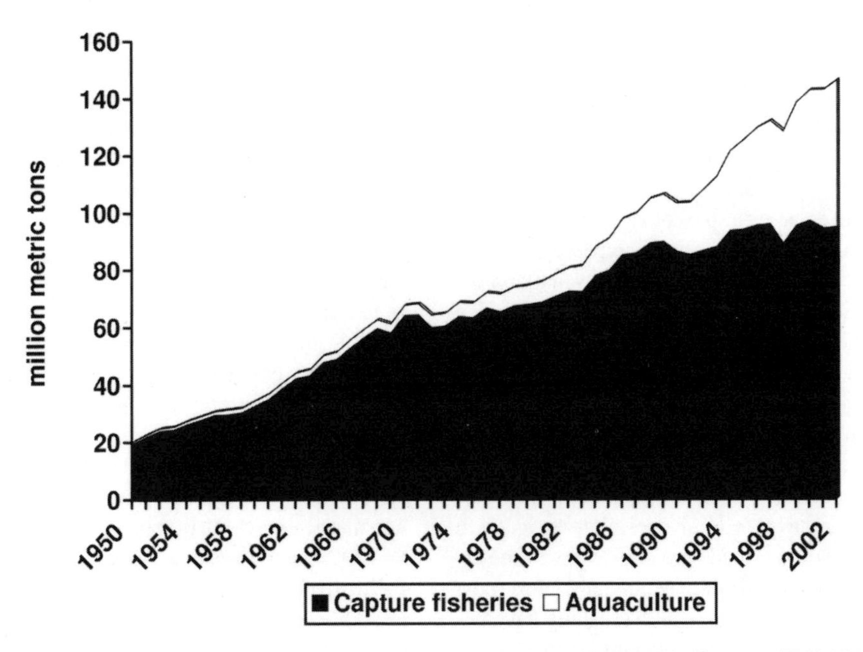

Figure 1.1. Volume of capture fisheries and aquaculture production, 1950–2002. (Source: FAO 2004.)

The relative costs of capture fisheries have increased over time while those of aquaculture production have decreased. The 200 nautical miles (370 km) Exclusive Economic Zone (EEZ) limits established by the Magnuson Fishery Conservation and Management Act, combined with declining abundance of many types of fish stocks, require trawlers to travel greater distances to find supplies of fish. Costs of capture fisheries are likely to continue to increase over time. At the same time, aquaculture costs have declined as new technologies have been developed and refined. For example, although feed currently represents 60% of the cost of raising shrimp (Csavas 1994; Valderrama and Engle 2001), improvements in feed efficiency over time and the development of cheaper feed ingredients will continue to decrease costs of shrimp production.

Where Are Most Aquaculture Crops Produced?

Asia is the birthplace of early aquaculture production technology and continues to be the world's leading aquaculture region. Production in Asia reached 46.7 million MT in 2002, accounting for 92% of the world's output (Fig. 1.2). Although aquaculture's contribution to world aquatic production averaged 35% in 2002, it reached 60% in some of the top aquaculture-producing countries. Next to Asia, Europe was the second leading aquaculture-producing region, but with only 4% of total world production. Europe was followed closely by Latin America and the Caribbean, North America, and the Near East. Africa, the Southwest Pacific, and other regions produced even less.

The nation that leads the world in aquaculture production is China (Fig. 1.3). Of the top 10 countries in aquaculture production, nine are located in Asia (China, India, Japan, the Philippines, Indonesia, Thailand, Bangladesh, Republic of Korea, and Vietnam). Chile is the only non-Asian country included in the top 10 (ranking ninth in terms of quantity produced). In terms of value, Norway and Chile replace the Republic of Korea and the Philippines in the top 10, but the other countries remain in the same order of rank.

Much of the aquaculture production in the world occurs in lesser-developed nations (FAO 2004). Of the top 20 aquaculture producing nations, only two, Japan and the United States, are considered developed nations by the FAO. Moreover, much of the increase in aquaculture production has been from low-income, food-deficit countries, such as China.

As mentioned earlier, global aquaculture production has grown at an annual rate of approximately 10% (FAO 2004). This average growth rate has remained fairly constant over time even though there is a great deal of variation from year to year. Aquaculture production in China has grown at an annual rate of about 14%, substantially higher than the world average rate of growth. Given that aquaculture production levels are very high in China, the continuing high growth rate results in large annual increases in aquaculture production each year. By comparison, the Southwest Pacific region of the world has had a slight decline in aquaculture production since 1999, and production levels in Europe appear to have leveled off since 1999 as well. However, the other regions of the world have continued to demonstrate growth in aquaculture production at rates ranging from 5% to 25%/yr.

What Are the Major Species Cultured Worldwide?

Worldwide, the greatest volume produced of an aquaculture product in 2001 was that of Japanese kelp (*Undaria* sp.) (Fig. 1.4). Kelp production is fol-

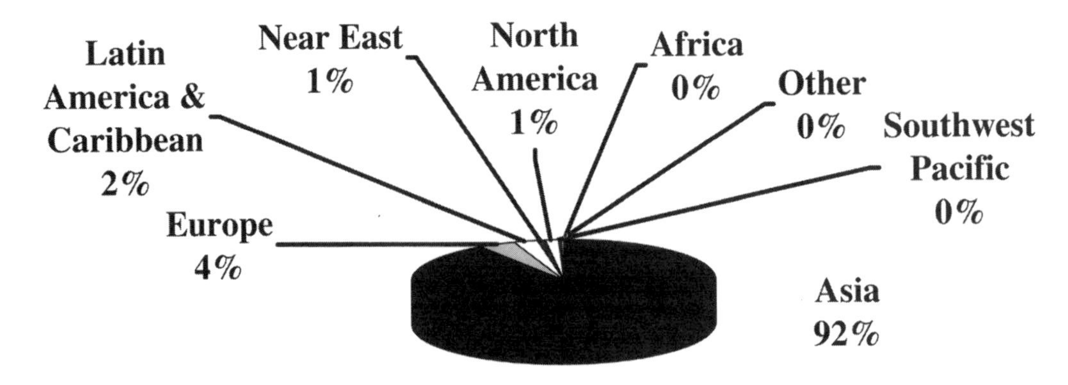

Figure 1.2. World Aquaculture Production by Region, 2002. (Source: FAO 2004.)

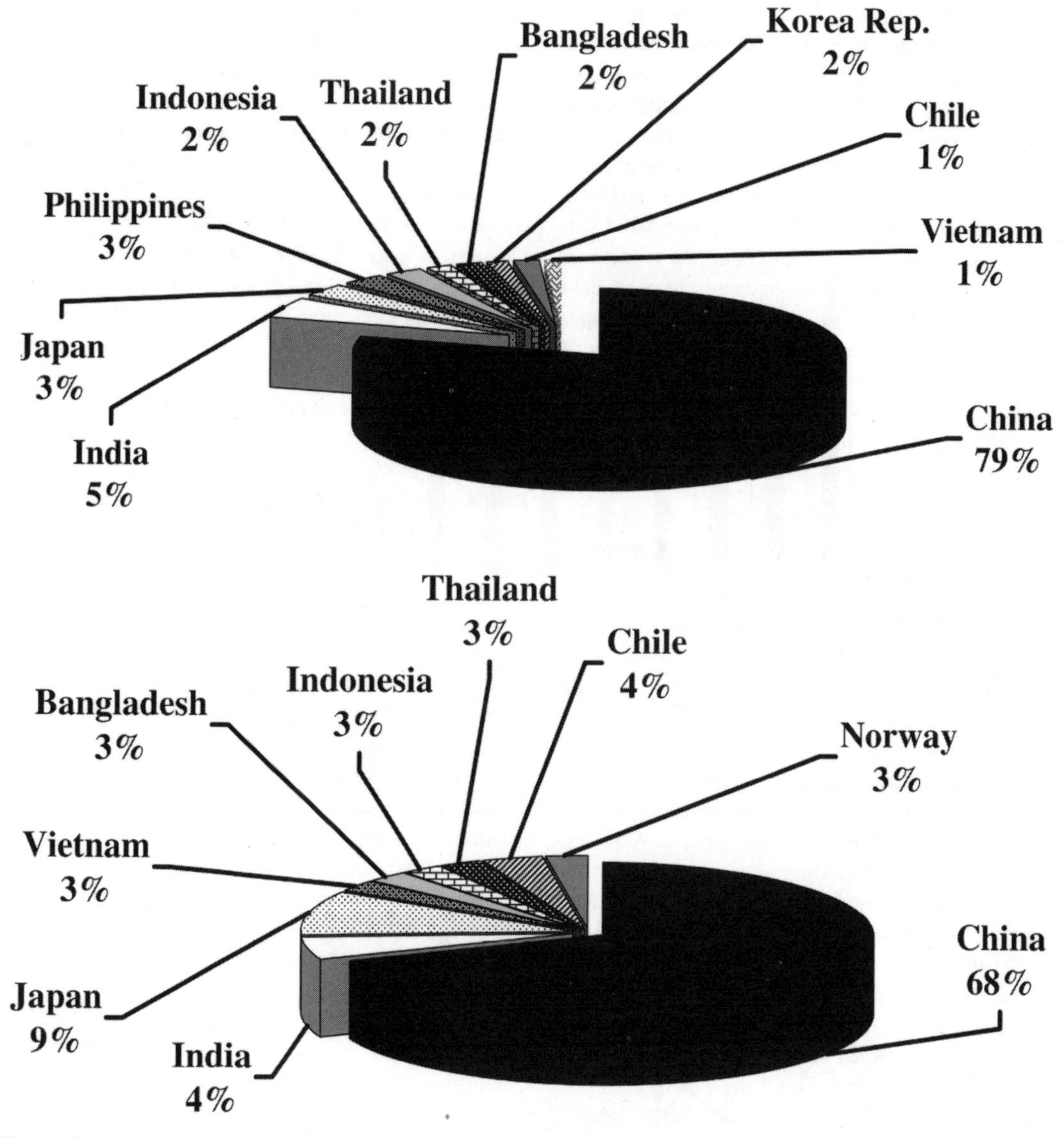

Figure 1.3. a. Quantity of world aquaculture production by country, 2002. **b.** Value of world aquaculture production by country, 2002. (Source: FAO 2004.)

lowed, in descending order, by Pacific cupped oyster (*Crassostrea gigas*), silver carp (*Hypophthalmichthys molitrix*), grass carp (*Ctenopharyngodon idellus*), common carp (*Cyprinus carpio*), Japanese carpet shell (*Ruditapes phillipinarium*), bighead carp (*Hypophthalmichthys nobilis*), Crucian carp (*Carassius carassius*), nori (*Porphyra*), Nile tilapia (*Oreochromis niloticus*), Yesso scallop (*Patinopecten yessoensis*), Atlantic salmon (*Salmo salar*), and the giant tiger prawn (*Penaeus monodon*). The various carp species combined represent the major volume of finfish harvested, by several orders of magnitude. The top five finfish species harvested, by volume, are all different species of carp, and carp are the only finfish included in the list of the top ten aquaculture products (by volume). Nile tilapia places tenth by volume.

The aquaculture species that generated the greatest value was the Pacific cupped oyster in 2002. The second highest value of aquaculture production was that of the silver carp, followed in descending order by the giant tiger prawn, common carp, Atlantic

salmon, grass carp, kelp, Japanese carpet shell, Yesso scallop, and bighead carp. Thus, the second, fourth, sixth, and tenth top aquaculture species and two of the top five finfish species by value were all carp species.

Figure 1.5 compares the value of the most important aquaculture species worldwide over time. The values used in Figure 1.5 were obtained by dividing the total value by the total quantity produced from FAO data (FAO 2004), and they provide general

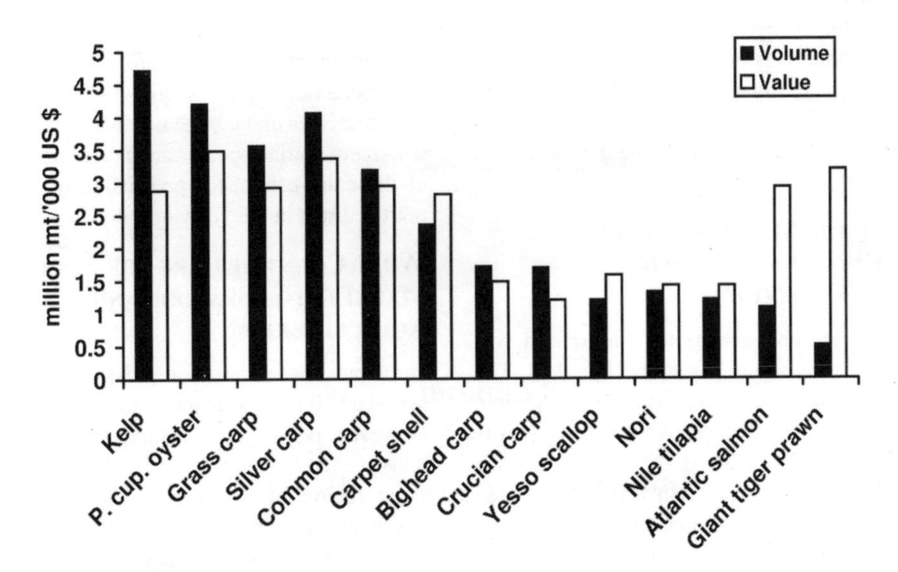

Figure 1.4. World aquaculture production by species, 2002. (Source: FAO 2004.)

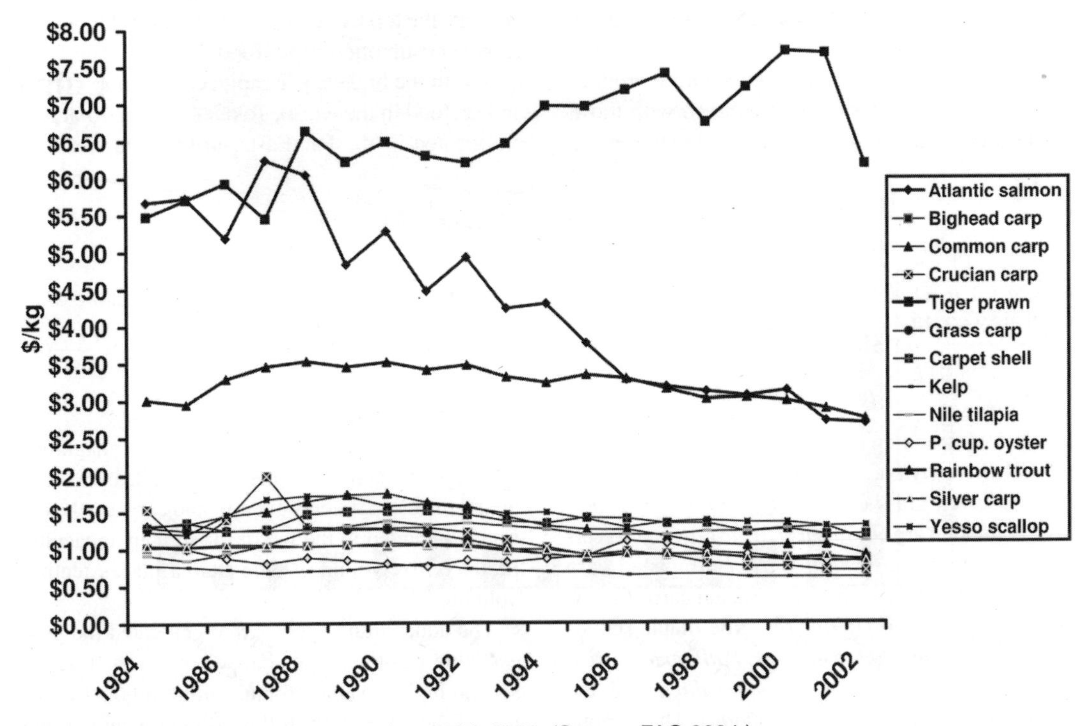

Figure 1.5. Value of aquaculture species, 1984–2002. (Source: FAO 2004.)

trends. The values of the majority of species have remained relatively stable over time, with slight decreases over time. There are two clear exceptions: tiger prawns and Atlantic salmon. The value of Atlantic salmon has declined dramatically over time since the mid-1980s. This is likely due to the rapid increases in aquaculture production of salmon worldwide. The value of tiger prawns has generally increased over time. It is the only species, of the top aquaculture species, that has exhibited increasing value over time. It is unclear whether the decline in 2002 represents annual variation or a longer-term trend.

What Are the Major Finfish Species Caught and Supplied to World Markets?

The Peruvian anchovy constitutes the greatest volume of worldwide capture fisheries (Fig. 1.6). The primary use of anchovies is for fishmeal production, not as a food product. The second greatest catch is that of pollock. Pollock is used commonly in fish sandwiches, fish sticks, and other popular frozen and breaded preparations. It is also used for production of surimi in many countries. Following pollock are several other types of mackerel, herring, and the Japanese anchovy. Both skipjack and yellowtail tuna, whiting, capelin, sardines, cod, and squid are included in the list of the top 15 capture species worldwide by volume.

If the volumes of worldwide aquaculture production (Figures 1.4 and 1.6) are compared with those of worldwide capture fisheries, it is clear that more grass or silver carp are produced worldwide than any single marine species used for direct food consumption by humans[1]. There was also more common carp produced from aquaculture (3.2 million MT) than of the next largest volume of wild-caught foodfish, the Chilean jack mackerel (1.75 million MT).

Nevertheless, total worldwide production from capture fisheries is still much larger than the total production from aquaculture. Culture techniques have been developed for only a small percentage of finfish species and a large number of different freshwater and marine species are caught and sold. Many of these are caught for production of fishmeal and not for direct human consumption.

What Countries Are the Major Markets for Seafood and Aquaculture?

Per capita consumption of seafood by world region[2] averaged 10 to 48 kg/capita (Table 1.2) (FAO 2004). However, per capita consumption varied tremendously, even from 0 to over 100 kg/capita within the same region of the world. The North American region averaged the highest average per capita consumption rate, but this is due to inclusion of Greenland, St. Pierce and Miquelong in the FAO classification system. Oceania ranked second, followed by the Near East and then the Caribbean. Table 1.3 presents the top five countries in terms of highest per capita consumption of seafood for 2001. The countries with the highest per capita consumption of fish and seafood in the world, Tokelau and the Maldives, are located in the Far East and Oceania world re-

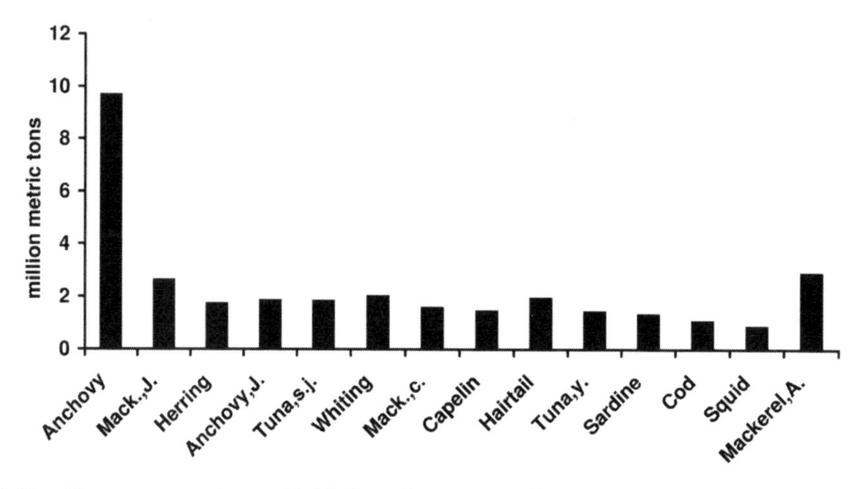

Figure 1.6. Top 15 capture species worldwide by volume, 2002. (Source: FAO 2004.)

Table 1.2. Average Per Capita Consumption of Fish and Shellfish by World Region, 1997–1999.

Region	Mean + SD	Maximum	Minimum
	kg/capita		
Africa	12 ± 13	63	0.2
Caribbean	21 ± 12	44	0.7
Europe	17 ± 21	90	0.1
Far East	30 ± 40	203	0
Latin America	13 ± 15	60	1.4
Near East	10 ± 8	26	0.1
North America	48 ± 31	84	21.2
Oceania	42 ± 41	200	2.9

Table 1.3. Top Five Countries Worldwide with Highest Per Capita Consumption and Highest Total Consumption of Fish And Seafood, 2001.

Country	Per capita consumption (kg)	Total population (million people)	Total consumption of fishery products (metric tons)
Highest per capita consumption			
Maldives	203.3	274	55,692
Iceland	90.2	275	24,763
Faroe Islands	86.5	45	3,891
Greenland	84.3	56	4,719
Kiribati	75.1	81	6,084
Countries with highest consumption of fish and seafood			
China	24.4	1,231,935	30,061,530
Japan	65.2	126,502	8,251,424
USA	21.2	277,534	5,872,138
India	4.7	976,346	4,542,780
Indonesia	19.0	206,412	3,916,836

SOURCE: Holliday 2002.

gions. However, these same regions included countries such as Guam and Mongolia, with extremely low per capita consumption rates of 2.9 and 0 kg/capita, respectively. In terms of the percent of countries within a region that consumed more than 25 kg/capita, there were 40% of the countries in the Far East region, 65% in Oceania, and 22% in Europe.

Table 1.3 presents the top five countries in terms of total volume of consumption of fish and seafood. The total amount is clearly related to the combination of per capita consumption and total population. Topping the list is China, which has both a high per capita consumption rate and the highest population in the world, resulting in consumption of more than 30 million MT. Japan follows, with total consumption of more than 8 million MT; the United States comes in third with 5.8 million MT. Although per capita consumption in India is among the lowest in the world, it still ranks fourth in total consumption due to its large population. Indonesia completes the top five countries in total consumption.

Trade in Seafood and Aquaculture

Approximately 38% (live weight equivalent) of world fish production was traded internationally in 2001 (Vannuccini 2003). However, this percentage increased by only 3% (from 34% to 37%) from 2001 to 2002. The percentage increases obscure the total increase in volume and value of the international trade in fish and seafood. The continued increase in aquaculture production results in continued increases in the total supply of fishery products that results in lower percentage increases.

ARE AQUACULTURE PRODUCTS DIFFERENT FROM AGRICULTURE PRODUCTS?

CHARACTERISTICS OF AQUACULTURE PRODUCTS

Aquaculture is a unique form of food production. Most cultured species of fish are not substantially different from wild-caught species. Common carp, with 2000 years of culture, has been bred selectively into strains of fish recognizably different from wild-caught fish. This is not the case for most other cultured aquatic species. Genetic advances may change this situation rapidly, but unlike animal and row crop agriculture, aquaculture growers find themselves competing in the marketplace with wild-caught seafood products. In many cases, wild-caught product still dominates the market and has a major effect on price. Some segments of the aquaculture industry have been more successful than others in differentiating their product from wild-caught supplies.

Aquaculture products offer distinct advantages in terms of control over the product. Many aquaculture products can be supplied year round. In contrast, most wild-caught seafood is characterized by seasonal fluctuations related to weather and fishing regulations that can result in dramatic price swings. The domination of seafood markets by wild-caught species has resulted in a tendency toward high volatility. Although aquaculture products offer the advantage of controlled year-round supply, these products must compete within the volatile seafood market.

Controlled production techniques also allow the aquaculture grower to produce a consistent product. Consistency in supply refers to size, quality, and other product characteristics in addition to consistency in the quantity supplied. Consistently supplied aquaculture products would be expected to lend some stability to the seafood market as the market share of aquaculture products continues to grow over time. Enhanced reliability and regularity in supply of farmed product should enable producers to negotiate better prices (Asche 2001). Theoretically, buyers would be willing to pay higher prices to compensate for reduced financial risk that results from supply problems. Market sectors, such as the retail sector, that prefer fresh product might be expected to prefer farmed supplies (Young et al. 1993). Fresh product requires a short re-order period. Supply chains of captured fisheries products are more fixed due to seasonality of supply and cannot respond readily to changes in retail demand.

Consumers in many countries and for many years have exhibited strong preferences for fresh seafood. By contrast, one rarely hears an emphasis on the freshness of beef, pork, or chicken. This strong consumer preference for fresh seafood likely derives from the perishability of seafood as compared to other products. Technological advances enable processors to produce quality frozen and preserved seafood products. However, the preferences for fresh seafood have driven some retail grocers to purchase frozen product, thaw it, and sell it as fresh.

It is easier to trace farmed product than wild-caught product back to its original source. The complexity of market channels for wild-caught product may obscure steps in the supply chain and make tracing products to their source difficult (Asche 2001). Some wild-caught seafood is marked, logged, and stored separately, but this is the exception. The greater traceability of aquaculture products should become increasingly advantageous as the new country-of-origin labeling laws take effect in the United States. The new labeling laws will require certification of product origin. Aquaculture suppliers should find compliance less onerous than will suppliers of wild-caught seafood.

The potential to control attributes and their levels in a product can offer an opportunity for farmers to target specific consumer segments (Asche 2001). For example, producing the exact fat content to produce a particular smoked flavor or production of fish of a given size may provide aquaculture growers a significant marketing advantage over capture fisheries. In most cases, additional research will be required to develop cost-effective means of producing these attributes.

Fish and other aquaculture production allows for reliable delivery schedules to comply with contractual agreements to supply fish of a given size and quality grade. The uncertainty of what species, size, and, to some extent, quality of fish will be caught is an important characteristic that can be used to differentiate farm-raised from wild-caught seafood.

The management required for successful aquaculture businesses can be used to reassure consumers of the safety of the product. Consumers increasingly desire assurances that products are free of chemicals, pesticides, and other undesirable additives. This concern can include assurance that the product has not been modified genetically.

Food scares and concerns over the use of pesticides in terrestrial agriculture have raised consumer concerns about pesticide use in other food products, particularly new ones (Smith et al. 1999). These concerns have been extended to seafood. The particular concerns for seafood are related to concentrations of dioxin and mercury in seafood products, and the status of menhaden and other pelagics used for fishmeal in fish diets (Millar 2001).

There has been growing resistance to aquaculture products by some activist groups, who consider aquaculture to be unnatural and detrimental to the environment. In some areas of the United States, for example, farmed salmon is considered to be less desirable than wild-caught salmon. On the other hand, some consumers may be convinced to pay a premium price for environmentally sustainable products. Farm-raised catfish is preferred to wild-caught catfish in southern states for a variety of reasons, but primarily for the consistency of flavor, quality, and the certainty that it is free of contaminants. Catfish and tilapia are frequently listed as environmentally acceptable seafood choices by the Monterrey Bay Aquarium and other groups (such as Audubon Living Oceans).

A major disadvantage of aquaculture products as compared to wild-caught seafood is the price. Costs of production have frequently been higher for aquaculture products than for wild-caught seafood. However, as wild fish stocks have declined and boats have had to travel farther on fewer fishing days, costs of capture fisheries have increased. At the same time, research and development have reduced costs of producing a number of aquaculture species. Thus, the costs of a number of farmed species are now more competitive with costs of wild-caught species than before. However, the consistent production and supply of aquaculture products results in consistent costs and prices. Buyers who are accustomed to waiting for periods of abundant supply and low price of wild-caught seafood may be reluctant to pay a consistently higher price for aquaculture products.

Market opportunities have developed for aquaculture species when declining stocks of the wild-caught species resulted in higher prices. This has been the case for hybrid striped bass in the United States, cultured turbot, halibut, and other species even though turbot and halibut are considered inferior to wild-caught product (Asche 2001).

Market Competition Between Wild-Caught and Farmed Finfish

Prices for several aquacultured species such as Atlantic salmon, rainbow trout, sea bass, and sea bream have fallen as production has increased. These finfish species have grown in importance in seafood markets in the European Union and in the United States (Asche 2001). Atlantic salmon, rainbow trout, sea bass, and sea bream were high-value species before aquaculture production became significant. The increased supplies from aquaculture have been accompanied by lower prices.

A farmed product that competes in a large market will face limited price effects from increased aquaculture production. As long as supplies of the farmed species are low in comparison with wild-caught species, the impact of the farmed quantity supplied on price will be small.

When the supply of the farmed species is high, farm-level production is likely to determine market price because of the greater control that farmers have over the production process (Asche 2001). Salmon (Asche et al. 1999) and catfish (Quagrainie and Engle 2002) are examples of seafood markets that are dominated by farmed production. With few or no substitutes, it may be more difficult for the industry to grow because farmers will then have to create and promote the market for their product.

U.S. catfish was a low-value species prior to development of the catfish farming industry. Although price in recent years has been low, there is no clear long-term trend. The U.S. catfish industry has successfully moved its product into new markets, thus sustaining price even with consistent growth in volumes produced and sold. New market development was predicated upon changing consumer attitudes toward what had been regarded as an inferior, scavenging fish.

Most seafood demand studies show that the seafood market is highly segmented. Farmed species seem to compete mainly with similar, wild species, but not with other species (Asche 2001). However, trout was found to be a low-fat, healthful substitute for meat in Europe (Gabriel 1990). Aquaculture growers are capturing market share even though demand studies have not determined clearly what market is being captured. Aquaculture products may create new market segments and may win parts of market shares from a variety of goods such that the effects on individual goods are not measurable (Asche 2001).

Consumption Trends in Seafood and Aquaculture Markets, Expenditures, Effects of Income, and At-Home Versus Away-From-Home Purchases

Until relatively recently, the majority of seafood consumed was wild-caught. Until the development of advanced transportation and refrigeration and freezing technologies, the only seafood available was what could be caught locally. There remains a strong tendency for consumers to prefer species that live in nearby waters (See Inset 1.1 for an example of regional preferences in the United States). European research showed that fish were associated with the natural environment in which they were found, that is, the sea, rivers, lagoons, and ponds; this leads to regional preferences for fish in Europe as in the United States (Gabriel 1990). Kinnucan et al. (1993) supported this by showing that preferences for fish products were influenced to a large degree by source availability.

Gabriel (1990) showed that the image of fish in Europe was changing. Traditionally, fish was consumed once a week at home as an inexpensive protein source that was prepared in simple forms. However, fish is increasingly viewed as a more expensive item to be prepared in more refined culinary dishes.

Preparation methods also vary by region and the associated culinary traditions. Northern Europeans, for example, prefer fish fried, in breadcrumbs, soused, smoked, and cooked in foil (Gabriel 1990). In Central Europe, French cuisine dominates, and fish are steamed, poached, fried, smoked, simmered, and wrapped in foil. In southern Europe, fish is most often fried, grilled, simmered, or eaten dried.

Consumer tastes and preferences change over time. In the United States, for example, beef consumption has declined whereas poultry consumption has increased. This is presumably due to increasing health concerns and choices of lower-fat protein

---

People Prefer to Eat What They Have Traditionally Been Able to Catch Close to Their Homes

Many people are conservative and traditional about the fish and seafood that they eat. Consumer preferences typically are based on what they, their family, and their friends have been able to catch or gather from their hometown areas. For example, Engle et al. (1990) asked consumers nationwide what their most preferred type of finfish was. From the following table, the regional nature of seafood preferences can be seen clearly. For example, the preferred finfish on the Pacific Coast of the United States was salmon. Consumers in the Mountain region preferred trout that is caught widely in the mountain streams in the region. Catfish was the most preferred in the West South Central and East South Central regions where catfish are abundant in the Mississippi River and its tributaries in the south. Catfish was also most preferred by consumers in the West North Central region through which the Mississippi River flows, but which also has a large number of inhabitants who have moved there from the south. The East North Central region has a tradition of Friday night fish fries that are based on the catch of the locally available yellow perch. The Middle and South Atlantic regions have provided consumers with an abundant flounder fishery, and the 1989 survey showed preferences by Middle and South Atlantic consumers for flounder. Haddock was most preferred by consumers in the New England region.

Top-Selling Fish/Seafood by Region

	Consumers		
Region	Most preferred species	Second most preferred	Third most preferred
Pacific	Salmon	Halibut	Catfish
Mountain	Trout	Halibut	Salmon
West North Central	Catfish	Trout	Cod
West South Central	Catfish	Flounder	Trout
East North Central	Perch	Catfish	Whitefish
East South Central	Catfish	Flounder	Bass
Middle Atlantic	Flounder	Haddock	Salmon
South Atlantic	Flounder	Trout	Red snapper
Northeast	Haddock	Cod	Swordfish

sources. However, dramatic declines in the cost of producing chicken in the United States no doubt have contributed to increased consumption of chicken. Pork and seafood consumption patterns, on the other hand, have changed little. Quality and flavor perceptions often have the greatest impact on preferences (Kinnucan et al. 1993). Other variables such as price, household size, coupon value, household income, geographic region, urbanization, race, and seasonality have been shown to explain the variation in household expenditures on fresh and frozen seafood commodities (Cheng and Capps 1988). Moreover, household size was found to be more important than household income. Kinnucan et al. (1993) also found that preferences for fish and seafood in the United States were generally not related to income, with the exception that lobster was preferred by high-income consumers and catfish by low-income consumers. The most promising target for efforts to increase overall seafood purchases in restaurants was higher-income, white, well-educated consumers in families with no young children present. These individuals were likely to be responsive to the nutritional benefits and other favorable attributes of fish.

The most promising customers for at-home sales were shown to be older, well-educated (four or more years of college), higher-income (more than $30,000), nonwhite urban-suburban residents in families without young children (age 10 or under present) (Rauniyar et al. 1997). New England households were significantly more likely to be frequent purchasers for at-home use as compared to households in the West North Central and West South Central regions.

Frequent purchasers at restaurants were more likely to have annual incomes above $20,000, and especially above $40,000 (Hanson et al. 1994). The role of income, race, seasonality, few small children and adherence to the Catholic faith were found to be important to restaurant consumption. The recognition in all consumer profiles of fish as a nutritious and healthful product represented an advantage for future marketing strategies in aquaculture.

AQUACULTURE MARKET SYNOPSIS: TILAPIA

Tilapia (*Oreochromis* sp.; *Tilapia* sp.) are some of the most widely cultured species of fish worldwide. Tilapia have been introduced from their native ranges in Africa and have been spread widely across the world (FAO 1997). The early introductions of tilapia (1950s–1970s) were part of development projects targeted toward increasing the availability of animal protein in subsistence farming areas. Surplus tilapia were sold as a means of generating cash income.

Large commercial tilapia ventures began to emerge in the 1990s. These businesses developed techniques that led to the production of export-quality fresh and frozen tilapia fillets. The availability of supply of high-quality fillets and marketing expertise has resulted in the successful introduction of fresh and frozen tilapia fillets into the U.S. and European markets.

World tilapia production has climbed steadily over the last half a century, with a marked increase in the rate of growth in the 1990s (Fig. 1.7). Total worldwide production of tilapia and cichlids reached

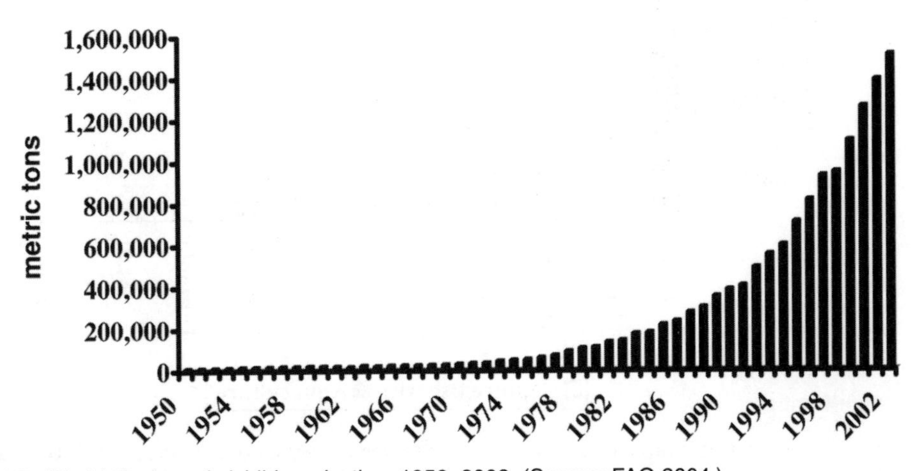

Figure 1.7. World tilapia and cichlid production, 1950–2002. (Source: FAO 2004.)

1.4 million MT in 2001. Average annual growth in tilapia production averaged 12.5% from 1991–2001.

Much of the growth in tilapia aquaculture is a result of the development of improved production practices and both domestic and export market development (Engle, in press). The development of export markets has resulted in a change in the major tilapia production centers and a shift from a dominance of tilapia from capture fisheries to tilapia produced on farms. In 1971, for example, the five leading tilapia-producing countries (Tanzania, Uganda, Mali, Madagascar, and Senegal) were all African countries with endemic tilapia populations (Fig. 1.8a). All this supply came from capture fisheries. Only Indonesia and Nigeria registered measurable amounts of tilapia production from aquaculture, and these were negligible. By the year 2001, only one of the five leading tilapia-producing countries (China, Egypt, Thailand, Philippines, and Indonesia) was an African country (Fig. 1.8b). Of these countries, only Egypt and Indonesia have endemic populations of tilapia; tilapia were introduced into the other countries. Moreover, the supply of tilapia had shifted heavily to production from aquaculture.

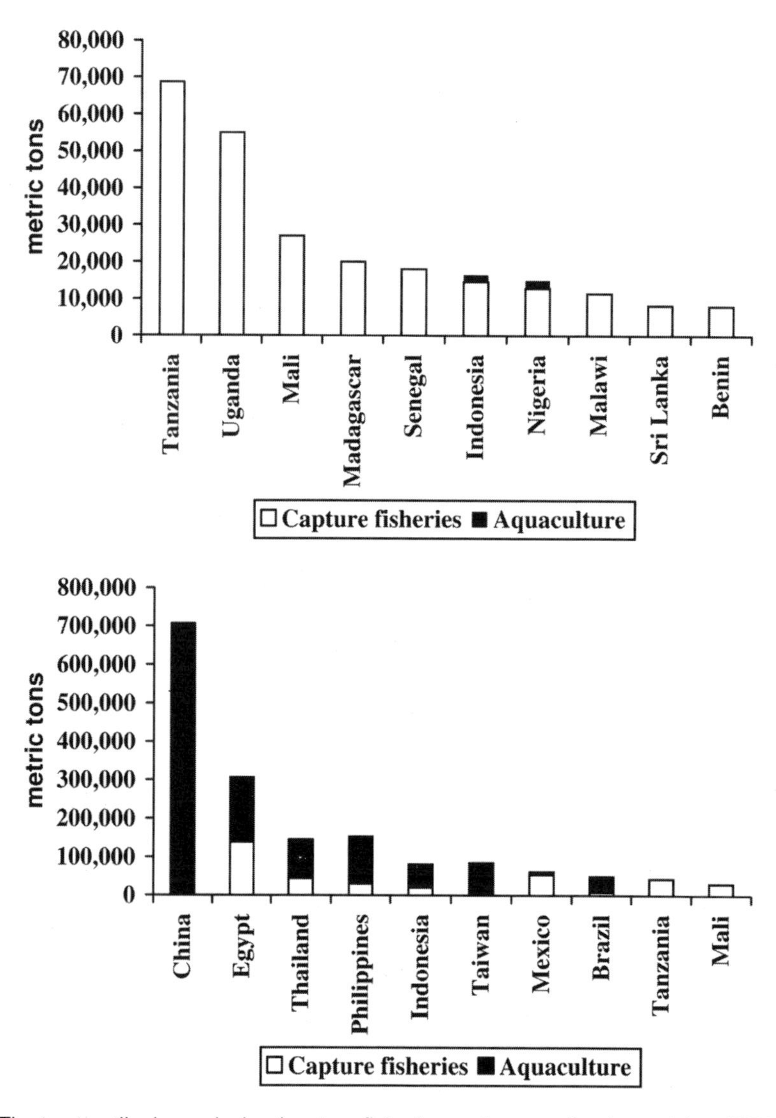

Figure 1.8. a. The top ten tilapia producing (capture fisheries and aquaculture) countries, 1971. **b.** The top ten tilapia producing (capture fisheries and aquaculture) countries, 2002. (Source: FAO 2004.)

China emerged as the dominant world producer of tilapia in the late 1990s (Fig. 1.9). Over the 10-yr period from 1992 to 2001, tilapia production increased by 327%, with an average annual increase of 19%/yr. Some of this production is exported, whereas other portions of the production are consumed in the domestic market.

Tilapia continue to be raised for subsistence purposes. In subsistence farming areas, tilapia are consumed whole, gutted, scaled, and either fried or roasted. Tilapia is now accepted in many national dishes around the world and is popular in many forms, including smoked, as sashimi, and even as fried tilapia skins. Whole-dressed tilapia are common in many open-air markets around the world. Export markets, however, require primarily filleted products although there is also international trade in frozen whole tilapia. Frozen whole tilapia imported into the United States is targeted toward Asian grocery stores throughout the United States. Taiwan has dominated the supply of frozen whole tilapia to the United States for many years, but China increased the export volume of frozen whole tilapia to the United States in the early 2000s.

The United States is the major export market for tilapia. Imports of tilapia into the United States have grown rapidly, particularly since 2000 (Fig. 1.10). The majority of this growth has been in the form of

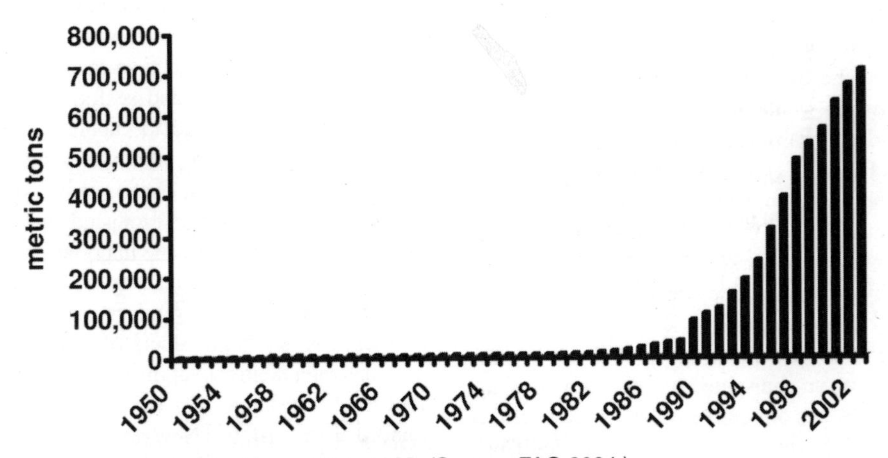

Figure 1.9. Tilapia production in China, 1950–2002. (Source: FAO 2004.)

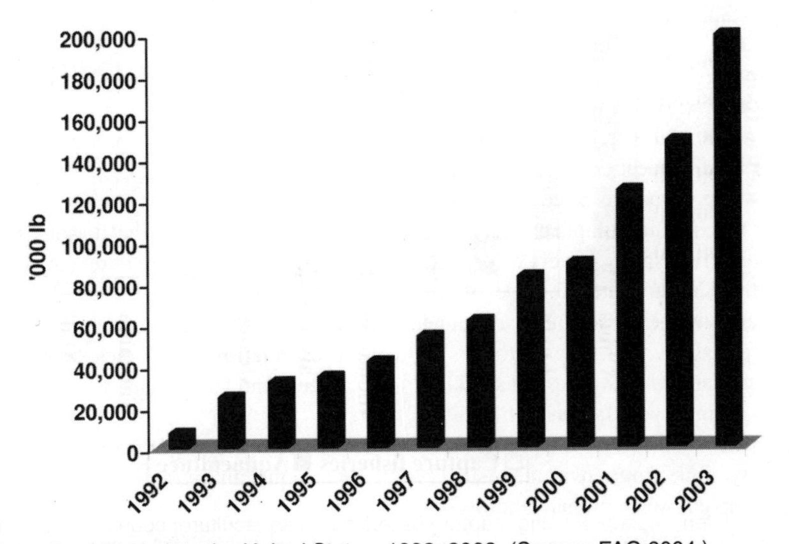

Figure 1.10. Imports of tilapia into the United States, 1992–2003. (Source: FAO 2004.)

imported fresh and frozen fillets. Tilapia are also imported as frozen whole fish, but these volumes have not increased as rapidly as the imported volumes of fresh and frozen tilapia fillets.

The major suppliers of fresh tilapia fillets to the United States in 2003 were Costa Rica, Ecuador, and Honduras. Tilapia from Costa Rica and Honduras originate primarily from farms designed to specialize in tilapia production, whereas in Ecuador shrimp farmers have begun to diversify into tilapia production. The pond and processing infrastructure in Ecuador allowed shrimp farmers to move quickly into tilapia production as shrimp disease problems escalated.

Indonesia has been the major supplier of frozen tilapia fillets into the United States for many years. In more recent years, though, Taiwan has begun to increase exports of frozen fillets in addition to the export of lower-priced, frozen whole tilapia. Taiwan continues to be the major supplier to the United States of frozen whole tilapia. The U.S. tilapia production industry has targeted sales of live tilapia to Asian and Hispanic grocery stores. Large cities such as New York, Toronto, Chicago, and San Francisco have historically been the major targets for the U.S. industry, but other markets have been developed successfully in smaller cities throughout the United States.

Although the growth of the global market for tilapia has been an undisputed success story in aquaculture, challenges are emerging that may begin to threaten the high rate of growth of tilapia sales. Controversy emerged in the late 1990s over the use of carbon monoxide by some tilapia processing plants (Seafood Business 2001–2003). Carbon monoxide treatment results in a deep-red color to the fillets, which is considered desirable. Second, tilapia fillets have a lower dress-out ratio (fillet weight: live weight of fish) than do fillets of other fish species. This results in a higher relative meat cost at the processing plant for the same farm-gate price of fish that dress out at higher ratios. Third, tilapia growers have recently come under criticism by buyers of organic supermarkets in the United States for use of the hormone methyltestosterone to sex reverse young tilapia. Sex reversal has allowed tilapia growers to achieve higher yields and growth rates by stocking the faster-growing all-male populations of tilapia.

A more significant challenge to tilapia production worldwide may come from environmentalists' groups. Some commercial-scale tilapia ventures depend upon high flows of surface water for the discharge of waste products. Increased awareness of environmental effects of effluent discharges may result in additional regulations. Also, concern globally over the introduction of exotic species is growing rapidly. Tilapia have become established in natural waters in many countries with tropical climates and are increasingly being labeled as an invasive species.

The tilapia industry can likely adapt to these challenges as they have over time. These are the types of challenges that arise as an industry matures and attracts increasing attention. The success in market development that has led to the growth of the tilapia industry will require it to continue to adapt to the new challenges that result.

SUMMARY

Much of the increased total fishery production is from aquaculture. Aquaculture costs of production have declined as the cost of capture fisheries has increased. The result has been an increase in the relevant proportion of fish and seafood supplies from aquaculture as compared to capture fisheries. The majority of aquaculture products in the world are produced in Asia. Kelp, oysters, and carps are the major aquaculture species produced and sold. Japan and the United States are the major seafood markets worldwide, whereas the leading seafood exporter is Thailand. Aquaculture products, as compared to wild-caught fisheries products, offer advantages such as: (1) greater control over the product and its consistency; (2) freshness; (3) traceability; and (4) enhanced food safety. Nevertheless, some activist groups consider farmed product undesirable and unsustainable; others prefer farm-raised product for its positive attributes.

STUDY AND DISCUSSION QUESTIONS

1. How do aquaculture products differ from wild-caught products?

2. What are some of the reasons that aquaculture has grown so rapidly in recent years?

3. Describe the major aquaculture-producing countries in terms of volumes, types of products produced, and target markets.

4. Describe the major world markets for seafood and aquaculture.

5. Discuss the controversies related to aquaculture and the various points of view.

Aquaculture Technology as an Alternative to Wild-Caught Species to Supply Demand for a Seafood Product

The supply of shrimp has continued to expand rapidly over time primarily due to the unprecedented success of the shrimp culture industry worldwide. Farmed shrimp production grew from a minuscule amount in the 1970s to exceed 27% of the total $8.4 billion shrimp supply by 2001.

Several technological breakthroughs in the 1980s laid the foundation for the exponential growth of shrimp farming. The development of hatchery seed and feed manufacturing technologies provided mechanisms for consistent and reliable production at a reduced cost. Wild seedstocks were not sufficient to sustain the demand for shrimp. The "PL wars" in Ecuador in the 1980s provide a graphic illustration of the competition over diminishing natural resources. For many years, hatchery-reared seed could not compete with collected wild seed of marine shrimp in terms of either price or performance. However, today hatchery technologies reliably produce high-quality and superior stocks such as the SPF (Specific Pathogen Free) stocks. The bulk of shrimp seed is now produced in low-tech, medium-sized or small hatcheries. The initial feeds used in shrimp production were by-products of poultry feeds. The use of extrusion in shrimp feed technology improved the water stability of pellets. Because shrimp nibble slowly at feed, improved water stability of the pellets allows shrimp to consume a greater percentage of the pellets. As shrimp production from aquaculture increased and supply improved, prices dropped and the quantity demanded increased.

The U.S. Seafood and Aquaculture Market

The United States imports more than 60% of its fish and shellfish and, after Japan, is the world's second largest importer of seafood. Fish and seafood imports into the United States are the largest agricultural and the second largest (after petroleum) natural resource component of the U.S. trade deficit. The $21.3 billion value of fisheries products imported in 2003 has increased by 64% since 1996 (National Marine Fisheries Service 2004).

Within the United States, natural commercial fisheries stocks are threatened by overfishing, and catches of many species have leveled off or are declining. Fish farming now accounts for 10–15% of U.S. fish and shellfish harvests, leaving clear potential for future growth.

U.S. aquaculture producers range from corporations employing several hundred workers to small family farms. Products raised include: finfish (catfish, trout, salmon, striped bass, tilapia, baitfish, ornamental fish); crustaceans (crawfish, shrimp); shellfish (oysters, clams, mussels); and aquatic plants (seaweeds for food, pharmaceutical, and biomedical industries). The aquaculture industry is supported by an infrastructure of feed mills, processing plants, equipment manufacturers, and suppliers of specialty services and products. Aquaculture also supplies substantial quantities of juvenile fish and shellfish for stocking recreational fisheries.

The top aquaculture species in the United States, by value, according to the most recent Census of Aquaculture in 1998, were, in order of importance: catfish , trout, mollusks, ornamental fish, baitfish, and crustaceans (USDA 1998). The total value of aquaculture production in the United States was $978 million in 1998. Of this, catfish contributed 46% of the total value.

6. How does consumption of seafood compare with that of other protein products in the United States?

7. Describe some important consumption trends related to seafood and aquaculture products.

NOTES

1. Grass carp volume was 3.6 million MT in 2001, and the volume of Alaskan pollock was 3.1 million MT.
2. FAO defines world regions as Africa, the Caribbean, Europe, the Far East, Latin America, the Near East, North America, and Oceania.

REFERENCES

Asche, F. 2001. Testing the effect of an anti-dumping duty: the U.S. salmon market. *Empirical Economics* 26:343–355.

Asche, F., H. Bremnes, and C.R. Wessells. 1999. Product aggregation, market integration and relationships between prices: an application to world salmon markets. *American Journal of Agricultural Economics* 81:568–581.

Audubon Living Oceans. http://seafood.audubon.org

Bardach, J.E., J.H. Ryther, and W.O. McLarney. 1972. *Aquaculture: the Farming and Husbandry of Freshwater and Marine Organisms.* Wiley-Interscience, New York.

Cheng, H. and O. Capps, Jr. 1988. Demand analysis of fresh and frozen finfish and shellfish in the United States. *American Journal of Agricultural Economics* 70:533–542.

Csavas, I. 1994. Important factors in the success of shrimp farming. *World Aquaculture* 25(1):34–56.

Diamond, J. 1999. *Guns, Germs, and Steel: The Fates of Human Societies.* W. W. Norton & Company, New York.

Engle, C.R. In press. Marketing and economics. In: Webster, C. and C. Lim, *Tilapia Nutrition, Feeding, and Culture.* The Haworth Press, Binghamton, NY.

Engle, C.R., O. Capps Jr., L. Dellenbarger, J. Dillard, U. Hatch, H. Kinnucan and R. Pomeroy. 1990. The U.S. Market for Farm-Raised Catfish: An Overview of Consumer, Supermarket and Restaurant Surveys. Arkansas Agricultural Experiment Station, Division of Agriculture, Bulletin 925.

FAO. 1997. FAO database on introduced aquatic species. Food and Agricultural Organization of the United Nations, Rome, Italy. Accessed at //www.fao.org

FAO. 2004. FISHSTAT+. Food and Agriculture Organization of the United Nations. Available at http://www.fao.org.

Gabriel, R. 1990. A market study of the portion-sized trout in Europe. Final Report, Federation of European Salmon and Trout Growers, Belgium.

Hanson, G.D., G.P. Rauniyar, and Herrmann. 1994. Using consumer profiles to increase the U.S. market for seafood: implications for aquaculture. *Aquaculture* 127:303–316.

Holliday, M.C. 2002. Fisheries of the United States, 2001. National Oceanic and Atmospheric Administration and National Marine Fisheries Service, U.S. Department of Commerce, Washington, D.C.

Kinnucan, H.W., R. G. Nelson, and J. Hiariay. 1993. U.S preferences for fish and seafood: an evoked set analysis. *Marine Resource Economics* 8:273–291.

Millar, S. 2001. Salmon farmers braced for clampdown on toxins. The Observer, 7th January.

National Marine Fisheries Service. 2004. Imports and exports of fishery products annual summary, 2003. Can be found at http:www.st.nmfs.gov.

Quagrainie, K. and C.R. Engle. 2002. Analysis of catfish pricing and market dynamics: the role of imported catfish. *Journal of the World Aquaculture Society* 33(4):389–397.

Rauniyar, G.P., R.O. Herrmann, and G.D. Hanson. 1997. Identifying frequent purchasers of seafood for at-home and restaurant consumption. *The Southern Business & Economic Journal* 20(2):114–129.

Seafood Business. 2001–2003. News archives. www.seafoodbusiness.com/archives

Smith, A.P., J.A. Young, and J. Gibson. 1999. How now, mad cow? Consumer confidence and source credibility during the 1996 BSE scare. *European Journal of Marketing* 33:1107–1122.

USDA. 1998. Census of aquaculture (1998). 1997 Census of Agriculture Volume 3, Special Studies, Part 3, United States Department of Agriculture, Washington, D.C.

Valderrama, D. and C.R. Engle. 2001. A risk analysis of shrimp farming in Honduras. *Aquaculture Economics and Management* 5(1/2):49–68.

Vannuccini, S. 2003. Overview of fish production, utilization, consumption and trade. Fishery Information, Data and Statistics Unit, Food and Agriculture Organization of the United Nations. Accessed at: http://fao.org

Young, J.A., S.L. Burt, and J.F. Muir. 1993. Study of the Marketing of Fisheries and Aquaculture. Products in the European Community. EC DG XIV, Brussels.

2
Demand and Supply, Basic Economic Premises

This chapter provides a definition and examples of the fundamental concepts of demand and supply and their relevant elasticities. The factors and determinants of demand and supply are explained along with the effects of changes in the levels of these factors. The chapter concludes with a discussion of special conditions related to demand and supply. The shrimp market synopsis describes one of the highest-valued aquaculture markets worldwide.

WHAT IS ECONOMICS?

Many people have their own definition of what economics involves, and it frequently revolves around money. Although economists do spend a great deal of time estimating monetary values, economics is much more than a study of money.

Most people understand that an economy includes both production and consumption. Consumers "demand" goods and services, and producers "supply" the goods and services that consumers "demand." However, if it were that simple, there would be no need for a field of study such as economics. The problem is that there is no end to the wants and desires of human beings and people never have everything that they would like to have. Yet, resources are often scarce.

Because resources are scarce and the wants and desires of human beings are unlimited, the problem is how to allocate scarce resources to meet the unlimited wants and needs of human beings. This is the fundamental economic problem. Thus, economics can be defined as: the allocation of scarce resources to meet the unlimited needs and wants of human beings. Demand represents the needs and wants of human beings and supply represents the scarce resources that have been converted into goods that are needed and wanted by human beings (Fig. 2.1). The allocation process takes place in what is called a "market." Although many people are familiar with markets as places for consumers to purchase goods, the concept of a market is a much broader concept. It encompasses the entire demand-and-supply relationship and everything that happens in between as scarce resources are converted into goods that are exchanged with consumers to meet their wants and needs. Marketing includes the transformations of the good or service from the time it leaves the producers until it is consumed by the end consumers.

How do producers and consumers come to an agreement about the value of the product and the terms of exchange? This is where money comes in. In the majority of markets today, money is the medium of exchange. The price of the good is the signal that sends information between producers and consumers about the extent and relative scarcity of resources used to produce that particular good and the extent to which consumers need and want that particular good. Thus, demand represents what people want and can afford to pay at different prices, and supply represents what producers can make available to the market at different prices. These forces of demand and supply interact in the marketplace until an "equilibrium" is achieved at which buyers and sellers agree to exchange a particular quantity at a particular price.

DEMAND

Consumer demand is the various quantities of a commodity that consumers are willing and able to purchase as the price varies, when all other factors that affect demand are held constant (*ceteris paribus*). *Ceteris paribus* is a term that means "all else being equal," or "all else held constant." In economics, the values of the various parameters that are important continue to change. Changes in certain values will cause other values to change. To understand

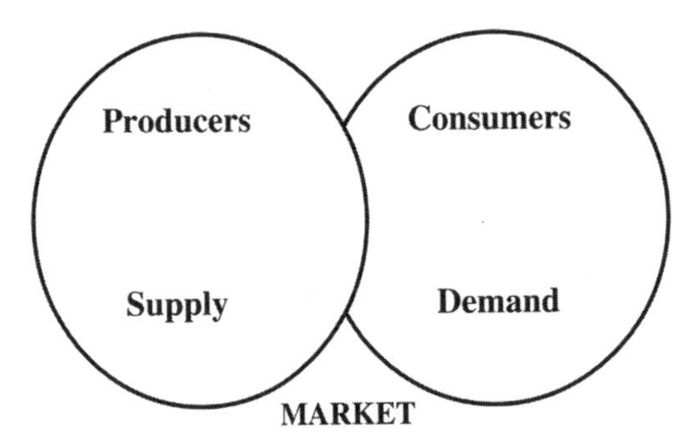

Figure 2.1. A market includes the supply obtained from aquaculture growers and all the functions, transactions, transformations, and exchanges required to meet the demand of consumers.

the fundamental relationships among different key parameters, it is important to first hold many of them constant to look at the relationships one at a time. This process assumes that each market is independent and self-contained. This type of analysis is referred to as a partial budget equilibrium analysis because changes in price in the market under consideration do not have dramatic effects on prices in other markets. This concept is similar to that used by aquaculture researchers who control for all but one or two variables in aquaculture experiments. After the individual relationships between specific variables are understood, more complex models can be built that include several variables at a time in order to begin to understand how these variables interact with one other. These analyses would be classified as general equilibrium analyses in which changes in price may affect other prices. These changes in other prices may then have an impact on the market under consideration.

An analysis that seeks to identify a profit-maximizing management strategy—say, the optimal stocking or feeding rate for a fish farm—may assume a market price. This would constitute a partial equilibrium analysis because an assumption is made that changes in the fish price will not result in changes in the prices of other goods associated with fish production, such as tractors or land. However, an analysis that would evaluate the impact of new environmental regulations on fish farming would likely take into consideration that increased costs on fish farms may affect the quantity of product supplied and the market price of the product. As the price of that particular type of fish product changes, the quantity demanded of other, similar types of fish products is likely to change. This type of analysis would require a general equilibrium analysis.

The word *demand* is often misused, and it is important to understand the different contexts within which the term is used. For example, existing demand represents the quantities of a particular product that would be purchased for a range of specific prices. In other words, existing demand measures the amount of fish taken from the market under all the imperfections that exist in any specific situation. However, many people speak about potential demand when they use the term "demand." Potential demand represents what consumers would do if the product were available. Potential demand estimation is often very important in the world of aquaculture. Farm-raised aquatic products are often introduced as new products in the marketplace. Demand cannot be measured if there are no historic sales data from which to measure prices and quantities involved in the transactions.

Formally, demand is represented as a relationship between the quantities of product that consumers are willing and able to take off the market at all alternative prices of that product. Economists typically represent this relationship as a graph (Fig. 2.2a). As price (P) goes up, the quantity demanded (Q) by consumers of that product typically goes down. Likewise, as the price (P) goes down, the quantity demanded (Q) by consumers of that product goes up. The resulting rate of change in the quantity demanded is represented by the slope of the curve in the graph (Fig. 2.2a). Products that perform this way in the market are called "normal goods." In aquaculture markets, this "Law of Demand" for normal goods can be seen in both shrimp and salmon markets. As the price of salmon and shrimp dropped in the 1990s and 2000s, the quantity demanded by U.S. consumers increased. Because quantity demanded will always move in the opposite direction from the price of normal goods, the demand relationship between the quantity demanded and the price of the good is said to be negative, or inverse. A negative relationship is depicted graphically as a line that slopes downward to the right. Figure 2.2a depicts a classic style of demand curve. An exception to this is a product termed a "giffen good." For giffen goods, if the price goes up, the quantity demanded also goes up. If the price goes down, the quantity demanded of giffen goods goes down.

Demand can be viewed as the maximum quantity of a product that is desired by consumers who are able to purchase the good at a given price. Demand can also be viewed as the maximum price that peo-

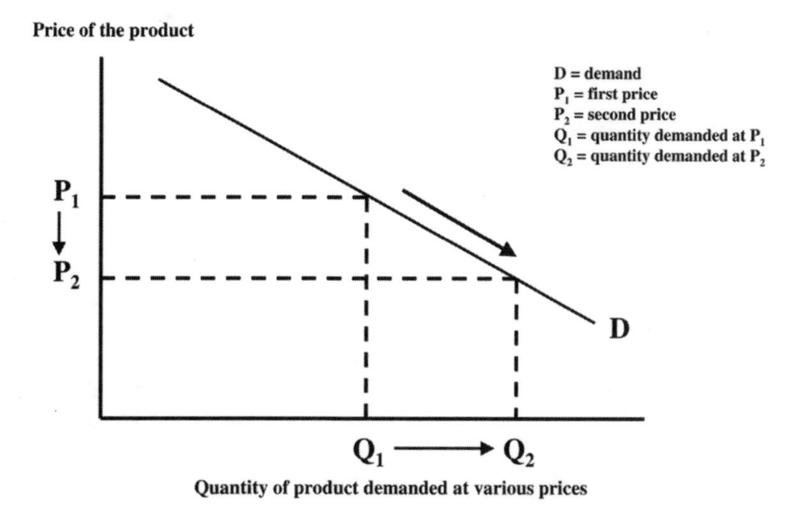

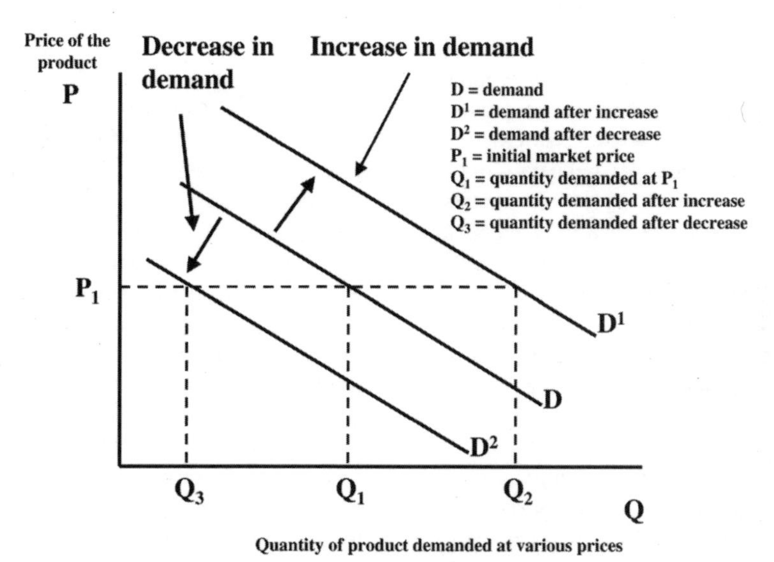

Figure 2.2. a. Demand and change in quantity demanded. **b.** Change in demand.

ple are willing and able to pay for a given quantity of the good.

As the price of a good changes, then, the quantity demanded changes. To identify the change in quantity demanded, one would track this change along the same demand curve, moving either up or down depending upon the direction of the price change. Thus, a change in price results in a movement along the demand curve and generates a change in the quantity demanded. It is important to note that this is a change in quantity demanded, not a change in demand. Because demand represents the total relationship between price and quantity demanded, a change in demand would require the entire demand curve to move.

Any given demand relationship between price and quantity demanded is based on holding a series of factors constant. These factors are also called determinants of demand and include:

- Population size and distribution
- Consumer income and distribution
- Consumer tastes and preferences
- Prices
- The availability of substitutes

When these factors are held constant, *ceteris paribus*, then we can discuss and analyze the relationships and interactions of price and quantity demanded. However, when any of these factors changes, then the demand curve itself will change, as shown in Figure 2.2b. As demand increases, the curve shifts to the right. When this happens, the quantity demanded is increased for all prices. If demand decreases, the demand curve shifts to the left. Consequently, for every price, then, the quantity demanded is less.

We can represent demand algebraically as follows:

$$Q_d = f(P_d, Pop_d, I, T, P_R)$$

where

Q_d = quantity demanded,
P_d = price of the product under investigation,
Pop_d = size of the population,
I = income,
T = tastes and preferences, and
P_R = prices of related products.

Traditional demand models use the quantity demanded as the dependent variable in a multiple regression analysis using econometric (statistical methods developed for economics analyses) methods. The independent variables frequently used are those listed above as factors or determinants of demand. Thus, the product's own price, population levels, consumers' incomes, consumer tastes and preferences, and prices of related products are frequently selected as independent variables.

Economists have specialized meanings for their terms, even those that have common, everyday meanings. So, it is important to be careful when using economics terms such as "demand" and "supply". The difference between a change in demand and a change in quantity demanded is very important in economic analysis. Thus, it is important to use the terms correctly to avoid misunderstandings that could result in mistaken recommendations and poor business decisions.

Shrimp consumption in the United States provides an example of the contrast between increases in quantity demanded and increasing demand. Shrimp was once a product consumed only occasionally and in certain consumer segments. Increasing quantities of shrimp supplied resulted in lower shrimp prices that resulted in a shift along the demand curve and increases in quantity demanded. In recent years, environmentalist groups opposed to shrimp production have developed advertising campaigns to convince consumers not to purchase shrimp due to the alleged environmental and social injustices related to shrimp production. If these advertising campaigns successfully change consumer tastes and preferences for shrimp, a decrease in demand could result. This decrease in demand would cause the demand curve to shift to the left, and quantities demanded would decrease at all prices.

The factors affecting demand are further described below.

Population

The world's population is projected to grow to 8.5 billion by 2025 from 5 billion in 2003. Even with stable per capita seafood consumption, world demand would increase from 95.6 to 162 million metric tons (MT) between 1990 and 2025 from population increases alone. Aquaculture must increase seven-fold, from 11 to 77 million MT, to supply this projected increasing demand.

Income

The demand for a good is affected by income levels of consumers. However, the demand for different types of goods can be affected in different ways by changing income levels. For example, in many developing countries, fish such as tilapia are considered "poor peoples' food" (Neira et al. 2003). In such cases, as peoples' incomes rise, their consumption habits may change by substituting a higher-priced source of protein. This might be large tilapia fillets instead of small, whole tilapia, or a change to shrimp or other seafood product that is considered to be of higher value. However, it could also mean that consumers would switch to fillet mignon or some other type of protein altogether. A product that is considered to be "poor peoples' food" would be classified by an economist as an inferior good. An inferior good is defined as one for which demand would decrease with an increase in income levels.

In other situations, aquaculture products may be considered to be superior goods. Superior goods are those for which demand increases as income levels increase. An example might be shrimp consumption in the United States. As income levels go up, many consumers will desire to eat more shrimp, thus increasing demand.

This relationship between demand and income levels of consumers can change over the life of the product. Several decades ago, for example, catfish was considered to be an inferior good that was consumed by poor people in rural southern areas. Demand for catfish has increased over time and the relationship has changed such that increasing income levels now typically result in increased demand for U.S. farm-raised catfish.

Consumer Tastes and Preferences

Consumer tastes and preferences affect demand over time, and demand for different products will change as tastes and preferences change. A good example of this in a seafood market is the demand for fresh salmon and tuna. With an increasingly health-conscious consumer population and information on the benefits of eating fish with a high content of omega−3 fatty acids, the demand for tuna steaks and salmon products has increased. Demand for whole-dressed finfish has decreased over time as consumers' preferences have turned more toward fresh and frozen fillets than whole-dressed fish.

Consumer Behavior

Consumer expectations of the future will also affect demand for different goods. If consumers expect the economy to grow and to enjoy increasing salaries and wages over the next year or so, they are likely to spend more money on luxury goods. A good is classified as a necessity or luxury depending on how expenditures on the good change with changes in income. For a necessity, the change in expenditure is less than proportionate to the change in price, but for a luxury good, the change in expenditure is greater than the change in income. Because many seafood products in the United States are considered by consumers to be luxury products, expectations of a healthy, growing economy often will result in increasing demand for seafood products. However, if consumers expect poor economic growth, or are concerned about the safety of seafood products, demand may decrease.

Products that are related to each other, in the minds of consumers, may affect the demand for one of those products. Some products, termed substitutes, are ones that can be considered to be competing or rival products. For example, most consumers are unlikely to purchase both sea bass and mullet to prepare for dinner. They are likely to choose either sea bass or mullet to prepare. If the price of mullet has recently gone up, someone considering mullet might choose to purchase sea bass instead, and the overall quantity demanded of sea bass will go up as the price of the substitute product (mullet, in this case) goes up.

Products can also be complements. Complementary products are those products that consumers tend to consume together at the same time. Many people serve lemon wedges with fish. Lemons could be considered to be a complement to fish. So, if the price of fish goes down such that consumers increase the quantity demanded of fish, the quantity demanded of lemons would be expected to increase as well.

The insets in this chapter provide examples of demand and supply at work in the marketplace. The

The World Shrimp Industry: The Influence of Taiwan, China, and Thailand

The Asian region dominates the supply of shrimp worldwide just as it does other forms of aquaculture. Thailand is the world's leading producer of shrimp and is followed by China, Indonesia, Ecuador, India, Vietnam, Taiwan, The Philippines, and Bangladesh. However, Thailand has not always been the world leader. In 1987, Taiwan led the world in shrimp production (Csavas 1994). Taiwan had pioneered the development of many of the technologies for farming marine shrimp and continued to lead the world in shrimp production through the 1980s. However, disease problems resulted in a near-collapse of Taiwan's industry in 1988. China moved very rapidly into shrimp production in the early 1990s and within just a few years went from minimal production to totally dominating the shrimp market. Disease problems resulted in serious decline of the Chinese industry, too. Thailand learned from the Taiwanese and Chinese experiences with disease problems and has been able to implement improved management practices to avoid the catastrophic losses that were characterized by the shrimp industries in other countries.

Penaeus monodon dominates the total shrimp market. For a number of years, this tiger shrimp was considered unmarketable in the United States due to the large black stripes. However, its size is a major positive characteristic in the U.S. market and, when cooked, it turns pink as do other shrimp. Test marketing clearly showed that it had considerable potential in the United States, and it is now sold here as well as throughout Asia. *P. vannamei*, the species most important in the Western hemisphere, is the second leading species produced, followed closely by *P. japonicus*. There has been a rapid increase in production of *P. vannamei* in Asia. Shrimp farmers report greater disease resistance than with *P. monodon* and have begun to substitute it. Much of the *P. vannamei* raised in Asia is destined for export to the United States.

Salmon Market Research

In the marketplace, "salmon" includes Atlantic salmon, coho salmon, and salmon trout. These species seem to be highly substitutable (Asche 2001). In 1999, about 80% of farmed salmon were Atlantic salmon. Production of farmed salmon increased substantially, beginning in the early 1980s. Norway, Chile, the United Kingdom, and Canada are the leading producers worldwide of farmed salmon, producing about 90% of the total quantity produced. The increase in production has been accompanied by a substantial reduction in prices. However, production costs have decreased and the expansion has been possible due to productivity growth (Bjorndal & Tveteras 1998; Tveteras 1999). This suggests that a large part of the growth in salmon aquaculture has been a move along the demand schedule. However, there is also evidence that there has been market growth, partly due to generic advertising programs (Bjorndal et al. 1992; Kinnucan and Myrland 1998).

A great deal of work has been done on the market for salmon. Demand for salmon has been analyzed by Bjorndal et al. (1992), Asche (1996, 1997), Asche et al. (1997), Eales et al. (1997), Salvanes and DeVoretz (1997), Asche et al. (1998), Johnson and Wessells (1998), Kinnucan and Myrland (1998), Eales and Wessells (1999), and Asche and Steen (2000). Other studies have investigated interactions between prices, including Gordon et al. (1993), Asche and Hannesson (1997), Asche et al. (1997), Asche et al. (1999), Clayton and Gordon (1999), Jaffry et al. (2000), Asche (2001), and Asche et al. (2001).

Results of these studies indicate that different salmon species and product forms are close substitutes. The Law of One Price holds, indicating that relative prices are stable (Asche 2001). This means that increased production of farmed salmon has had a substantial impact on the markets and prices for wild Pacific salmon and that the increased supplies of Pacific salmon is evidence of moving down along a backward-bending supply schedule. This would mean that salmon farming has been beneficial for wild salmon stocks. However, landings are at historically high levels. Substantial hatching programs implemented in the 1980s may be the underlying cause for increased landings. Although the number of fishermen and their revenues have decreased, salmon farming may not have had any impact on landings.

Salmon does not in general seem to compete with the large volume species of fish that constitute the global whitefish market (Asche 2001). Asche and Steen (2000) report that several fish species in the EU are complements for salmon. This implies that expansion of the salmon market has made more room for other species as well. With reference to competition with other forms of meat, Eales and Wessells (1999) show evidence of a shift from no interaction to some competition in the mid-1990s.

development of aquaculture technologies has resulted in increased supply of shrimp and salmon. The increased supplies have resulted in lower prices. As prices have decreased, the quantity demanded of these goods has increased.

SUPPLY

Supply is the amount of goods and services that producers are willing and able to offer in the marketplace at specific prices. Formally, supply is represented as a relationship between the quantity of product that producers are willing and able to place on the market at all alternative prices of the product. Economists typically represent this relationship as a graph (Fig. 2.3a). The data used are referred to as a "schedule." The supply schedule includes the alternative quantities (Q) offered for sale at different prices (P). As price goes up, the quantity supplied by growers typically goes up (Fig. 2.3a). Likewise, as the price goes down, the quantity supplied by growers goes down (Fig. 2.3a). This classic relationship can be seen in the hybrid striped bass market in the United States in the 1980s–1990s. As the price of hybrid striped bass increased, the quantity supplied by U.S. growers increased. Because quantity supplied will typically move in the same direction as price, the relationship between the quantity supplied and price of the good is said to be positive, or direct. A positive relationship is depicted graphically as a line that slopes upward to the right. Figure 2.3a depicts a classic style of supply curve.

As the price of a good changes, then, the quantity supplied changes. To identify the change in quantity supplied, this change would be tracked along the same supply curve, moving either up or down depending upon the direction of the price change. Thus, a change in price results in a movement along the supply curve and generates a change in the quantity supplied. It is important to note that this is a change in quantity supplied, not a change in supply.

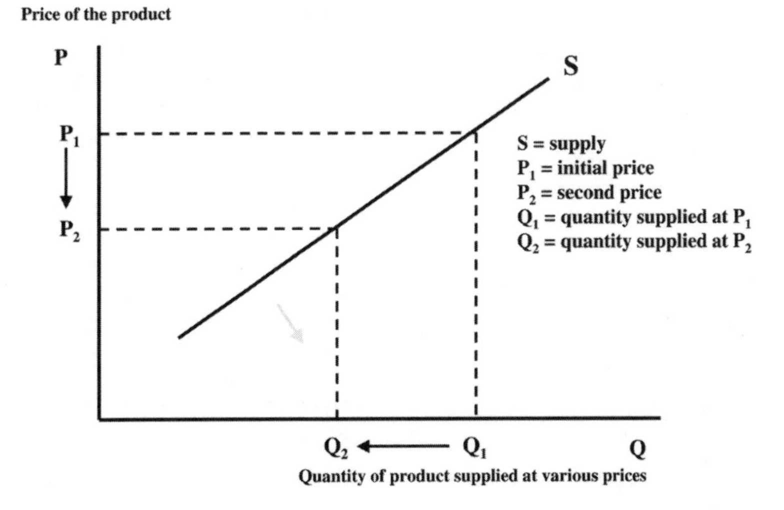

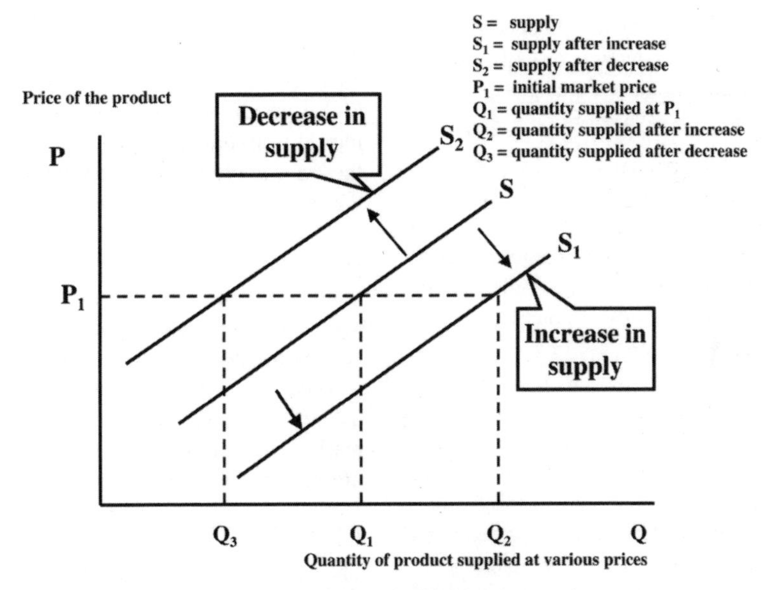

Figure 2.3. a. Supply and change in quantity supplied. **b.** Change in supply.

Because supply represents the total relationship between price and quantity supplied, a change in supply would require the entire supply curve to move.

Figure 2.3b shows that a shift to the right of the supply curve will result in increased quantities supplied at all prices. A decrease in supply will shift the supply curve to the left and will result in lower quantities supplied at all prices.

Any given supply relationship between price and quantity supplied is based on holding a series of factors constant. These factors are also called factors affecting supply and include:

- Changes in prices of inputs
- Changes in profitability of related commodities
- Changes in production technology
- Changes in prices of joint products
- Institutional and environmental changes—government regulations and programs

When these factors are held constant, *ceteris paribus*, then we can discuss and analyze the relationships and interactions of price and quantity supplied. However, when any of these factors changes, then the supply curve itself will change.

We can represent supply algebraically as follows:

$Q_s = f(P_1$, prices of inputs, prices of related commodities, production technology, price of joint products, and institutional and environmental changes)

where

Q_s = quantity supplied

P = price of the product under investigation

Prices of inputs = prices of feed, labor, electricity, and so on

Prices of related good = prices of other types of finfish or shellfish that consumers would consider as own price fluctuates

Production technology = change in the way the product is produced,

Price of joint products = price of a product that is produced in the same production system, and institutional and environmental changes.

The farm-raised shrimp industry can also provide examples of the differences between changes in the quantity supplied versus changes in supply (See market synopsis at end of this chapter). Improved feed formulations and the development of hatchery techniques to consistently and reliably supply shrimp seed (post-larvae) were major technological breakthroughs that resulted in increases in supply. Thus, the supply curve shifted to the right, and greater quantities were produced at all prices. However, when market prices fall, shrimp farmers produce less, and some farmers go out of business if prices fall below their costs of production. In these situations, the quantity supplied decreases, representing a movement downward along the supply curve.

The factors affecting supply are further described in the following sections.

Costs of Production

Increases in production costs will cause supply to decrease. For example, increased costs of feed, labor, or utilities will decrease the quantity that producers can supply for any given price. Kouka and Engle (1998) showed that food-size supply of catfish would decrease by 2% with a 20% increase in the cost of feed.

Technology

Improved technologies can cause supply to increase. Technologies that improve productivity of use of inputs will result in increases in supply. The development of improved diet formulation and more water stable pellets for shrimp contributed greatly to expansion of the shrimp industry (Csavas 1994). The improved feeds allowed farmers to increase feed efficiency and increase yields. Production costs declined as yields increased and the result contributed to an increase in the supply of farm-raised shrimp. The development of efficient aerators in the 1980s resulted in a similar increase in catfish supply. With a consistent, reliable, and low-cost source of oxygen, farmers could stock and feed at higher rates and increase yields with lower yield risk. Supply increased as a result.

PRICE DETERMINATION

The amount and price of a product are determined in the marketplace by the interactions between demand and supply (Fig. 2-4). If producers seek too high a price, there will be fewer buyers who are willing and able to purchase the product at that price. Thus, the quantity demanded in the market will be less at higher prices. Not all the product offered will be removed from the market and some producers will either have to offer lower quantities or some producers will go out of business. In order to move their products, sellers will have to drop the price. At lower prices, fewer sellers will be able to sell at a profitable level and some will go out of business. However, at lower prices, the quantity demanded by consumers increases. At some point, an equilibrium is reached in which the quantity demanded equals the quantity supplied for a given price. This is called market equilibrium and it is described by the equilibrium price and the equilibrium quantity.

ELASTICITY

The concepts of demand and supply are fundamental to any discussion of market and economic forces and effects. However, it is essential to understand the demand and supply concepts to be able to read and understand the marketing literature. Characteristics of the specific demand and supply relationships depend on how prices behave in the marketplace. In particular, characteristics relate to the elasticity of demand and of supply. Elasticity concepts measure the degree to which the quantity demanded or supplied will change with a given change in market price.

Demand Elasticity

Elasticity of demand is the degree of responsiveness of quantity demanded to a given change in price. Thus, elasticity is a measure of changes relative to a single demand curve, not changes in the determi-

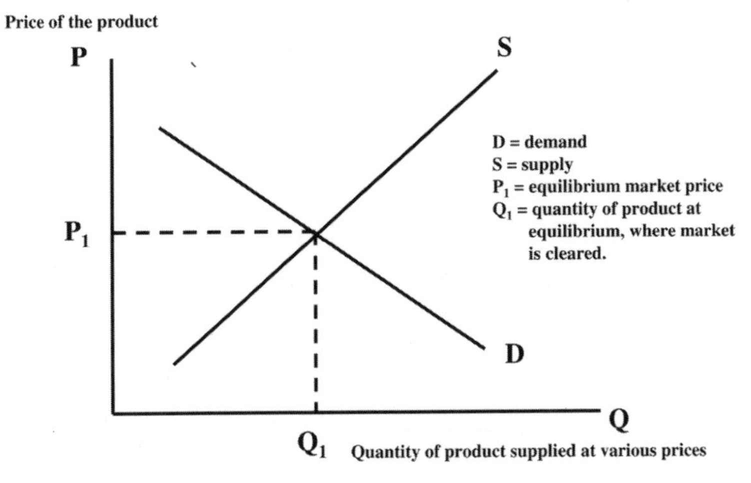

Figure 2.4. Price determination.

nants of demand that result in a shift in the demand curve itself. In discussing elasticity of demand, we are measuring the change in quantity demanded as a result of a given change in price or the percentage change in the quantity demanded that results from a 1% change in one of the independent variables. If the product's own price is used to calculate elasticity, this measure is also referred to as price elasticity of demand. It is measured as the percentage change in quantity demanded due to a percentage change in price, *ceteris paribus*.

Price elasticity of demand (E_d) is measured by the following equation:

$$Ed = \frac{\text{percentage change in quantity demanded}}{\text{percentage change in price}}$$

$$= \frac{\%\Delta Q}{\%\Delta P}$$

It is important to note that, at this point, we are discussing the percentage change in price of the good itself that results from a 1% change in its "own" price. Later in this chapter, elasticities related to changes in price of a related good are discussed. Note that the price elasticity of demand is not the same along the entire length of the demand curve. Price elasticity of demand is more elastic for demand relationships with higher prices and lower quantities, whereas price elasticity of demand becomes more inelastic as price decreases and quantity demanded increases. There is a mid-point along a demand curve that corresponds to unitary elasticity. Normal goods have a negative elasticity and a downward sloping demand curve.

Elasticity does not measure the slope of the curve. If it did, it would not change along the length of the demand curve (straight lines have constant slopes). It does not determine the shape of the demand curve, but there are some shape relationships. For example, if the coefficient for the price elasticity of demand is greater than the absolute value of 1, then demand is considered to be elastic. This means that the percentage change in quantity demanded is greater than the percentage change in price. Thus, the quantity demanded is very responsive to price changes, or the relationship is very "elastic." The quantity "stretches" a great deal with a small change in price. Goods that are price elastic tend to be goods with many substitutes. If price goes up and there are other types of fish that can readily be substituted by consumers, buyers are likely to switch quickly to the other types of fish. Thus, the quantity demanded will decrease quickly as price increases if there are many substitutes.

However, if the absolute value of the coefficient of the price elasticity of demand is less than 1 ($E_d < 1$), then demand is considered to be inelastic. In other words, the percentage change in quantity demanded is less than the percentage change in price and the quantity demanded is not responsive to changes in price.

Figures 2.5a and 2.5b present the two extremes, perfectly elastic demand and perfectly inelastic demand, respectively. Perfectly elastic demand would result in a market that absorbs any increase in supply at the same price level (Fig. 2.5a). The cod fishery in the 1800s exhibited characteristics of a product with highly elastic demand. The abundance of cod resulted in lower prices that enabled cod to be

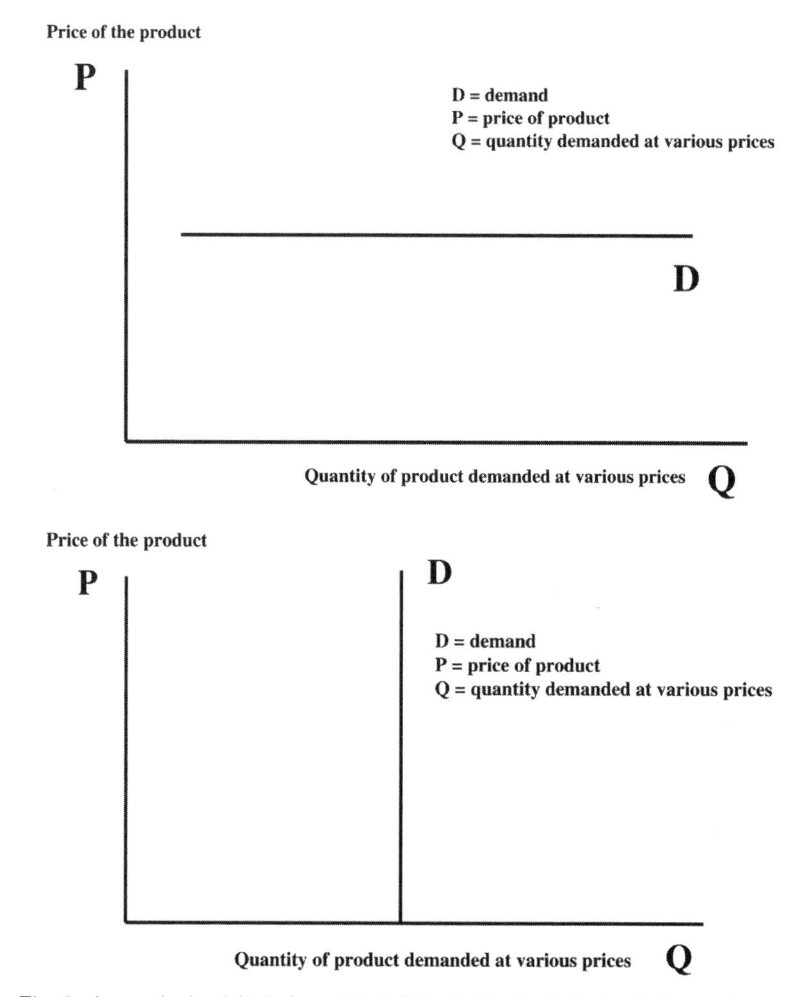

Figure 2.5. **a.** Elastic demand: elasticity is from 0 to infinity. **b.** Perfectly inelastic demand.

transported all over the world and to become a staple commodity for over a century. However, as cod stocks diminished and price increased, buyers began to substitute other types of fish for cod and quantities decreased. At the other extreme, if the quantity demanded of a product does not change regardless of changes in price, then demand is perfectly inelastic (Fig. 2-5b). Carp served for traditional Christmas Eve dinner in European countries would be expected to have highly inelastic prices. If people believe that they need to pay whatever price to have the appropriate meal for an important holiday or ceremony, they are likely to purchase the same quantity regardless of its price.

The price elasticity of demand determines the extent to which the good is considered a necessity. Goods that are basic necessities will have inelastic demand. If a particular item is truly necessary for survival, consumers will purchase it regardless of changes in price. Thus, the quantity demanded will change very little even if price changes by a great deal. More formally, a 1% change in price results in less than a 1% change in quantity demanded. The opposite will also be true. If a good is a luxury good, then the price elasticity of demand will be highly elastic. The quantity demanded of a luxury good will vary greatly with relatively small changes in price. Fish in some developing countries are considered a necessity; therefore demand for fish in those markets is inelastic. Shrimp at one time were a high-priced seafood product with highly elastic demand. As price has decreased, shrimp demand has become less elastic because the product has become less of a luxury item.

Cross Price Elasticity

Cross price elasticity measures the responsiveness of quantity demanded in one good to changes in price of a related good. It is measured by the following equation:

$$E_{x,y} = \frac{\textit{percentage change in quantity demanded of good x}}{\textit{percentage change in price of good y}}$$
$$= \frac{\%\Delta Q_d}{\%\Delta P_y}$$

where

$E_{x,y}$ = cross price elasticity of good x with respect to changes in the price of y
Q_d = the quantity demanded of good x
P_y = price of good y

Cross price elasticity is used to measure the degree of substitutability of goods, or the degree to which goods compete in the same market. It measures the effect on quantity demanded of one product as a result of changes in price of another product. Completely unrelated goods have a zero cross elasticity. A negative sign shows an inverse relationship that indicates that the two goods are complements. Substitute goods would have a positive relationship between the change in price of one and the quantity demanded of the other.

The availability of substitute goods will also determine the price elasticity of demand. The more substitutes that are available, the easier it will be for consumers to switch to another good when price increases. Thus, demand tends to be more price elastic because more substitutes are available. However, if there are no close substitutes for a good, those consumers who really wish to purchase it will find it necessary to pay whatever the market price is. Thus, if the price increases, all else being equal and if substitutes are available, consumers will switch to the cheaper good. However, if seafood is not considered to be a good substitute for red meat and poultry consumption, then the demand for seafood will be more price inelastic.

The literature on substitutes among types of seafood is not clear. However, it would be reasonable to suppose that different species of marine fish fillets would substitute for each other. Thus, an increase in the price of orange roughy fillets might result in an increased quantity demanded of red snapper as an example. Shrimp and cocktail sauce could be considered complements. Decreasing prices of shrimp would be expected to increase quantity demanded of shrimp and also of cocktail sauce to accompany the shrimp.

Economists refer to income elasticity as a measure of the response of the quantity demanded to changes in income *ceteris paribus*. Specifically, income elasticity is measured by the following equation:

$$E_I = \frac{\textit{percentage change in quantity demanded}}{\textit{percentage change in income}}$$
$$= \frac{\%\Delta Q}{\%\Delta I}$$

Income elasticity will vary with the proportion of income spent on the product. Goods are classified as a necessity or luxury depending on the income elasticity. For a necessity, a 1% change in income results in less than a 1% change in quantity demanded. The value of the coefficient of income elasticity is between 0 and 1. For a luxury good, the income elasticity is greater than 1. The larger the proportion of income spent on the good, the greater the elasticity. Elasticity varies with the type of fish and the form in which it is presented. The sign of the coefficient is important. A negative sign indicates an inferior good because an increase in income results in decreasing quantities demanded. However, a positive sign indicates a direct relationship and a normal good. As incomes increase, quantities demanded also increase.

Price Elasticity and Total Revenue

There is an important relationship between price elasticity of demand and total revenue. With an elastic demand, a decrease in price will result in a proportionately greater increase in quantity demanded. This is because the demand curve is relatively flat. Thus, a decrease in price will increase total revenue ($TR = P \times Q$; where TR equals Total Revenue, P = price, and Q = quantity). Similarly, a price increase will result in a proportionately greater decrease in quantity demanded that will decrease total revenue if demand is elastic.

Conversely, with an inelastic demand, a decrease in price will result in a smaller proportionate increase in quantity demanded. Thus, a decrease in price will result in lower total revenue. An increase in price will result in a smaller proportionate decrease in quantity demanded. Thus, an increase in prices will result in increased total revenue.

Elasticity of Supply

Elasticity of supply is a similar concept to that of demand elasticity. It measures the degree of responsiveness of the quantity supplied to changes in the price of the good.

The price elasticity of supply expresses the percentage change in quantity supplied in response to a 1% change in price, *ceteris paribus*, and is calculated as follows:

$$E_s = \frac{\textit{percentage change in quantity supplied}}{\textit{percentage change in price}}$$
$$= \frac{\%\Delta Q}{\%\Delta P}$$

Q = quantity supplied of good
P = price of the good
Δ = very small change

A value of 0 means that supply is perfectly inelastic, or fixed. In other words, quantity supplied will not change, irrespective of price changes. An elasticity value greater than 1 indicates that supply is elastic. In other words, a 1% change in price will result in a percentage change in the quantity supplied that is greater than 1.

Some shrimp growers in Ecuador have diversified their farm production by co-culturing tilapia with shrimp in ponds. These species occupy different niches in the pond, offer more market opportunity, and can be used to reduce market risk. Supply becomes more elastic as farmers have greater flexibility to respond to prices by holding crops for a better price or by switching to marketing a higher-priced product. However, freshwater prawns might represent a substitute product for marine shrimp.

MARKET STRUCTURES AND IMPLICATIONS FOR COMPETITION AND PRICING

Economists use the term "market structure" to describe the factors that determine the competitiveness of the industry (Carlton and Perloff 2000). The degree of competitiveness is determined by (1) the number of firms (businesses); (2) the type of product (homogeneous, differentiated, or unique); (3) whether there is control over the price; and (4) the degree of freedom of entry and exit. The resulting classifications of market structures are discussed in greater detail in Chapter 5, and Table 5.2 illustrates the differences among market structures. The market conduct of businesses within the industry affects market performance.

Fish farmers often discuss issues related to market power, or whether a particular level of the marketing chain has greater or lesser control over the price. Processors or other middlemen are often thought to exercise "unfair" control over prices. Whether or not market power occurs in an industry is measured through market performance. These include the rate of return, the price-cost margin, and Tobin's q (value of the market value of a firm to its replacement cost) (Carlton and Perloff 2000).

SPECIAL CONDITIONS

If production of farmed fish is increasing relative to wild-caught product, this implies that the productivity for farmed fish production increases faster than for wild-caught product (Asche 2001). If the two products are close substitutes, farmed fish can then win market share from the other product. Moreover, if demand is not perfectly elastic, the price will decline as will the income of the producers of wild-caught fish. However, if the goods are not substitutes there are no market effects, and the increase in the supply of the farmed fish will lead only to a move down the demand schedule for farmed fish. Hence, for the producers of farmed fish it is easier to expand when the farmed fish has substitutes with established markets.

This situation changes if the potential substitute is fish from a fishery located on the backward-bending part of the supply schedule (Anderson 1985). Captured fish has a backward-bending supply schedule as a consequence of the biological growth curve (Fig. 2.6) (Anderson 1985). A backward-bending supply curve indicates that as price continues to increase, the quantity supplied begins to decrease. This can happen when overexploitation of fish stocks results in scarcity that causes price to increase. However, at some point, the scarcity results in decreasing spawning and recruitment to the fishery. Thus, the decline in stocks leads to decreased quantity supplied, and the supply curve bends in a backward fashion. Moreover, as many of the world's fish stocks are reported to be fully or overexploited, it is likely that the market equilibrium for many of the world's fish stocks is on the backward-bending part of their supply schedule (Asche 2001). The increased supply of farmed fish will then actually lead to a higher supply of wild-caught fish and thus to sharp competition. Price will decline and fisherman's revenues may increase or decrease depending on the slope of the backward-bending supply schedule. However, stock size will increase as one moves down the backward-bending supply curve. If a fishery is on the "normal" part of the supply schedule, the effects will be as for conventionally produced goods, and the reduction of supply that is caused by decreased price will also tend to enhance stock size.

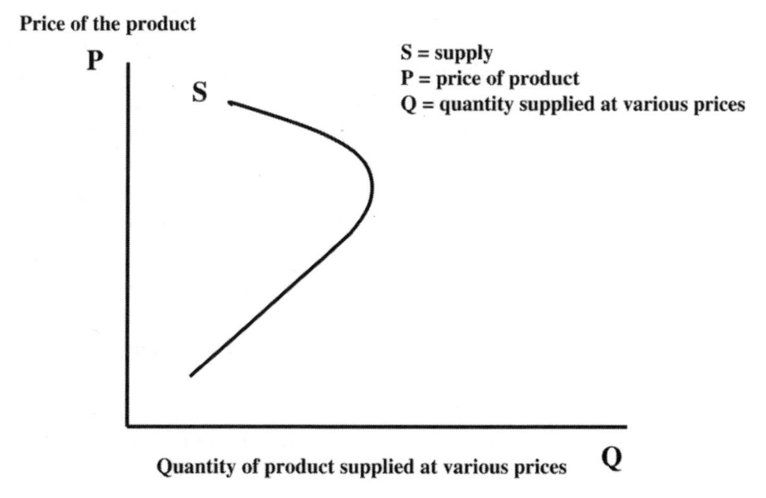

Figure 2.6. Backward-bending supply curve.

AQUACULTURE MARKET SYNOPSIS: SHRIMP

The early development of the shrimp market was the result of Japanese trading companies and American importers urging developing nations to develop culture techniques in the 1950s and 1960s. Currently, U.S. shrimp production fills only one-third of the demand. The increase in demand for shrimp is three times faster than for all other meats. The downturns in price are usually in a four-year cycle.

Shrimp is a globally traded commodity, principally in the United States, Japan, and Europe. It is not considered exclusively a commodity because it is marketed by brand, packaged at the source of production, and in the final container in which it will be sold. The number of colors, sizes, and species also set it apart from other commodities.

Shrimp is the most important seafood species entering international trade, with about one-third of global shrimp production entering the international trade. Worldwide production is approaching 3 billion MT. Sources include 40 countries, of which 15 produce 80% of the shrimp. Principal nations include India, China, the United States, Indonesia, Thailand, Taiwan, and Mexico. The three principal markets are the United States, Japan, and Europe, which together consume more than 50% of the final product.

At least 25 species of shrimp enter world trade. In Latin America, the best prices are found for 31/35 and 36/40 count shrimp. Prices depend on the size of individual shrimp, the quality, and the source. Marketing is complicated because of the number of countries involved, the range of sizes sold, the number of species sold, the number of product forms, and the types of markets.

There are literally hundreds of species of saltwater shrimp in the world. The industry divides them into two categories: (1) cold-water/northern (*F. Pandalidae*) and (2) warmwater/tropical (*F. Penaeidae*). However, only eight species dominate the market in the United States. The product mix changes when aquaculture technologies change.

There are a variety of species of shrimp that are typically traded. The Pacific white shrimp (*Litopenaeus vannamei*) is one of the most common species of shrimp that is farmed. It is imported into the United States from South America, Central America, and Mexico. It is similar to the Gulf whites and is seasonal. In the past, *L. vannamei* was cultured only in the Western hemisphere, but in recent years it has become widely cultured in Asia. Although this species is smaller and therefore attracts a lower price, Asian growers have found it to be more resistant to diseases and have begun to culture it for this trait.

The major species raised in Asia are the black tiger prawns (*Penaeus monodon*). The black tigers have dark dorsal/lateral stripes, hence the name. There is some market adversity to the stripes in the United States. Tiger shrimp are usually sold peeled. Large tiger prawns are sold heads-on to the Japanese market. In recent years, increased incidence of White Spot Syndrome Virus in black tiger shrimp has resulted in many shrimp farmers in Asia converting to culture of the exotic *L. vannamei*.

The Chinese whites (*Penaeus [Fenneropenaeus] orientalis*), or mandarin whites, are medium-sized and generally sell for lower prices. The Gulf brown shrimp (*Farfantepenaeus aztecus*) has varied shell color and is popular in Texas (50% of the wild harvest). It is usually a little cheaper than the whites and is caught offshore. The Gulf Pink (*Farfantepenaeus duorarum*) is a very sweet shrimp with firm flesh and tails that turn red when cooked. Most Gulf pinks are packed shell-on. Gulf white shrimp (*Litopenaeus setiferus*) are often confused with brown shrimp. However, browns have a groove in the last tail segment. The flesh is often colorless, has good flavor, and turns pink when cooked.

Shrimp are traded daily as a commodity by a multitude of exporters. Most of the trade is from developing nations to industrialized countries. This trade is characterized by price cycles from fluctuating supply and demand, marginal producers, capital requirements, contracts and futures, fluctuating exchange rates, short and long positions, and overproduction and speculation. Shrimp is actually considered a medium of foreign exchange in some areas.

Shrimp prices are cyclical and subject to a variety of influences. Shrimp price cycles tend to match price levels in prosperous and recession years and appear to follow consumer discretionary income. Price breaks tend to follow sizeable accumulations of secondary and substandard quality product.

The processor is a key between the producer or shrimper and the market. Processing has two stages: (1) turning the shrimp into a form in which it can be traded as a commodity, and (2) changing it from a commodity into a value-added product form (for example, peeled, cooked, or individually quick frozen [IQF]). Packaging is improving in developing nations where most of the production occurs and refrigerated vessels are more readily available. Air shipments of fresh and live product are becoming more common.

International trade takes place almost exclusively between importers and exporters. Financing is provided by the importer, who typically opens an irrevocable letter of credit (LC) in favor of the exporter. Importers are marketers themselves and usually sell to wholesalers, distributors, reprocessors, restaurant chains, and supermarket chains. Financing within the producing country is often provided by exporters, who finance the processor, who in turn finances the farmer.

An exporter may be a processor, farmer, or an independent third party that takes financial responsibility and communicates with the importer. Many governments require that prices be set before shipment. Others set minimum sales prices and quality parameters.

Importers may purchase shrimp outright from foreign traders, paying for the purchase at full invoice value either at the time of shipment or upon passing through FDA. Alternatively, the importer may work on a consignment arrangement whereby an advance is made to the exporter by means of an LC (letter of credit). In some cases, the importer acts as a sales agent (broker) for the exporter and collects a commission. Importers can also make preseason advances to producers (therefore tying up their production). This is how Japanese typically maintain a strong grip on Asian sources. The availability of supply determines the direction shrimp markets will take. The level of supply determines prices. Prices, as they relate to competing products, determine quantity demanded.

Japanese and U.S. shrimp markets are interdependent; prices prevailing in one market tend to affect the other. Fluctuations in each country's rates of exchange can cause a reaction in both markets and elsewhere. If the U.S. dollar is strong, people will want to export shrimp to the United States. Low U.S. dollar and yen values favor sales to Europe. European markets prefer heads-on product. Increased flow to the major markets of the United States, Japan, and Europe often result in decreased supplies in minor markets. The result is that prices stay firm in the minor markets. U.S. importers, however, are not as concerned with foreign exchange markets as are Japanese importers because the U.S. dollar is a major medium of currency exchange.

The majority (75%) of shrimp is consumed outside the home in the United States. Shrimp are sold in units of counts per kg (lb). For example, 16–20 means that there are 16–20 shrimp per pound. These sizes range from under 10 (giant) per lb to more than 300–500 (canned).

Product forms and packs are generally the same from species to species. Most shrimp are sold raw with the head off (green headless) and the shell on. Raw, without shells are referred to as "peeled." Heads-on, cephalothorax included, appear as the entire shrimp (known as "enteros" in Spanish). PUD (peeled, undeveined) shrimp have the vein, or digestive tract, intact. This vein varies in color from dark to light. P & D refers to peeled and deveined. PDI means peeled, deveined, and individually packed. Tail-on peeled refers to a product form in which the tail fin and an adjacent shell segment are left on. Tail-on round refers to shrimp with the tail on and are undeveined. Butterfly shrimp have been cut along the vein. These are also called split or fan-tail. "Western-style" refers to splitting the shrimp through the first four segments.

Cooked shrimp are usually sold IQF (individually quick frozen). These are often sold as P&D tail-on or P&D tail-off and shell on. Other forms include minced, canned, dried, and value-added (marinated, flavored, or breaded). The U.S. Department of Commerce has established standards for green headless and breaded shrimp.

Breading consists of two components: wet, adhesive batter and a dry, crunchy breading. The percentage of breading by weight is critical and is regulated by the Food and Drug Administration of the United States. Labeling standards require that breaded shrimp be more than 50% shrimp; lightly breaded, more than 65% shrimp; and imitation breaded products, more than 50% shrimp.

There are a variety of forms of breaded shrimp. Whole breaded can be tail-on or tail-off, are usually headless (although called "whole"), and are deveined if fewer than 70 count. Butterfly breaded are split partway on the vein side (dorsal) and spread open. Split breaded are completely bisected ("western" or "cowboy" style). Hand-breaded are labor intensive, expensive, but more attractive and are usually prepared tail-on. Machine-breaded is done either tail-on or tail-off. If the tail remains, it may or may not be breaded. Not breaded is referred to as "pinched."

Green headless shrimp are usually packed in 2.3 kg (5 lb) blocks (net weight). With ice, the total weight of the box is often 2.7–3.2 kg (6–7 lb). Blocks are packed in two styles: (1) layer or finger packed and (2) random jumble or shovel packed. IQF are individually quick frozen, usually in bags (1–30 lb), labeled to net weight, and are without glaze. Breaded shrimp are packed in boxes with a moisture-resistant barrier and are completely sealed.

Shrimp are usually dipped in solutions to either add weight or to serve as a preservative. Sodium tripolyphosphate (STP) is added on peeled and breaded shrimp to reduce drip loss (which maintains weight). The label must advise of use of STP. Sodium bisulfite is used primarily for shell-on shrimp to prevent melanosis ("black spot"). The limit is 100 ppm but is higher in Europe. "Ever-Fresh" (4-hydroxyresorcinol) is a naturally occurring, Generally Regarded as Safe (GRAS) compound but is expensive.

Over 0.5 billion kg (1.1 billion lb) of shrimp were imported into the United States in 2003 (Fig. 2.7), an increase of 18% over 2002 (Rosenberry 2004). The frozen shrimp category showed the greatest increase in sales, nearly 90% of the increase. The primary exporting countries of shrimp to the United States, in decreasing order of importance in 2003, were Thailand, Vietnam, China, India, Mexico, Ecuador, Indonesia, Brazil, Bangladesh, and the Philippines. Countries showing steady increases in export volumes to the United States include: Vietnam, China, India, Ecuador, Indonesia, and Brazil.

Although shrimp imports into the United States have been increasing, particularly since the mid-1990s, the average price of shrimp imported has been decreasing (Fig. 2.8). In fact, the average price in 2003 ($7.45/kg) was 33% lower than the 1994 average price ($11.13/kg). The lower prices of shrimp have resulted in greater sales and higher per capita consumption of shrimp in the United States. (Johnson 2003). The decreases in prices were greatest in product from Thailand and China, whereas prices were higher for shrimp from Mexico and India (Rosenberry 2004). Mexico and India ship a higher percentage of larger shrimp that bring a higher price. However, trade disputes and discontent on the part of both commercial shrimpers and shrimp farmers over the lower prices may lead to tariffs. Imposing tariffs on imported shrimp would likely put upward pressure on shrimp prices in the United States.

The shrimp industry continues to face challenges from environmentalist groups. Some groups allege that shrimp farms have had negative environmental and social impacts. It is unclear to what extent the actions of the environmentalist groups have had an effect on the overall market for shrimp. Nevertheless, it is likely that there will be continued pressure for the industry to continue to adopt environmentally friendly production practices.

The shrimp industry will also need to learn to adapt to a market position in which shrimp is regarded less and less as a high-value luxury good and more of a lower-priced good to be consumed more frequently. The challenge will be for growers to improve efficiencies to maintain profitability with higher volumes and lower prices.

SUMMARY

Economics is the study of how resources are allocated to meet the unlimited wants and needs of

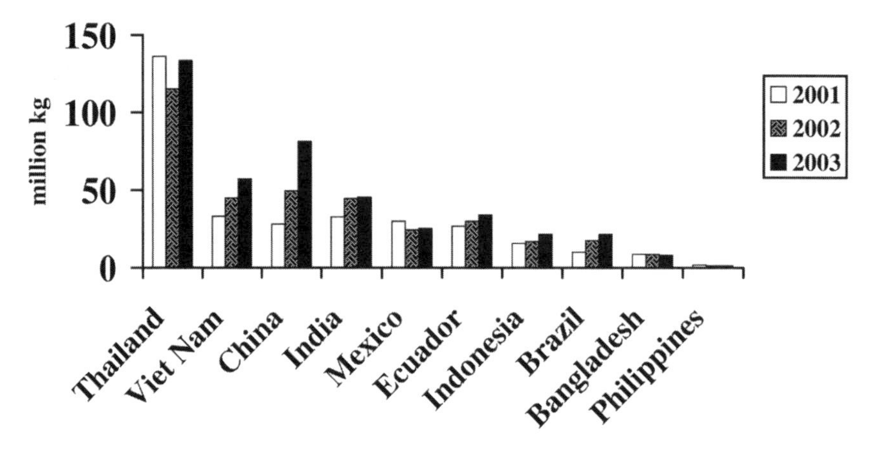

Figure 2.7. Increase in volume of shrimp imports into the United States, 2001–2003. (Source: NMFS 2004.)

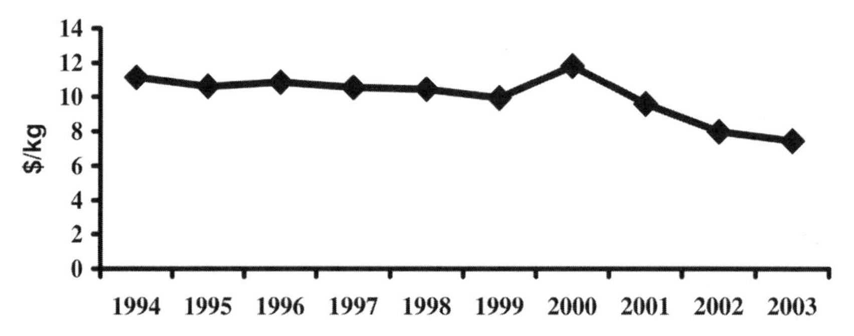

Figure 2.8. Average annual prices of frozen shell-on shrimp imports into the United States, 1994–2003. (Source: NMFS 2004.)

consumers. The forces of demand and supply interact to determine the equilibrium price and quantity. Demand and supply curves can change as the determinants of either demand or supply change. Elasticities provide further insight into the demand and supply relationships.

STUDY AND DISCUSSION QUESTIONS

1. Draw the graphs in Figure 2.4. Now draw graphs showing an increase in demand and then a decrease in demand and discuss what would happen to price and quantity with each. Provide an aquaculture example (different from those in the text) for each.

2. Draw the graphs in Figure 2.4. Now draw graphs showing an increase in supply and then a decrease in supply and discuss what would happen to price and quantity with each. Provide an aquaculture example (different from those in the text for each).

3. Draw graphs of what you think the following would look like:

   a. Perfectly inelastic demand
   b. Perfectly elastic demand
   c. Elastic demand
   d. Perfectly inelastic supply
   e. Perfectly elastic supply
   f. Inelastic supply

4. Draw a backward-bending supply curve and explain how the biological growth curve and exploitation levels can result in this type of supply curve.

REFERENCES

Anderson, J.L. 1985. Market interactions between aquaculture and the common-property commercial fishery. *Marine Resource Economics* 2:1–24.

Asche, F. 1996. A system approach to the demand for salmon in the European Union. *Applied Economics* 28:97–101.

Asche, F. 1997. Dynamic adjustment in demand equations. *Marine Resource Economics* 12:221–237.

Asche, F. 2001. Testing the effect of an anti-dumping duty: the U.S. salmon market. *Empirical Economics* 26:343–355.

Asche, F. and R. Hannesson. 1997. On the global integration of the markets for whitefish. SNF Report 98/97.

Asche, F. and F. Steen. 2000. Is salmon a substitute to whitefish? Paper presented at IIFET 2000, Oregon State University, Corvallis, Oregon.

Asche, F., T. Bjørndal, and K.G. Salvanes. 1998. The demand for salmon in the European Union: the importance of product form and origin. *Canadian Journal of Agricultural Economics* 46:69–82.

Asche, F., H. Bremnes, and C.R. Wessells. 1999. Product aggregation, market integration and relationships between prices: an application to world salmon markets. *American Journal of Agricultural Economics* 81:568–581.

Asche, F., D.V. Gordon, and R. Hannesson. 2001. Searching for price parity in the European whitefish market. *Applied Economics* 34:1017–1024.

Asche, F., K.G. Salvanes, and F. Steen. 1997. Market delineation and demand structure. *American Journal of Agricultural Economics* 79:139–150.

Bjørndal, T. and R. Tveterås. 1998. Production, competition and markets: the evolution of the salmon aquaculture industry. Discussion Paper No. 7/1998. Centre for Fisheries Economics, Norwegian School of Economics and Business Administration.

Bjørndal, T., K.G. Salvanes, and J.H. Andreassen. 1992. The demand for salmon in France: the effects of marketing and structural change. *Applied Economics* 24:1027–1034.

Carlton, D.W. and J. M. Perloff. 2000. Modern industrial organization. Addison-Wesley Longman, Inc., Reading, Massachusetts.

Clayton, P.L. and D.V. Gordon. 1999. From Atlantic to Pacific: price links in the U.S. wild and farmed salmon market. *Aquaculture Economics and Management* 3:93–104.

Csavas, I. 1994. Important factors in the success of shrimp farming. *World Aquaculture* 25(1):34–55.

Eales, J. and C.R. Wessells. 1999. Testing separability of Japanese demand for meat and fish within differential demand systems. *Journal of Agricultural and Resource Economics* 24:114–126.

Eales, J., C. Durham, and C.R. Wessells. 1997. Generalized models of Japanese demand for fish. *American Journal of Agricultural Economics* 79:1153–1163.

Gordon, D.V., K.G. Salvanes, and F. Atkins. 1993. A fish is a fish is a fish: testing for market linkage on the Paris fish market. *Marine Resource Economics* 8:331–343.

Jaffry, S., S. Pascoe, G. Taylor, and U. Zabala. 2000. Price interactions between salmon and wild caught fish species on the Spanish market. *Aquaculture Economics and Management* 4:157–167.

Johnson, H.M. 2003. 2003 annual report on the United States seafood industry. H.M. Johnson & Associates, Jacksonville, Oregon.

Johnson, A., C.A. Durham, and C.R. Wessells. 1998. Seasonality in Japanese household demand for meat and seafood. *Agribusiness* 14:337–351.

Kinnucan, H.W. and O. Myrland. 1998. Optimal advertising levies with application to the Norway-EU. In: IIFET 98 Proceedings (eds. A. Eide and T. Vassdal), p. 701–711. University of Tromsø, Tromsø.

Kouka, P.J. and C.R. Engle. 1998. An estimation of supply in the catfish industry. *Journal of Applied Aquaculture* 8(3):1–15.

Neira, I., C.R. Engle, and K. Quagrainie. 2003. Potential restaurant markets for farm-raised tilapia in Nicaragua. *Aquaculture Economics and Management* 7(3/4):1–17.

Rosenberry, B. 2004. World shrimp farming. *Shrimp News International*, San Diego, California.

Salvanes, K.G. and D.J. Devoretz. 1997. Household demand for fish and meat products: separability and demographic effects. *Marine Resource Economics* 12:37–55.

Tveterås, R. 1999. Production risk and productivity growth: some findings for Norwegian salmon aquaculture. *Journal of Productivity Analysis* 12:161–179.

3
Aquaculture Marketing Concepts

The purpose of this chapter is to help readers understand some key marketing concepts used throughout the rest of the book. This chapter is particularly useful as a quick review of marketing concepts. Detailed explanations and illustration of these concepts are provided in the following chapters, including specific examples and case studies that illustrate the application of the concepts to aquaculture business situations. The market synopsis on salmon outlines the development of an industry that has grown to have major impacts on the world supply of that species and has had a major effect on world trade.

WHAT IS MARKETING?

In simple terms, marketing covers all the processes that occur between the moment the product leaves the farm and when it is consumed by the end user. Aquaculture products must be harvested, transported, and assembled in adequate volume for resale. Many products are processed in some fashion before resale and consolidated by product form to provide volumes that are large enough to be traded and negotiated. Advertising programs are designed to increase demand for the product by communicating the attributes of the product and are included in the marketing process. Sales are also a part of marketing, but especially in today's complex economy, sales will not occur in the absence of the other marketing processes.

MARKETING PLAN

Every aquaculture business should have a well-defined marketing strategy defined in a written marketing plan. A comprehensive marketing plan includes an assessment of the current market situation, identification of opportunities and threats to the business, and a clearly defined marketing strategy. The marketing strategy developed includes a market summary, description of market demographics, market trends, market growth, analysis of strengths and weaknesses of the company, product offerings, specific objectives, financial analysis of the relative costs associated with these market objectives, and a monitoring and control plan. The plan should define the product, identify the buyers and sellers, and articulate the market rules. Chapter 12 includes more detail on developing marketing plans.

MARKET PRODUCTS

Selection of the specific product or products to be marketed is a key decision for the business. The company needs to effectively articulate what is new or different about the product and understand how consumers view their product(s) when compared to the competition. When products available on the market are virtually identical, the products are said to be homogeneous, but if products are significantly differentiated, they are said to be heterogeneous. For example, if the only tilapia product available on the market were a frozen fillet, then tilapia would be considered a homogenous product. However, if there are diverse tilapia products with different characteristics, such as battered, breaded, stuffed, dried, marinated, and canned, it is possible to distinguish or differentiate each specific product from the other. Any product or service offered to the market must be defined clearly and compared to competing products available on the market.

PROCESSORS

Processors add what is called form utility to raw farm products by processing live fish into more convenient product forms such as fillets, steaks, or nuggets. Processors may also add value to aquaculture products by providing transportation from farms to processing plants, and storing processed product in coolers and freezers until it is sold. Some processing companies even extend credit to farmers to help them finance their production operations. Processing plants may also provide transportation and may store the product until it is sold. Chapter 5 presents additional details on seafood and aquaculture product processing.

Some food processing industries have a dominant core of a few large firms that produce well-known brands, advertise, and have a strong influence on product price. National processors concentrate their selling efforts on innovation, quality, and other forms of nonprice competition. Others consist of a large number of smaller firms that process products under wholesaler and retailer private labels. These competitive fringe food processors rely largely on price competition for their success. Brands allow processors to differentiate products and certify product quality. Food processing firms are among the nation's leading advertisers of food products, and food products are the most heavily advertised consumer products.

The trend in food processing has been to consolidate into fewer, but larger, processing companies. This concentration is typically expressed as the share of the market controlled by the top food processing firms. As an example of recent trends, the top 20 food processing firms in the United States increased market share from 36% in 1987 to 51% in 1997 (ERS 2002).

Processing involves significant investment in facilities and equipment, and thorough planning is critical to select the most efficient size levels of processing plants. However, it is often difficult to determine the optimal number and size of plants. Plants will run efficiently when running close to full capacity because of the high proportion of fixed costs in building and equipment infrastructure. Supplies from fish farms may fluctuate due to the time of the year, changing feed prices, or availability of fingerlings or seedstock. Market demand for seafood affects the price, as will prices of other similar types of fish that may be caught from the wild or imported at lower prices. Fluctuations in supply will affect the plant's ability to operate near its capacity. Replacing a single, large plant with several smaller ones reduces some assembly and transport costs but may require sacrificing the operational efficiencies of large-scale centralized plants.

The major market channel for U.S. catfish farmers has been through processing plants. Between 1990 and 2001, 93% of catfish produced in the United States was sold to processing plants (U.S. Department of Agriculture 2001). U.S. catfish processing plants are specialized in processing catfish and have well-automated systems designed to handle large quantities of catfish. Although higher prices can be obtained in other marketing channels, processing plants are the only channel option that can absorb the production volume of the majority of catfish farms (Kinnucan et al. 1986).

MARKET OR DISTRIBUTION CHANNELS

Market channel decisions are some of the most important decisions made by a company. Marketing channels can be thought of as customer value delivery systems in which each channel member adds value for the customer. Examples of companies that have successfully identified and utilized a market and distribution channel as a key component of their overall business strategy include: (1) Fed Ex, in small package delivery; (2) Dell Computer, in sales directly to consumers; (3) Charles Schwab, in delivering financial services on the Internet; and (4) Caterpillar, in network, powerful support, and partnership with dealers. Their respective market channel strategies have led these companies to become dominant in their respective product categories. Additional information on market channels can be found in Chapter 6.

Most producers use some type of intermediary to get their product to market, thus forging a distribution channel. The use of intermediaries is most common when growers or fishermen do not have the resources to develop marketing capabilities in transportation and storage. The importance of intermediaries lies in enhancing efficiency of the distribution system and thereby reducing marketing costs. An efficient organization of the distribution system requires the performance of some key distribution functions. These include:

- Gathering information and conducting market research and intelligence that are important for marketing planning
- Promoting and advertising products
- Searching and contacting by finding and communicating with prospective buyers
- Matching products to buyer needs through grading, assembling, and packaging
- Negotiating price and other contractual arrangements in the marketplace
- Distributing products through transportation and storage
- Financing the costs of distribution
- Assuming some commercial risks by holding stocks

All the preceding functions need to be undertaken in any marketplace. Intermediaries perform these functions to create a supply chain and a total distribution system that serves customers. In some cases the food processor sells directly to the consuming public. Many food companies market directly to consumers during holiday seasons such as Christmas and Thanksgiving in addition to marketing through traditional intermediaries. Although the use of intermediaries requires the grower to give up some control, the contacts, experience, specialization, and scale of operation often allow intermediaries to offer the firm more than it can achieve on its own (Armstrong and Kotler 2003).

There are three basic forms of marketing channels for delivering products from the producer to the consumer (Fig. 3.1): conventional distribution channels, vertical distribution channels, and horizontal marketing systems. Conventional distribution channels typically consist of one or more independent producers, wholesalers, and retailers. Each is a separate business looking to maximize its own profits, not those of the system as a whole.

Vertical distribution channels are composed of producers, wholesalers, and retailers that are part of one marketing system. Typically, one member is strong enough to either own the others or wield enough power to insist on cooperation (Armstrong and Kotler 2003). Vertically integrated corporations are single business entities that control all phases of the distribution channel. Some vertically integrated systems are based upon contracts among independent firms at different levels of production and distribution. Franchises, retailer cooperatives, and wholesalers that organize voluntary chains of independent retailers integrate vertically through contracts. Members of a top brand, such as Wal-Mart, can administer or exert strong influence on suppliers through the company's size and power.

Horizontal marketing systems are those in which two or more companies at one level join together to pursue a new marketing opportunity by combining capital, production capabilities, and marketing resources. Kentucky Fried Chicken franchises in a Shell gas station create joint benefits from colocation of retail outlets. Companies may use different types of distribution channels to target different market segments.

E-commerce and other technological advances have led to a degree of disintermediation (bypassing intermediaries to sell directly to final buyers). Web-based sales allow sellers to capture the profit margins of the entire marketing chain. However, the seller must assume all the customer service, shipping, and advertising functions that are critical for success. Moreover, some types of products are less amenable to Internet sales than others. Examples of marketing channel systems for specific commodities are presented in Chapter 6.

TRANSPORTATION

Transportation is a marketing function that provides what is called "place" utility. In order to sell a product, the buyer and seller need to be able to physically make the exchange. The availability of adequate transportation alternatives can affect the type of product that can be sold, the quality of the product, the timeliness of deliveries, and the volume of product that can be moved, among others. The lack of transportation can result in lack of access to specific markets, can reduce competitiveness in preferred markets, and can restrict the growth and development of the business. The types of transportation available can affect storage requirements, inventory costs, and the location of processing plants. Transportation costs will affect marketing margins and food prices.

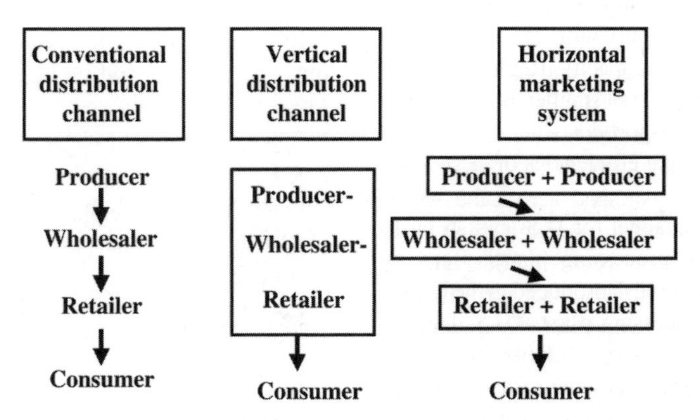

Figure 3.1. Various forms of distribution channels.

Improvements in transportation technologies have allowed for the emergence of complex, global markets for a wide variety of food products. Much of the fish and seafood consumed around the world is transported to other countries, other continents, or the other side of the world while maintaining freshness and quality.

Transportation is especially critical to marketing fish and seafood because of its perishable nature. Moreover, the diversity of types of aquaculture products and markets has resulted in a wide variety of methods of transportation used throughout the world for fish and seafood markets. Even today, while shrimp, salmon, and tilapia are air freighted around the world, fish and seafood are still transported by canoe (Ababouch no date) or boats with live holds (FAO 1978), bicycle (Kada 1997; Jagger and Pender 2001), and pickup truck (Jagger and Pender 2001).

Many fish and seafood products continue to be transported live. Within the United States, catfish and trout are hauled live on large hauling trucks equipped with liquid oxygen to processing plants or pay lakes. Baitfish species are hauled to wholesale distributors and reloaded on to smaller trucks for transportation to retail bait shops. Fingerlings of many species are hauled by truck from hatcheries to growout facilities that may be only a short distance away or a several-hour drive.

Fish fry are often shipped in plastic bags with oxygenated water. The bags are typically packed in Styrofoam boxes as insulation for temperature control.

Boats with wells are used in the salmon and other industries to transport live fish to processing plants (Schoemaker 1991). Mechanical aeration is used to maintain adequate oxygen levels. Boats with hulls with holes in the sides have been used in China (FAO 1978). The holes allow for water exchange. Live fish were reported to sell at prices 30–60% higher than fish on ice.

Shellfish must be kept shaded and cool in a humid environment (Schoemaker 1991). Prechilled oysters are shipped by air in Styrofoam boxes.

Processed fish may be transported to local grocery stores or restaurants or shipped across the world. Fresh fillets are chilled before packaging and packaged to avoid direct contact between fillets and ice. Refrigerated transport is required for shipping frozen fish products with temperatures maintained at −30 to −25° C for most products.

Fresh tilapia fillets are flown from Central America to markets in the United States, whereas frozen tilapia fillets are imported into the United States from China, Taiwan, and Indonesia. Careful coordination of harvesting, processing, and shipping times has been a key to the success of tilapia companies that export to the United States.

In developing countries, access to markets can be a critical problem, especially if there are few transportation alternatives. Poor road conditions, unreliable vehicles, and lack of ice can prevent aquaculture products from reaching those markets with the greatest demand for their product. Leyva (2004) developed an analysis of optimal markets for different sizes of tilapia farms located in different locations in Honduras. The mixed-integer transshipment mathematical programming model explicitly accounted for varying costs associated with different truck sizes, varying distances to various markets and seasonality of demand. The models were used to suggest recommendations for farmers on the most profitable cities and outlets to target.

WHOLESALING

The wholesaling process includes everything involved in selling products to companies that resell the products. Typically, wholesalers buy from producers and resell to retailers or other wholesalers. They operate buying offices, warehouses, trucking and delivery services. Businesses will choose to sell to a wholesaler often because the wholesaling company may be more efficient at selling and carrying out other marketing functions. Wholesalers promote products, build variety to meet customer demand, break bulk quantities into smaller lots for customers at lower prices, warehouse to offer adequate inventory, and also transport the product. Some wholesalers finance customers and suppliers with credit and bear risk related to title and theft, damage, spoilage, and obsolescence. A wholesaler may provide information to suppliers and customers on competitors, new products, and price development. They may help retailers train sales clerks, improve store layouts and displays, and set up accounting and inventory control systems. Retailers are focused on servicing their customers and cannot search out and deal with producers and processors to source all products. Processors are not in the best position to take care of the needs of retailers. Successful wholesalers step into this gap to facilitate coordination between processors and retailers. Additional information on wholesaler marketing can be found in Chapter 8.

Recent market trends have made it more difficult to distinguish between retailers and wholesalers. Wholesaler clubs and hypermarkets may be operated by retailers but perform wholesale functions.

Some large wholesalers such as Super Valu may perform retail functions. Rising costs combined with demand for increased services squeeze wholesale profit margins and require wholesalers to find ways to deliver more value to customers. Many large wholesalers are now expanding to operate on a global level.

BROKERS

Brokers and agents primarily buy and sell products and earn a commission on the selling price. Brokers and agents often specialize in a particular product line, but do not take title to the goods (Armstrong and Kotler 2003). Brokers are paid by the party that hired them and do not carry inventory. Agents, on the other hand, represent either buyers or sellers on a more permanent basis. Some agents represent two or more manufacturers of complementary product lines. Selling agents have contractual authority to sell the entire output of a manufacturer while purchasing agents have a long-term relationship with buyers and make purchases for them. Commission merchants take a load of commodities to a market, sell it for the best price, deduct a commission, and send the balance back to the growers. Commission merchants are most often used by farmers who do not belong to a growers' cooperative.

RETAILING

Retailing includes everything involved in selling products to the end consumers. Thus, retailing includes grocery stores, restaurants, and direct sales to consumers. Food retailing is one of the most expensive parts of the food marketing chain, but retailers have considerable market power in the food industry. Retail businesses take many different forms, and aquaculture marketers should carefully understand the differences to identify potentially profitable marketing alternatives. Shopping patterns and demographics change rapidly. Additional detail on retail market trends can be found in Chapter 9.

There are a number of examples of successful retail fish marketing concepts. For example, fish and chip shops in the United Kingdom (UK) face severe competition from other fast food retailers but continue to be popular.

Supermarkets have become the primary form of food grocer in the United States and in many other countries. For example, in many European countries the share of fish being sold by supermarkets has increased from less than 20% in the late 1980s to more than 60% in the late 1990s (Guillotreau 1998). Supermarkets tend to be fairly large grocery stores that sell high volumes at low cost and are organized as self-service businesses. A supermarket can be described as a full-line, departmentalized, cash-and-carry, self-service food store. Supermarkets were products of growth in suburban areas and became an American symbol of innovation, affluence, abundance, efficiency and the good life. Chain stores represent both a horizontal affiliation of retail stores and a vertical affiliation of food retailing, wholesaling, and sometimes processing businesses. Chain stores developed to take advantage of the efficiencies to be gained through large-scale buying and selling. The food chain store movement triggered competitive reactions on the part of independent retailer and service wholesalers, who developed their own joint action (retailer-owned cooperative, wholesaler, and wholesaler-sponsored voluntary retail chains).

FOOD GROCERS

Supermarkets have experienced slow sales growth in recent years with slower population growth and increased competition from convenience stores, discount food stores, superstores, and increased consumption away from home. Fresh seafood departments have been used to attract customers away from competing outlets. Marketbasket pricing provides the retailer latitude in pricing any one food. Loss-leaders can attract business without each individual item's being priced based on wholesale prices.

The growing market share of multiple retail stores (supermarket and hypermarkets) in food distribution has also changed patterns of production, supply, and distribution. Hypermarkets are stores with up to 200,000 square feet of selling space in groceries, sporting goods, auto supplies, and so on. A warehouse foodstore eliminates some services and frills to reduce retail costs and prices. Superstores are larger supermarkets (up to 60,000 sq. ft.) that seek to supply all the products, food and nonfood, that consumers want. Superstores are growing at the rate of 25%/yr whereas supermarkets are growing at only 1%/yr. Wal-Mart has more than 720 supercenters with 2/3 of the total supercenter volume. Warehouse or wholesale clubs sell annual membership fees, often $25 to $50, and then sell a variety of grocery and nonfood items at deeply discounted prices. Sam's Club and Costco are examples. The weakening economy of recent years has been accompanied by increased growth of warehouse club sales while traditional supermarket sales have slowed. Supermarkets, superstores, and wholesale clubs all handle a variety of aquaculture products. Convenience stores (small stores located near residential areas) and specialty stores (stores that sell a narrow product line) rarely handle aquaculture products. Other common types of retailers such as department stores, discount stores, and off-price retailers other than warehouse clubs do not generally sell food products.

There has been an increase in cooperative organizations among retailers. Voluntary chains such as IGA (Independent Grocers Alliance), for example, are retailers that have formed an association to purchase in bulk and to merchandise jointly. Chain stores are companies with two or more retail outlets. Larger in size than independent grocers, they can purchase in bulk to benefit from lower prices. Associated Grocers is a retailer cooperative that has established a central buying organization to conduct joint promotion efforts. Other, nongrocer examples of retail organizations include corporate chain stores (Pottery Barn), franchises (Subway, 7-Eleven), and merchandising conglomerates such as Dayton-Hudson.

Livehaulers

Livehaulers buy live fish from producers and function as middlemen. Livehaulers market fish to a variety of outlets including processing plants, fee fishing operations, community fishing ponds, retailers, or other outlets. Sales to livehaulers were 3.2 million kg/yr of food-sized channel catfish in 2002, and farmers can receive ± $0.11/kg as compared to the processor price (Heikes 2003).

Restaurants

Restaurants are also retail outlets that operate in an extremely competitive environment. The away-from-home food consumption market is very different from the home food preparation market. Prices typically are more stable in the restaurant trade, and there is a higher ratio of marketing services to food. In restaurants, 45–65% of the price charged is in nonfood costs as compared to supermarkets in which only 20% of the costs are nonfood costs. Food service managers are more concerned with standardization, portion control, and labor-saving foods.

DIRECT SALES

The complexity of market channels can be avoided by moving smaller quantities directly to the end consumer without any intermediaries (Palfreman 1999). However, for direct sales to be feasible, the grower must develop the capacity to transport and possibly store the product.

In the U.S. catfish industry, direct sales accounted for 1.2 million kg of food-sized catfish sold, or 0.4% of all food-sized fish sold (U.S. Department of Agriculture 2001). Much of these direct catfish sales occur through fee fishing operations. Fee fishing operations either charge a fee for customers to fish in their ponds or charge by the unit weight of fish caught. Most successful fee-fishing businesses provide picnic areas, concession stands, bait, on-site dressing, fishing piers, and ice (Cichra et al. 1994). Locations close to a large customer base and constant restocking of large fish are important to the success of fee fishing businesses (Engle 1997). Inhabitants of local and nearby towns accounted for 88% of customers in a Kentucky study (Cremer et al. 1984). Direct marketing to outlets other than processing plants results in higher prices, but retail outlets are relatively limited in size and not a feasible option for the entire crops of large-scale farms (Wiese and Quagrainie 2004).

Producers may form cooperatives to assemble and sell produce directly to consumers. There have been a number of attempts to develop aquaculture cooperatives. As with many other forms of business, the failure rate typically is high. A successful cooperative must have a strong, skilled manager who is viewed as being fair to all members and who has the marketing and business acumen to position the cooperative's products competitively.

PROFIT MARGINS

As a product moves through each market channel, the price increases at each stage in accord with the value added to the product. The amount added, or the marketing margin, is affected by the time of sale and the price paid for the raw product. Government price controls, producer organizations, types of products, and level of market concentration will all affect the amount of the marketing margin.

Although the intermediaries, or middlemen, of the market channels are frequently called such names as "coyotes" and viewed as abusive of growers, legitimate costs are incurred as value is added to the product by intermediaries. In addition to storage, packaging, and transportation costs, time spent by the intermediary to identify buyers and coordinate with suppliers also has a value. Fish farmers and fishermen often forget to budget a cost for their time spent in marketing functions. This cost is referred to as an opportunity cost. Opportunity costs are defined as the cost of an input in its next best alternative use. Owners should value their time spent in marketing activities at their true opportunity cost (what they could earn working for someone else) to ensure that prices charged reflect all costs.

Retailers typically seek either high markups or high volume but rarely both. Specialty stores typically select high markup on low volumes, whereas supermarkets have lower markups on higher volumes. That said, retail grocer markups for seafood tend to be higher than for other store products and have been even higher over the last several years. In a survey of seafood retailers conducted in 2002, 45% of the respondents indicated that retail margins for seafood were over 30%, and another 21% had margins of 25–30% (Seafood Business 2001–2003). These margins were higher than in previous years, and seafood was being used to compensate for lower margins in other food sales categories.

ECONOMIES OF SCALE IN MARKETING

Economies of scale refer to decreasing costs with increasing size of the business. This is particularly true with the growth of large supermarket and other chains that take advantage of the large economies of scale in food distribution (Asche 2001). These economies of scale allow for productivity growth to occur throughout the value chain for fish (Zidack et al. 1992).

Economies of scale in marketing seafood are one of the reasons for consolidation among seafood suppliers (Seafood Business 2001–2003). Mergers and acquisitions occurred frequently through the late 1990s and into the 2000s. Seafood supply companies such as ConAgra Seafood Companies, StarKist Seafood Company, and Marine Harvest all have annual sales exceeding $1 billion. The second tier of supply companies include the Red Chamber Group Company, Trident Seafoods Corporation, Nippon Suisan, and the Pacific Seafood Group, with annual sales that range from $550 to $680 million. Aquaculture supply companies must compete with these large conglomerates in the seafood marketplace and with the marketing economies of scale that come from the ability to supply a wide variety of seafood products.

SUPPLY CHAIN MANAGEMENT

Supply chain management is a term that has emerged to refer to the complexity of efficiently managing the flow of goods and information from suppliers to resellers and final users. Improved logistics associated with tracking inventories and moving product efficiently through market channels have provided a mechanism for managing the entire supply chain.

Supply chain management involves far more than just the marketing logistics, or physical distribution, of product to consumers. Supply chain management is more of a customer-centered approach that works backward from end consumers in the market to the producer and back to the resources that are used as inputs. Efficient supply chain management can result in better service to customers or lower prices that may offer a competitive advantage to the business. It may also result in cost savings to both the business and its customers. Moreover, retail trends toward increased product diversification have made supplying large customers more complex. Information technology provides tools to manage supply in ways previously unknown, such as with point-of-sale scanners, uniform product codes, satellite tracking, Web-based systems, electronic orders, and payments.

Although the early stages in the value chain tend to receive more attention (Asche 2001), in modern retail markets, buyers increasingly demand that products can be traced to determine origin and history. Hazard Analysis Critical Control Point plans are expected and required. HACCP regulations require the development of a HACCP plan that identifies potential food safety hazards in processing and develops procedures to minimize food safety problems (HACCP) (National Fisheries Institute 1997).

PRICING SYSTEMS

PRICE DETERMINATION

In a purely competitive market situation, prices typically are determined by the interaction between supply and demand in the market (see Chapter 2 for more details on equilibrium prices). One of the best examples of purely competitive pricing in seafood markets is the fish auctions that continue to operate in the United Kingdom and other countries. In auction markets, potential buyers bid for various lots of fish, and prices are bid either upward or downward, depending on the particular auction's guidelines, until buyer and seller agree on a price (Palfreman 1999).

In corporate settings, however, pricing decisions are made based on other processes. Administered pricing describes all pricing in which a seller or buyer announces a nonnegotiable selling (buying) price (Breimyer 1976). Prices paid by catfish processing plants to catfish farmers are administered prices. In the case of catfish processing plants, these prices are based on wholesale prices received by the plants from brokers and food service distributors.

Other companies may use cost-plus pricing in which a set margin is added to costs of production to determine selling price. In cost-plus pricing, an arbitrary amount of profit is added to the production costs. For highly valued products for which few substitutes exist, this pricing mechanism may be effective. Some companies base their pricing on competition-oriented pricing. Often this is done based on prices for similar and competing goods. In this type of strategy, there typically is one price leader with a dominant market share. Other companies set prices compared to the leader. Regardless of the pricing mechanism, the price established for a particular product should be established based on in-depth understanding of the targeted consumers, their attitudes and preferences, and where the product is to be positioned within the price-quality matrix.

Other companies use demand-oriented pricing. This is especially true for customers with different quality standards. Sales of higher-priced species are supported by advertising the quality attributes to those population segments willing and able to pay higher prices for a high-quality product. Lower-cost species are sold by emphasizing the corresponding lower price to market segments that seek out more inexpensive types of seafood.

Psychological pricing involves establishing prices that either look better or convey a certain message to the buyer. An example would be to charge $6.58/kg ($2.99/lb) instead of $6.60/kg ($3.00/lb) to make the product appear to be more of a bargain. Perceived-value pricing positions and promotes the product based on nonprice factors such as quality, healthfulness, environmental sustainability, or prestige.

Some temporary pricing strategies are used to increase sales and market share. Skimming involves introducing the product at a relatively higher price for more affluent, quality-conscious consumers and then lowering the price as the market becomes saturated. Discount pricing offers customers a reduction from advertised prices for specific reasons. Discount coupons in the newspaper or radio ads may attract new customers. With loss-leader pricing, a portion of the product is offered at a reduced price (below cost) to attract customers. This is used to attract new customers to farmers' markets or supermarkets. Price-penetration is a pricing strategy in which a low price is charged to gain increased market share.

Marketing Margins, Marketing Bill, and Farm-Retail Price Spreads

The marketing margin is that portion of the consumer's food dollar that goes to businesses engaged in marketing (Armstrong and Kotler 2003). Another way to view the marketing margin is as the difference between what the consumer pays for food and what the farmer receives. It must be remembered that this difference includes costs associated with all marketing functions performed. Thus, the price that the consumer faces includes both the farm price and the marketing price of food. These two prices may not always move in the same direction.

The size of the marketing margin cannot be used to measure efficiency. Small margins exist with shorter marketing chains, but these are not always the most efficient. A fish farmer who sells directly to the public may have a small marketing margin, but it may not be efficient for the farmer to make a large number of deliveries to satisfy his or her customers.

The size of the marketing margin reflects the marketing costs involved, not the number of intermediaries. Marketing costs include profit to each intermediary, but although middlemen may be eliminated, the costs of the required marketing functions minus the intermediary profit will still exist. Eliminating middlemen will not decrease the marketing margin if the farmer cannot perform the marketing functions as efficiently as the middlemen. Increased marketing margins also increase the retail value and price of food.

The food marketing bill is the difference between total consumer expenditures for all domestically produced food products and what farmers receive for equivalent farm products (Kohls and Uhl 1985). It provides an aggregate view of the division of consumer food expenditures between farmers and food marketing businesses.

The farm value share of the food dollar has declined continuously over time. The marketing bill is now more than four times larger than the farm value (Fig. 3.2) and includes all transportation, processing, and distribution of foods as well as foods consumed both at home and away from home. The decline in the farm value share does not necessarily mean that farmers' welfare has declined. Relative costs of production and returns are necessary to evaluate the economic and financial health of the farm sector. The increasing share of the marketing bill reflects market trends toward more complex processing, and distribution systems related to increasing food expenditures away from home and the growth in convenience foods for at-home consumption. Of the marketing bill, labor is the largest portion, followed by packaging, profits, transportation, rent, advertising, depreciation, business taxes, energy, interest, and repairs.

Figure 3.2. Food marketing bill, 2000. (Source: USDA-ERS 2004).

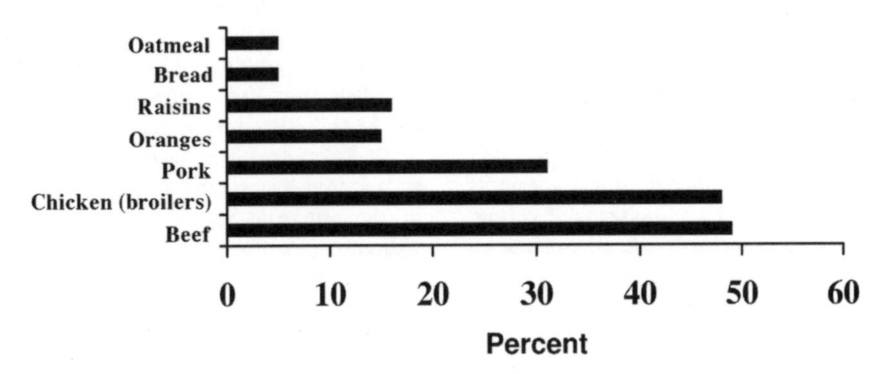

Figure 3.3. Farm-retail price spreads (calculated as the farm value share of the retail price), 2000. (Source: FAO 2004.)

Changes in the farmers' share of the marketing bill can occur due to changes in supply and demand at the farm or retail levels or changes in marketing costs. Commodities for which the marketing agencies provide a relatively large share of utilities occur for products that have a lot of processing, are highly perishable, are seasonal, have high transportation costs, and are bulky in relation to product value.

Assembly market functions tend to be a small portion of the marketing bill. Products such as seafood that are frequently marketed fresh tend to have larger farm values, with the reverse being true for highly processed products. Transportation and wholesaling costs tend to be higher for more perishable goods.

Whereas the marketing bill is concerned with expenditure margins, the farm-retail price spread is concerned with price margins for individual foods. It measures the gross return per unit to food marketing, or the profits and costs of all marketing functions. The spread is the difference between the retail price per unit and the farm value of an equivalent amount of food sold by farmers. There is wide variation by different food crops. Figure 3.3 shows that the farm-retail price spread (calculated as the farm value share of the retail price) for beef is nearly 50% whereas that of oatmeal and bread is only 5%. Farm-raised aquaculture products would fall somewhere in between, with values that would range somewhere between 15 and 20%.

Pricing at Different Market Levels

Elasticities can vary for different market levels. (Chapter 2 presents information on how elasticities are calculated and gives additional detail on interpretation.). Kinnucan et al. (1988) estimated demand to be elastic at the processor level but inelastic at the farm level for U.S. catfish. Elasticities have important implications for pricing strategies. If the demand for a product is inelastic, increasing price will result in greater total revenue. However, if demand is elastic, increasing price will result in lower total revenue.

Most estimates of elasticity of seafood products show elastic demand (Anderson 2003). For example, cod (Brooks and Anderson 1991), flounder (Brooks and Anderson 1991; Wessells and Wilen 1994), salmon (Kinnucan and Myrland 2002; Herrmann et al. 1993; and Wessells and Wilen 1994), and catfish (Lambregts et al. 1993) were shown to be price elastic at the retail level. Only imported shrimp (Keithly et al. 1993) and tuna at the retail level (Wessells and Wilen 1994) were shown to be price inelastic.

PRICE BEHAVIOR, TRENDS, AND FLUCTUATIONS

Prices react to a variety of different forces, shocks, and events, some long-term and others short-term. Fish and seafood prices tend to be more volatile and exhibit greater fluctuations than do some other, less perishable, types of products. Shortages of certain species of fish, whether in the off-season or due to overfishing, will tend to drive prices up, whereas increased supplies from aquaculture production or during the peak fishing season for that species will tend to drive prices down. Weather disasters may affect prices of products raised in affected areas. Food scares or reports of contamination in aquaculture growing areas may cause demand to decrease and prices to drop as a result.

Longer-term trends in prices will be affected by changing trends in demand as well as in supply. The increased interest in fish and seafood consumption is driven, in part, by recommendations to eat more fish and seafood as part of a healthful diet. In the United States, per-capita beef consumption has declined, whereas per-capita consumption of broilers has increased (Fig. 3.4). Increasing per-capita consumption, combined with a world population that continues to grow, will increase overall demand and put upward pressure on price. However, the increased supplies of shrimp and salmon that have come from aquaculture production have tended to drive prices of those products downward over time.

GEOGRAPHIC MARKETS

There are great differences in food consumption patterns among regions. In the United States, for example, catfish is a preferred fish species in areas along the Mississippi River, whereas trout is preferred in the Rocky Mountain states, yellow perch in the Great Lakes region, flounder in the mid-Atlantic, and carps in Asian communities. In Peru, consumers in the Amazon Region prefer freshwater fish, whereas consumers in Lima prefer marine fish (Neira et al. 2003).

INTERNATIONAL TRADE

All the factors that affect demand and supply in each country will play a role and interact in the international market. In addition, national regulations in each country and of international organizations will affect the international flow of goods and services.

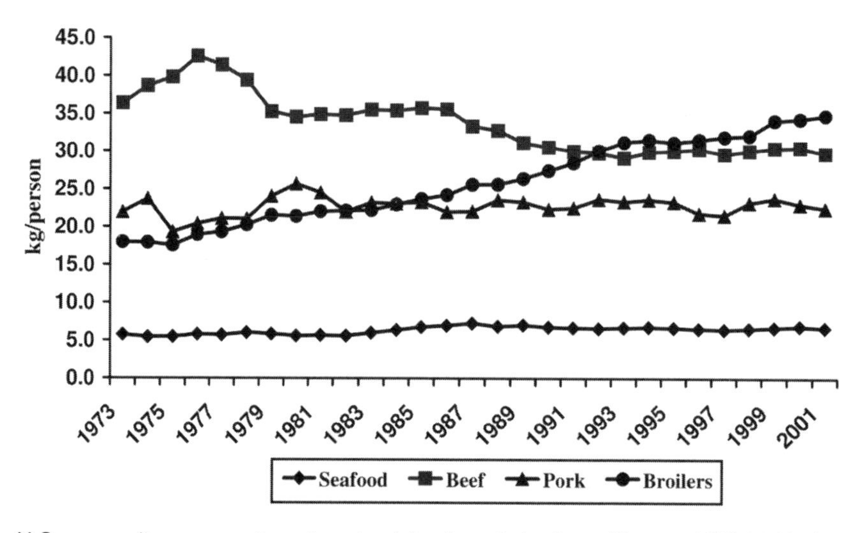

Figure 3.4. U.S. per-capita consumption of seafood, beef, pork, broilers. (Source: USDA 2004.)

The principle of comparative advantage is one of the most important principles underlying trade. Comparative advantage indicates that if free trade conditions apply, some countries will specialize in production of the commodities that can be produced relatively most efficiently in that country. Thus, price ratios among countries guide the flows of trade among those countries. A country generally will be better off importing a commodity that is produced in other countries at a lower cost than the commodity can be produced at home; that is, it is cheaper to import a commodity than produce it at home.

However, trade policies frequently are developed to "protect" domestic industries from competition from similar products imported into that country. A decision must be made based on the expected benefits and costs associated with free trade or protectionist policies. Protectionist policies may be based on raising tax revenue, supporting producers' income, reducing consumers' food costs, attaining self-sufficiency, or countering interventions of other trading partners. Protectionist policy instruments are numerous and include those restricting quantities that can be imported (quotas), increasing the price of the imported product (tariffs), encouraging export (export subsidies), controlling exchange rates, and supporting domestic prices through the use of price premiums, marketing boards, supply quotas, commodity programs, and so on. Export subsidies increase the share of the exporter in the world market at the cost of others, and tend to depress world market prices and may make them more unstable because decisions on export subsidy levels can be changed unpredictably (Pearson and Sharma 2003).

International trade research generally shows that the volume of international trade will be greater if trade policies are reduced or eliminated. This is referred to as trade liberalization. Free trade can raise aggregate economic efficiency (Suranovic 1997–2004). This increase in economic efficiency can include benefits from increased production efficiencies that result in producing more with the same amount of resources and providing more different types of goods and services that satisfy more consumer needs. Differences in how individuals seek profits along with differences in price will result in efficient trade under free trade conditions.

However, free trade will result in losses for some people, and protection from international competition may benefit some countries. Some groups may lose because it is difficult to quickly change investments from one industry to another in the short run, or in some cases even in the long run.

Although free trade appears to offer many economic benefits, people who are not trained in international economics tend to not favor free trade policies that relate to imports to their own nation. No major economic nation allows completely free trade. Companies that seek to export must learn the details of quotas, tariffs, subsidies, inspections, and certifications required by each country. Indeed, many trade barriers exist in many countries. Arguments often heard for supporting trade barriers include: (1) prevent countries from dumping products below production costs in other countries to gain market share; (2) protect farm programs; (3) food self-sufficiency; (4) manage the national economy; (5) maintain employment; (6) stabilize the industry; (7) the domestic industry hurt is an infant industry; (8) presence of international monopolies; (9) international politics; and (10) national security (Rhodes 1993; Suranovic 1997–2004). Additional detail on international trade is found in Chapter 10.

PRODUCT STORAGE

The timing of the production of food commodities does not always coincide perfectly with demand for those commodities. To meet consumer demand for products typically will require some type of storage.

Working inventory is necessary for an efficient marketing chain because, without stocks stored, there will be disruptions in supply. Storage is necessary for products that are harvested in a short time but are consumed throughout the year. Carryover stocks are those that are left from one marketing year to the next. Some farmers will store stocks of a product to wait for a higher price; these stocks are considered speculative stocks.

Processing plants often provide an important storage function in the supply chain of a product. With fish processing, unlike some other commodities, processors typically store processed volumes, and farmers "store" raw material, that is, live fish, in ponds until it can be sold.

There are a number of costs associated with storing products. The direct storage costs will include such items as repairs to storage facilities, depreciation, insurance, and utilities to maintain optimal temperatures. There is also a cost that represents the interest on the financial investment in the product while it is in storage. This cost is incurred whether the grower actually borrowed money or not. If the product deteriorates in quality during the storage period, another cost is incurred. If consumers prefer fresh product and will pay higher prices for fresh than frozen product, then storage incurs the costs of the price differences. There is also the risk that the price of the product may decline while in storage, and that shrinkage may occur that will increase the costs of the product.

MARKET POWER

Market power is the ability to affect the behavior and performance of exchanges in the marketplace to the advantage of the particular firm. It is often expressed as the ability to affect prices, but can represent influence over marketing functions, product flows, quality, or others.

Some growers choose to form marketing cooperatives to increase market power. Cooperatives are businesses that are owned and controlled by those working in it (Palfreman 1999). Membership is open to all employees and each member has one vote, irrespective of shareholdings. Profits typically are shared according to agreed-upon rules. In most countries, cooperatives must be registered. Palfreman (1999) points out that cooperatives may have lost ground in the UK because they resist efficiency-enhancing change such as computerized buying and selling. In Europe, Fish Producers Organizations (FPOs) are cooperatives but are also companies limited by guarantee. They are backed by a group of subscribers who guarantee their debts.

Few studies have been conducted on market power in aquaculture. Early research indicated that in certain areas, the U.S. catfish industry may exhibit monopsonistic (one buyer) control (Kinnucan et al. 1986). In west Alabama, for example, when only one processor existed, an imbalance in market power between catfish producers and processors may occur that could result in lower prices paid to producers. However, the number of new entrants into the industry might indicate trends toward a more competitive structure.

ADVERTISING AND PROMOTION

Promotion of a product is a form of communicating the product's attributes to prospective consumers. Ultimately, it is the product itself that communicates with the consumer, and the consumer decides whether the product meets his or her expectations. New products typically need to draw upon consumer perceptions through brand image, quality marks, labels, reputation of suppliers, and other point of sale information and support to introduce the new product to the consumer. For a consumer to reach a decision to purchase a new product, he or she must pass through the stages of: awareness, interest, favorable perception, and evaluation (Marshall 1996).

Paid promotions are referred to as advertising. Advertising includes various media such as print media, radio, T.V., and other forms. Advertising reaches large numbers of possible consumers and subjects consumers to repeat messages. The type of advertising selected depends upon the stage of the life cycle. A new product that is being introduced will require informative advertising. After competition increases, persuasive advertising becomes more important, to convince consumers of the benefits of one particular brand. Mature products require reminder advertising so that consumers do not forget about the product.

Advertisements should include a headline, picture, text, and where to buy it. Not all advertising programs must be costly. There are low-cost ways to advertise products. Bold, funny, and striking graphics on tee shirts or in stores can be very effective. Large quantities of the product can be donated for people to try to spread news by word of mouth. Using creative company and product names such as Ben and Jerry's Ice Cream can be effective advertising. Some businesses even invite tourists to visit and promote their business as a tourist attraction.

Sales promotions refer to short-term incentives to encourage purchase of a product. Point-of-purchase displays, premiums, discount coupons, specialty advertising, and demonstrations can all be used to promote sales. Sales promotions are short lived but attract consumer attention and may be used to boost lackluster sales. Sales refer to building customer relationships with the expressed purpose of making sales. Personal selling is most effective to create preferences and purchase actions. In sales, there are two fundamental rules of thumb: (1) never promise more than you can guarantee, and (2) never deliver less than you guaranteed. Obtaining favorable publicity for a product to build a favorable corporate image is called public relations and often will take the form of press releases and special events.

An increasingly common problem in marketing communications is that various groups within companies are not well organized or coordinated. Paid advertisements may send a message that is not well supported by a price promotion or by the label. Strong brand identification comes only from seamless coordination and reinforcing of images and messages.

PRODUCT GRADES, QUALITY, AND MARKETING IMPLICATIONS

Product attributes such as color, taste, aroma, texture, size, and shape can be combined in an infinite number of combinations. Product quality is a subjective evaluation of the value of the particular combination of attributes possessed by a specific product. Consumers perceive quality not only in terms of the sensory attributes such as taste but also in terms of appearance, nutritional value, and safety of the product.

Quality standards can be used to sort varied mixes of product categories into uniform categories. This grading process can result in homogeneous product categories, with a pricing structure that conforms to buyer and seller preferences for the various bundles of attributes represented by each product grade.

The establishment of standards (commonly agreed-upon yardsticks of measurement) to sort agricultural products into grades can simplify marketing and reduce marketing costs. Producers can charge price premiums for higher-quality products. Product grades provide clear product information to consumers, and consumers benefit as they select products more closely aligned with their needs. Although grading can be done at any stage of the marketing chain, food grades tend to operate mostly at the wholesale, not the retail, level.

One of the most critical steps in developing a food grading system is to select the criteria to be used to judge the adequacy of standards. Standards should be based on those characteristics considered most important by consumers. For standards to be successful, the standards should be those that can be measured and interpreted accurately. Individual grades that exhibit a great deal of variation in quality within a grade will reduce the usefulness of grading. The terminology used to identify grades must be understood clearly by consumers. Standards need to also capture a significant portion of the average production. Grading costs must be reasonable. The ultimate test is adoption in the market place.

There are a variety of problems associated with establishing grades. Key issues include types of tolerances and what terms should be used to identify grades. Positive terms are typically selected rather than those suggesting an inferior product.

Research has shown that consumers may not readily discriminate among different grades and may not be willing to pay price premiums for higher grades. This can be a problem particularly if grades were viewed as convenient for traders but were not consumer oriented. Confusion can occur between grades and with federally required inspections related to food safety.

Farmers that produce the highest-quality product gain the most with grading systems, sometimes at the expense of farmers that produce lower-quality products. Typically, the larger, more specialized producers are the most receptive to developing grades. Programs that establish product grades can result in raising standards and quality across an industry, as producers seek to gain higher prices.

Large chain retailers benefit from grading because it simplifies their procurement decisions. Smaller processors also benefit because grades allow them to supply larger market outlets. Larger plants may be opposed to grades because federal grades may compete with their own brands. Grades may also result in decreased market concentration because it allows smaller packers to compete in the market.

There are a number of examples of product grading in the seafood market. In the United States, the U.S. Department of Commerce has established standards for grades of fishery products that range from whole-dressed to frozen minced blocks to fillets to breaded products (U.S. Department of Commerce 2004). Individual tuna fish, for example, are assigned a grade that accompanies that fish through the market chain to its final sale (Kaneko et al. 1996). Different grades of tuna are sold to different market niches. Tuna grading is done subjectively by visually inspecting the appearance and directly sampling a small section of fish muscle. The price of fresh tuna can range from $1.10/kg ($0.50/lb) to $121/kg ($55/lb) depending upon the grade assigned. There are four basic grade distinctions for the Japanese and U.S. markets: Grade #1 has bright red muscle, firm texture, clear flesh, and little fat; Grade #2 is red, firm, with some translucency, and no fat; Grade #3 has some red but some brown muscle, is firm and opaque, with no fat; and Grade #4 is brown and gray, soft, and opaque (Ledafish 1996). The top grade (#1) is sold for high-end Japanese sashimi, Grade #2 is sold in lower-end Japanese and Hawaiian sashimi; Grade #3 is sold in lower-end restaurants in the United States; and Grade #4 is either canned or frozen. European markets use Grades #2 and #3.

Another example of product grades in seafood is for frozen, raw, breaded shrimp (U.S. Department of Commerce 2004). "U.S. Grade A" is a product that when cooked possesses good flavor and is rated at 85 points and above. "U.S. Grade B" is rated at 70–84 points, and "Substandard" product fails to meet the standard of "U.S. Grade B."

AQUACULTURE MARKET SYNOPSIS: SALMON

The global supply of salmon has increased rapidly over time, largely due to the increase in farm production of salmon (Fig. 3.5). Salmon culture technologies were originally developed in hatcheries in Europe, Japan, and the United States to produce salmon for restocking overfished natural populations.

Culture technologies began to be adopted successfully on a commercial scale in the 1960s (Anderson 2003). Since then, the farm-raised salmon industries primarily in Norway, Chile, the United Kingdom (mostly Scotland), and Canada have experienced rapid and dramatic growth (Figs. 3.6 and 3.7). By 2002, farm-raised salmon production constituted 1.8 million MT, 69% of the total world supply (in terms of weight).

Norway has continued to lead the world in production of farm-raised salmon (Fig. 3.8). Chilean production has demonstrated the most rapid growth in recent years and has replaced Norwegian product in the U.S. market (see Chapter 10 for details on the legal actions that led to this). Canada and the United Kingdom have also demonstrated growth over the last decade. Much of the growth in Chilean imports into the United States has been in the form of value-added fillets (Bjorndal et al. 2003). Salmon imported from Canada into the United States is mainly a round (headed and gutted) product. Although Denmark appears in statistics for salmon exports, these result from the re-export of fresh or smoked product produced from fresh salmon imported from Norway (Anderson 2003).

There are a number of different salmon species sold on the world market, and several of these are cultured on farms. However, the aquaculture indus-

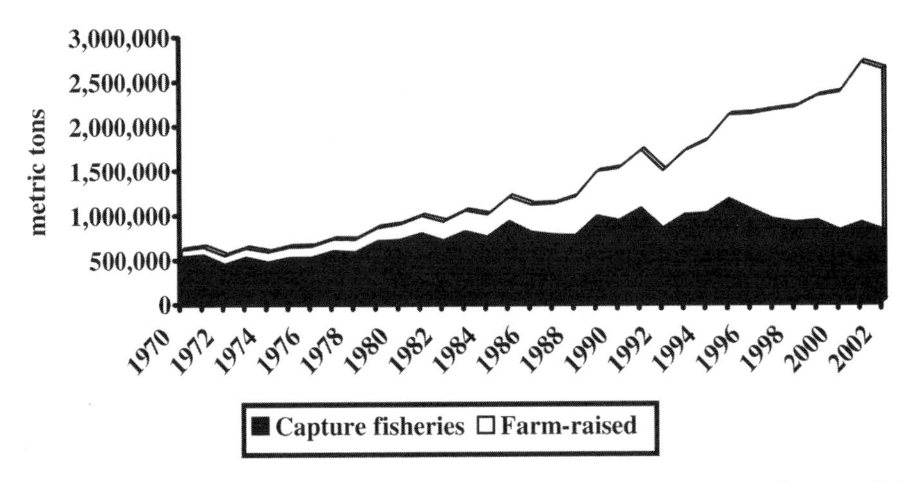

Figure 3.5. World aquaculture production and capture fisheries of salmon, 1970–2002. (Source: FAO 2004.)

Figure 3.6. Salmon farm in Scotland. Photo courtesy of Dr. Joann Sadler.

Figure 3.7. Salmon farm in Vancouver Island. Photo courtesy of Dr. Joann Sadler.

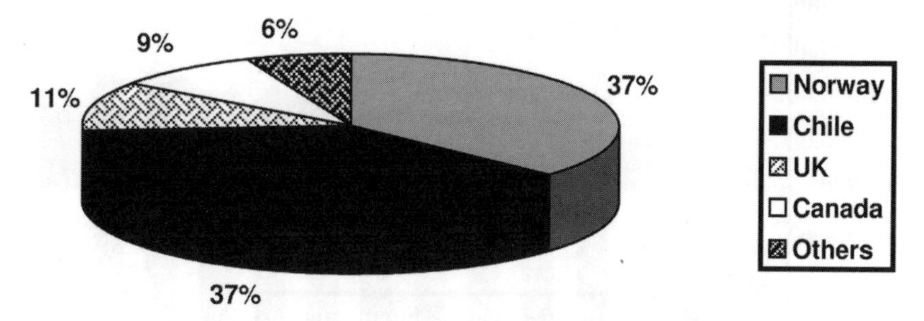

Figure 3.8. Major world suppliers of farm-raised salmon. (Source: FAO 2004.)

try primarily raises the Atlantic salmon (*Salmo salar*) with some additional production of coho salmon (*Oncorhynchus kisutch*). Wild catches of salmon are based primarily on pink salmon (*Oncorhynchus gorbuscha*), chum salmon (*Oncorhynchus keta*), and sockeye salmon (*Oncorhynchus nerka*).

The earliest sales of salmon were as a canned product prepared from wild-caught fresh salmon. The United States was a leading producer of canned salmon that were caught primarily in Alaskan waters. Japan and the Russian Federation historically have also been major suppliers of wild-caught salmon. The major salmon products sold in recent years have been fresh and frozen fillets. New packaging technologies, such as leak-proof Styrofoam packaging, was developed in the 1980s for farmed salmon (Anderson 2003). The new packaging provided a means to increase air shipment of fresh salmon.

The farmed salmon industry has tended to be dominated by large, integrated, agribusiness firms. A restructuring process began in the salmon industry in Norway in the 1990s as firms merged and larger firms were created (Bjorndal et al. 2003). In Norway, salmon firms have become increasingly consolidated (Bjorndal et al. 2003). The four largest firms controlled 28% of Norway's production capacity, and the largest 10% controlled 46% in 2001. Ownership structures have become more international. There are now Norwegian interests in both the Chilean and Scottish industries. Salmon farms have integrated vertically into processing facilities with sales offices in several countries. In Chile, the four largest firms accounted for 35% of exports in 2001, and the 10 largest accounted for 60% of exports (Bjorndal et al. 2003).

The increased consolidation provided a means to achieve economic efficiency and supply salmon at lower prices, putting smaller firms at an economic disadvantage. For example, Scottish farmers had trouble competing with Norwegian farmers on price

(Bjorndal et al. 2003). To compete in the salmon market, Scottish product has been oriented toward a higher-quality, lower-volume product produced under stringent health controls under the name "Scottish Quality Salmon" and "Tartan Quality Mark." Scottish salmon were the first fish to be awarded the highly regarded Label Rouge label in France, which is recognition of quality.

International markets have developed over the last decade, mostly due to aquaculture production (Bjorndal et al. 2003). The largest markets for salmon globally are the United States, Japan, and the European Union. However, new salmon markets are developing in Central and Eastern Europe, S.E. Asia, China, and South America. Per-capita consumption has risen as prices have declined. In the United States, for example, per-capita consumption of salmon has increased from 0.332 kg (0.7 lb) in 1990 to 1.009 kg (2.2 lb) in 2003 (Fig. 3.9).

The largest suppliers of farmed salmon are Norway and Chile, whereas the United States, Japan, and the Russian Federation are the major suppliers of wild-caught salmon. As supplies have grown and increased, the price for salmon has dropped over the long term (Fig. 3.10). Aquaculture production costs have decreased due to economies of scale, improved technology, and better feeds. Overall, price fell by 60% between 1990 and 2000 (Anderson 1997). Salmon consumption has increased and new products have been developed. However, as price has decreased with the increased supply from aquaculture, the salmon industry has turned to development of

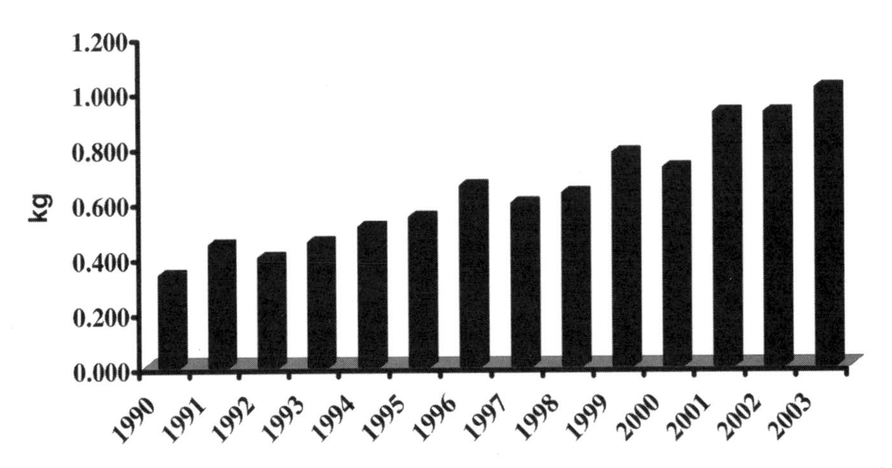

Figure 3.9. Per-capita consumption of salmon in the United States, 1990–2003. (Source: NFI 2004.)

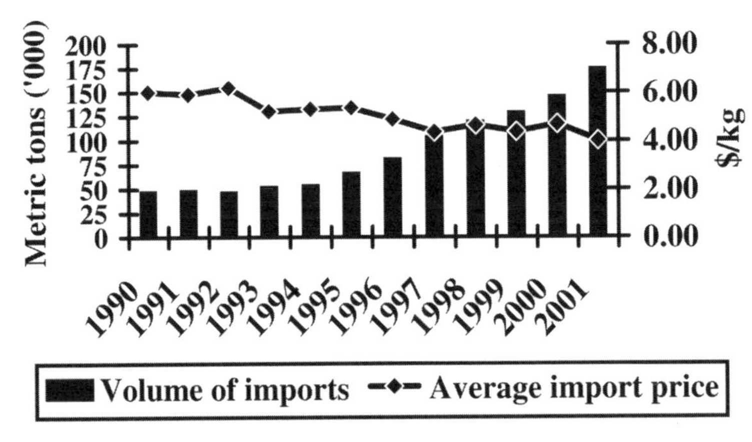

Figure 3.10. Average import price and volume of frozen Atlantic salmon in the United States, 1990–2001. (Source: NMFS 2004.)

a wide range of value-added products that range from gourmet smoked products to salmon jerky and salmon bits for salad toppings. In the United States, salmon is sold in a wide variety of market outlets, including restaurants, cafeterias, and grocery stores (Bjorndal et al. 2003). Salmon is estimated to be on the menu of 39% of all restaurant menus. This includes 71% of fine dining, 71% of hotel/motel, and 49% of casual/theme restaurant establishments.

The salmon industry has been at the center of numerous accusations from environmentalist nongovernmental organizations (NGOs). Salmon production has been labeled as unsustainable and environmentally unsound for the following alleged reasons: (1) use of Atlantic salmon in Pacific waters has potential for escaped fish to weaken the genetic pool in the Pacific Ocean; (2) discharge of waste products from the net pens where salmon are raised pollutes surrounding waters; (3) mercury and PCB (polychlorinated biphenyl) concentrations are higher in farm-raised than in wild-caught salmon; (4) the use of astaxanthin in salmon feeds is unnatural and should be labeled as an additive; and (5) the use of fish meal and fish oil in salmon will lead to overfishing of pelagic species upon which other species and fisheries depend. Many of these claims by NGOs have been exaggerated, and information has been used incorrectly out of context. Nevertheless, the very active opposition of some environmental NGOs to farm-raised salmon production has constrained sales and dampened market growth to some degree.

SUMMARY

Marketing is a broad term that encompasses all the interactions involved from the point of production to the end consumer. Marketing functions have grown in scope and complexity as consumer income levels, sophistication of consumer demand, and technology have grown. This chapter introduced terminology and fundamental concepts of marketing. It lays the groundwork in the terminology and conceptual understanding for the discussion that follows in subsequent chapters.

STUDY AND DISCUSSION QUESTIONS

1. Compare and contrast vertical and horizontal marketing systems.

2. Explain the marketing functions provided by wholesalers and why these are important.

3. Explain the differences between brokers and food service distributors.

4. List the different types of retail outlets.

5. How have economies of scale affected seafood marketing?

6. Describe supply chain management.

7. Explain the differences among the terms: marketing margin, marketing bill, and farm-retail price spread.

8. What is market power? What does it mean for the aquaculture industry?

9. Outline a marketing channel for an aquaculture product sold in your hometown. Calculate the marketing margins.

REFERENCES

Ababouch, L. No date. Transportation of fish and fish products. Accessed at http://www.oceansatlas.com/world_fisheries_and_aquaculture/htm

Anderson, J.L. 1997. The growth of salmon aquaculture and the emerging new world order of the salmon industry. In: E.K. Pikitch, D.D. Hubbert, and M.P. Sissenwine (eds.), *Global Trends: Fisheries Management*. American Fisheries Society, Bethesda, Maryland.

Anderson, J.L. 2003. *The International Seafood Trade*. Woodhead Publishing Limited, CRC Press, Boca Raton, Florida.

Armstrong, B. and P. Kotler. 2003. *Marketing: an Introduction*. Upper Saddle River, New Jersey.

Asche, F. 2001. Testing the effect of an anti-dumping duty: the US salmon market. *Empirical Economics* 26:343–355.

Bjorndal, T., G.A. Knapp, A. Lem. 2003. Salmon—a study of global supply and demand. Globefish Research Programme Volume 73. Food and Agriculture Organization of the United Nations, Fishery Industries Division, Rome, Italy.

Breimyer, H. 1976. *Economics of the Product Markets of Agriculture.* The Iowa State Press, Ames, Iowa.

Brooks, P.M. and J.L. Anderson. 1991. Effects of retail pricing, seasonality and advertising on fresh seafood sales. *The Journal of Business and Economic Studies* 1(1):77–90.

Cichra, Charles E., Michael P. Masser, and Ronnie J. Gilbert. 1994. Fee-fishing: an Introduction. Southern Regional Aquaculture Center, Publication No. 479, November.

Cremer, Michael C., Steven D. Mims, and Gregory Sullivan. 1984. Pay Lakes as a Marketing Alternative for Kentucky Fish Producers. Kentucky State University Community Research Service Program Report Number 8, Frankfort, Kentucky.

Engle, C.R. 1997. Marketing of the Tilapias. In: Costa-Pierce, B. and Rakocy, J. (eds.), *Tilapia: Aquaculture in the Americas, Volume 1*. World Aquaculture Society, Baton Rouge, Louisiana.

ERS. 2002. Economics Research Service. United States Department of Agriculture, Washington, D.C. Accessed at //www.ers.usda.gov

FAO. 1978. Marketing. Accessed at: http://www.fao.org

FAO. 2004. Fishstat Plus. Food and Agriculture Organization of the United Nations. Accessed at: //www.fao.org

Guillotreau, P. 1998. Foreign Trade and Seafood Prices: Implications for the CFP. Final Report. Len-Corrail, Nantes. France.

Heikes, D. 2003. Extension Specialist, University of Arkansas at Pine Bluff, personal communications.

Herrmann, M., R. Mittelhammer, and B.H. Lin. 1993. An international econometric model for wild and pen-reared salmon. In: Hatch, U. and H. Kinnucan, *Aquaculture: Models and Economics*. Westview Press, Colorado.

Jagger, P. and J. Pender. 2001. Markets, marketing and production issues for aquaculture in East Africa: the case of Uganda. Naga, the ICLARM Quarterly 24(102):42–51.

Kada, Y. 1997. Transporting fish by car. Accessed at: http: //www2.lbm.go

Kaneko, J., P. Bartram, and P. Garrad. 1996. Quality and product determination as price determinants in the marketing of fresh Pacific tuna and marlin. JIMAR Contributions 96-304, SOEST 96-06. www.soest.hawaii.edu

Keithly, W., K.J. Roberts, and J.M. Ward. 1993. Effects of shrimp aquaculture on the US market: an econometric analysis. In: Hatch, U. and H. Kinnucan, *Aquaculture: Models and Economics*. Westview Press, Colorado.

Kinnucan, H. and O. Myrland. 2002. Optimal seasonal allocation of generic advertising expenditures with product substitution: salmon in France. *Marine Resource Economics* 7(2):103–20.

Kinnucan, H., O. Cacho, and G. Hanson. 1986. Effects of selected tax policies on management and growth of a catfish enterprise. *Southern Journal of Agricultural Economics* 215–225.

Kinnucan, H., S. Sindelar, D. Wineholt, and U. Hatch, 1988. Processor demand and price-markup functions for catfish: a disaggregated analysis with implications for the off-flavor problem. *Southern Journal of Agricultural Economics* 81–91.

Kohls, R.L. and J.N. Uhl. 1985. *Marketing of Agricultural Products*. Macmillan Publishing Company, New York, NY.

Lambregts, J., O. Capps, and W. Griffin. 1993. Seasonal demand characteristics for U.S. farm-raised catfish. In: Hatch, U. and H. Kinnucan, *Aquaculture: Models and Economics*. Westview Press, Colorado.

Ledafish. 1996. Accessed at www.ledafish.com/tuna.htm

Leyva, C. 2004. Optimizing tilapia (*Oreochromis sp.*) marketing in Honduras. Master's thesis, Department of Aquaculture and Fisheries, University of Arkansas at Pine Bluff, Pine Bluff, Arkansas.

Marshall, D.W. 1996. *Food Choice and the Consumer*. Blackie Academic Professional, Glasgow.

National Fisheries Institute. 1997. NFI Seafood HACCP Manual (A Guide for Developing Seafood HACCP Plans). National Fisheries Institute, 1901 N. Fort Myer Drive, Suite 704, Arlington, VA 22209.

National Fisheries Institute. 2004. Top ten seafoods. Accessed at www.nfi.org

National Marine Fisheries Service. 2004. Fisheries of the United States. 2003. National Marine Fisheries Service, National Oceanic and Atmospheric Administration, Washington, D.C.

Neira, I., C.R. Engle, and K. Quagrainie. 2003. Potential restaurant markets for farm-raised tilapia in Nicaragua. *Aquaculture Economics and Management* 7(3/4):1–17.

Palfreman, A. 1999. *Fish Business Management*. Fishing News Books, Blackwell Science, Oxford, UK.

Pearson, R. and P. Sharma. 2003. Export subsidies. Agreement on Agriculture. www.fao.org

Rhodes, V.J. 1993. *The Agricultural Marketing System*. Fourth Edition. Gorsuch Scarisbruck Publishers, Scottsdale, Arizona. 484 p.

Schoemaker, R. 1991. Transportation of live and processed seafood. INFOFISH Technical Handbook 3. INFOFISH, Kuala Lumpur, Malaysia.

Seafood Business. 2001–2003. Diversified Business Communications, Portland, Maine, USA.

Suranovic, S. 1997-2004. International trade theory and policy. The International Economics Study center. http://internationalecon.com/v1.0

United States Department of Agriculture. 2001. Catfish and Trout Production. National Agricultural Statistics Service, United States Department of Agriculture, Washington, D.C.

United States Department of Agriculture. 2004. Economics Research Service. Accessed at: http://www.ers.usda.gov

U.S. Department of Commerce. 2004. Standards for Grades of Fishery Products. Seafood Inspection Program, U.S. Department of Commerce. Seafood.nmfs.noaa.gov/standards.htm

Wessells, C.R. and J.E. Wilen. 1994. Seasonal patterns and regional preferences in Japanese household demand for seafood. *Canadian Journal of Agricultural Economics* 42:87–103.

Wiese, N. and K. Quagrainie. 2004. Market characteristics of farm-raised catfish. Master's thesis. Department of Aquaculture and Fisheries, University of Arkansas at Pine Bluff, Pine Bluff, Arkansas.

Zidack, W., H. Kinnucan, and U. Hatch. 1992. Wholesale- and farm-level impacts of generic advertising: the case of catfish. *Applied Economics* 24: 959–968.

4
Aquaculture Growers and Their Marketing Choices

Aquaculture growers must make choices with regard to what type of markets to pursue with their products. One of the leading causes of failure of aquaculture businesses is the failure to spend time planning for the marketing component of their business. Successful businesses are market driven, and successful aquaculture businesses are those that have spent time analyzing their marketing options. Aquaculture businesses have an advantage over seafood supply businesses based on wild-caught species in that the businesses can be designed to be market driven. The greater level of control over stocking and harvesting schedules and the general control over the product possible in aquaculture allows aquaculture businesses to adjust production to meet market demands.

This chapter discusses some of the supply characteristics of aquaculture that are unusual and that pose special challenges to aquaculture growers. Requirements for sales to several particular types of markets are described, including sales to processors, livehaulers, and directly to consumers. The market synopsis on baitfish presents an example of a market that poses a contrast to markets for foodfish.

FISH SPECIES AND MARKETS

Aquaculture includes an astonishing diversity and complexity of cultured organisms. There are more than 210 species of aquatic finfish, crustaceans, mollusks and aquatic plants raised (Engle and Stone, 2005). Of these, the vast majority of species (99%) are raised for human consumption (FAO 2002). Other animal protein sources of food are far more limited and are composed of products from just a few species of animals (cows, swine, chickens) and a few other specialty livestock crops. Fish and seafood consumption has tended to be driven by the species that are available locally from the wild. Consumer preferences and the seafood markets that have developed over time are as diverse and complex as the number and types of species raised.

The diversity of species raised and the resulting diversity of specific markets developed for these products present a different type of challenge to aquaculture growers. Business growth and development requires market growth, but seafood consumer preferences have tended to be provincial and regional in nature. Thus, market expansion of aquaculture products often requires aquaculture companies to change attitudes and preferences of consumers located outside the region where their species has traditionally been sold.

Market development is complicated by the biological differences among the multitude of aquatic species cultured. For example, aquatic organisms are poikilothermic, meaning that they cannot control their body temperature. All the other animals that supply animal protein (cattle, swine, and poultry) for human consumption can control their body temperature. Although temperature levels play a role in production of terrestrial livestock, it is not a key production parameter. In aquaculture, cultured organisms are divided into warmwater and cold-water organisms. Trout, for example, are cold-water fish that cannot tolerate the average water temperatures under which channel catfish or most shrimp species are cultured. Although chickens can and are raised throughout the world in a variety of climates and temperature zones, trout can be raised only where there is a source of cold water. The majority of cultured shrimp are warmwater species and can be raised only where there is a source of warm water. There are also cold-water shrimp species, but cold-water shrimp cannot be raised in warm waters. Aquaculture supply will, thus, be more partitioned based on temperature than will terrestrial livestock production. Supply of particular aquaculture species will be concentrated in regions of the world that present optimal temperature conditions for that specific species.

Catfish farming in the United States, for example, is generally profitable in those regions in which ambient temperatures are suitable. Pond water temperatures (which closely follow air temperature) should be above 20° C (65° F) for at least 180 days of the year and above 25° C (75° F) for at least 125 days (Tucker et al. 2004). Fluctuations in water temperature trigger key phases of the reproductive cycle, making it difficult to raise channel catfish outside regions with temperate climates.

The majority of aquaculture species are also restricted to either freshwater or saltwater. Thus, carps, trout, and channel catfish must be grown in freshwater, whereas salmon, sea bass, yellowtail, and flounder require saltwater for culture. Some species, such as tilapia and shrimp (*Litopenaeus vannamei*), have been cultured successfully in both saltwater and freshwater although the majority of tilapia continue to be raised in freshwater and the majority of shrimp in brackish water. However, it remains to be seen whether culturing species under salinities different from those of their natural environments can be the basis for large industrial sectors.

TYPES OF PRODUCTS FROM AQUACULTURE

Just as there are a wide variety of species that can be raised in aquaculture, there are a wide variety of types of products raised and sold from aquaculture. The largest volume and value of aquaculture products are those sold as food. However, there are aquaculture products that are sold for other uses, including enhancing stocks of natural fish populations, eggs, fry, post-larvae, and fingerlings that are sold to growout businesses, fish and shrimp sold for use as bait in recreational fishing, and ornamental fish sold in the pet trade. Each of these types of products poses different and unique market requirements and challenges.

Selling fish for food requires delivery of live fish to either wholesalers, retailers, or processing plants. Conditions during transport must be adequate to keep the fish alive and in good condition. Other quality standards, as discussed in Chapter 11, need to be observed.

Hatchery segments have developed in many sectors of aquaculture. It is often more efficient economically for a growout operator to purchase eggs or fingerlings than to produce them on the farm. International trade in these intermediate hatchery products has developed. Trout eggs, for example, are shipped across the world. The development of hatchery technologies to produce post-larvae (PLs) of shrimp was a major factor that contributed to the emergence of a commercial shrimp industry (Csavas 1994). Wild stocks of post-larvae were not adequate to supply the demand for PLs as shrimp production expanded. The development of hatcheries allowed the supply of PLs to keep pace with the growth in food shrimp production.

Some of the earliest culture efforts in the United States were those designed by public agencies to produce fingerlings for enhancing stocks of trout, salmon, catfish, largemouth bass, hybrid striped bass, and other species. The hatchery programs of state and federal agencies developed technologies that were later adapted by the private sector to develop fish farming businesses. These hatchery businesses continue to be important. The trout synopsis in Chapter 6 illustrates that, in 2003, 12% of the 15- to 30-cm trout produced in the United States were sold to the government for stocking programs. This does not include the volume of trout produced in the public sector in the United States for stock enhancement purposes.

Other aquaculture products are sold live for use as bait or as pets. The market synopsis in this chapter presents an overview of the baitfish industry in the United States, and the market synopsis in Chapter 13 describes the global ornamental fish industry.

PRODUCTION SYSTEMS AND INTENSIFICATION

The earliest recorded aquaculture production was practiced in earthen ponds in China (Avault 1996). Much of the early expansion of aquaculture occurred in freshwater earthen ponds (common, grass, bighead, silver carps, and tilapia) throughout the world. Today, the majority of aquaculture production worldwide continues to come from earthen ponds.

From a marketing perspective, ponds are a convenient way to store and hold fish for long periods of time. However, production is heavily dependent upon the climate and ambient water temperatures. In temperate climates, for example, warmwater species such as tilapia will grow only over the warm months and must be brought indoors or sold before temperatures fall too low. Thus, supplies of warmwater aquaculture product are seasonal and highest prior to the onset of the cold season in temperate regions. Temperatures in tropical areas vary also, but not to the same extent as in temperate climates. Even so, there are important differences in salinity, sunlight intensity, and cloudiness imposed by the rainy and dry seasons of the tropics.

In many respects, the technologies underlying many aquaculture production systems have developed along with the corresponding markets. Fish that are produced primarily for home consumption frequently are cultured in polyculture systems that take maximum advantage of various ecological niches in the pond while supplying the family with a variety of different fish products for consumption in the home. Production of fish and crustaceans for commercial markets often requires production of a single species to capture economic efficiencies in utilizing production inputs. Most production of major commercial species such as shrimp, salmon, tilapia, and catfish in the United States are monoculture, or single-species production systems.

Tilapia are raised in ponds throughout the world at a wide variety of densities and levels of intensification (Pillay 1990; Pullen et al. 1987). Subsistence farmers in many countries stock tilapia at low densities with composted vegetative matter as the primary input to produce fish for family consumption (Boyd and Egna 1997). Near-subsistence farmers culture tilapia using manure and supplemental feeds to maintain a "savings account" for future cash needs or to generate some cash through local sales (Smith and Peterson 1982; Engle 1997; Setboonsarng and Edwards 1998; Little 1995) (Fig. 4.1).

Figure 4.1. Roadside sales of tilapia in Honduras, Central America. Photo by Dr. Carole R. Engle.

Tilapia are also raised in integrated agro-aquaculture farming systems (Pullin and Shehadeh 1980). Animal manure from livestock pens constructed on or near the water allows for manure to be washed into the pond as fertilizer. Filter-feeding fish such as tilapia and carps grow well in these systems. Other vegetable and tree crops can be grown on the levees, and pond water can be used for other uses on the farm (Lovshin et al. 1986).

Commercial small-scale tilapia production requires regular feeding but produces higher yields that allow growers to supply domestic markets in larger urban areas (Popma and Rodriguez 2000; Green and Engle 2000; Fitzsimmons 2000; Hanley 2000). Tilapia are also produced intensively in several countries to meet the high quality standards required to export to markets in the United States and the European Union (Engle, in press).

The majority of shrimp produced in the world are produced in earthen ponds along brackish water estuaries. In Honduras (the leading producer of *Litopenaeus vannamei* in the Central American region [Rosenberry 1999]), for example, pond shrimp production is classified as semi-intensive; farmers raise two crops a year with stocking densities that vary from 5–20 post-larvae (PLs) per m^2 (Dunning 1989; Stanley 1993; Valderrama and Engle 2001). Typical yields of shrimp from semi-intensive production systems in Honduras range from 400 to 2,000 kg/ha/yr of shrimp tails, whereas feed quantities range from 1 to 8 MT/ha/yr (Valderrama and Engle 2001). However, shrimp are also produced extensively on artisanal farms with low stocking densities and feeding rates. About one-third of the world's pond-raised shrimp continue to be produced in extensive systems (Rosenberry 1999).

Shrimp production in Asia has developed more intensive production methods, probably due to the higher costs of land in areas with greater population pressure. For example, in Thailand, shrimp are stocked at rates up to 30–35 PLs per m^2 with yields of 7,000–8,000 kg/ha/yr (Lin 1995). The higher yields are necessary to spread the higher fixed costs of land over greater amounts of production to be price competitive.

More than 98% of the catfish produced in the United States are grown in earthen ponds. Production costs are generally lower for catfish grown in ponds than in other culture systems. More than 98%

of levee-style ponds use groundwater pumped from shallow wells (less than 391 m). These wells yield abundant water at low cost. The most common stocking strategy in the catfish industry is to stock the fish in multiple batches to supply processing plants year round. In this system, fingerlings are stocked each year in the spring, but multiple harvests are made throughout the year to sell market-sized fish. Because the production cycle is approximately 18 months, varying sizes of fish are present in the pond at the same time. Single-batch stocking strategies are more profitable, but farmers stock in multiple batches due to the necessity of spreading the market risk of delayed sales associated with the presence of off-flavor (Engle and Pounds 1993). Multiple-batch stocking strategies provide the cash flow pattern necessary to meet financial obligations on catfish farms.

Cage production has grown in importance in certain areas of the world. A number of countries with high population densities, small land areas, and ample access to marine resources have developed successful cage-based aquaculture industries. Cage and net pen technologies developed rapidly in the latter decades of the 1900s. Large-scale net pen operations were developed to culture marine species such as salmon (Norway, Chile, Canada), sea bream and bass (Greece), yellowtail (Japan), cobia (Taiwan), cod (Norway, Canada), and others.

Marketing fish from cage operations presents unique transportation challenges. Net pen companies have developed technologies that include use of helicopters, barges with fish pumps, and others, to transfer fish from net pens to hauling tanks to markets. Significant market coordination is required to minimize losses when transporting large quantities of fish from net pen operations.

Atlantic salmon (*Salmar salmo*) are the fastest-growing of the salmon species and typically reach market size in two years. However, the high costs of the marine cage systems used for production create large economies of scale that have resulted in large, vertically integrated companies. These companies compete globally, largely on price. As salmon prices have decreased over time, largely due to the increased supply from aquaculture, the industry has moved into new product development. There currently is a wide variety of salmon products available on the market, which include gourmet, smoked and canned products, salmon burgers, salmon jerky, salmon bits (as a substitute for bacon bits), and many others.

Abundant flowing surface waters have been used in some areas to develop flow-through raceway production systems. Raceways are the predominant production system for cold-water trout and in more recent years have become the main production system used for large-scale warmwater production of tilapia in tropical regions. Raceways provide for ease of harvest, which provides flexibility to accommodate processing and sales schedules and programs. Size classes of trout can be maintained in separate tanks within a raceway. This size separation facilitates marketing specific sizes to specific markets.

Indoor recirculating aquaculture systems have been developed with advances in engineering technologies and have become a reliable, but high cost, production system.

Recirculating systems, if constructed indoors, provide a means of controlling water temperatures that allows for year-round growth of fish in temperate climates. Moreover, indoor recirculating systems can be constructed closer to major seafood market areas than can pond-based production systems. One of the major disadvantages of recirculating systems, however, is the continuing high cost of production in spite of the extent of improvement in the reliability of the systems.

The primary species raised successfully in indoor recirculating systems has been tilapia. This is primarily due to the ability of tilapia to withstand adverse water quality fluctuations. Other species, such as turbot and sole (in Spain), yellow perch (in the United States) and even more recently, shrimp (*Litopenaeus vannamei* in the United States and Israel) have been raised successfully in indoor recirculating systems.

SIZES OF PRODUCERS

The diversity and complexity of aquaculture species, production systems, and markets is accompanied by an equal diversity in the size of aquaculture businesses. A tilapia "farm" in Rwanda may consist of one pond with a surface area of only 0.01 ha (0.03 ac), whereas a company that exports tilapia fillets may have 60–70 ha (150–175 ac) of land.

Most of the shrimp produced on a commercial scale come from large-scale, private farms (>50 ha; 125 ac). Valderrama and Engle (2001) identified (Honduran survey data) groupings of commercial shrimp farms clustered around farm sizes of 73 ha (182 ac), 293 ha (732 ac), and 966 ha (2,415 ac). However, there were also a total of 68 artisanal producers operating 239 ha (598 ac) of ponds in Honduras in 1997 (ANDAH 1997). A typical artisanal farm is operated by a family group and is composed of from 1 to 30 ha (25–75 ac) of ponds. Although most of the semi-intensive farms in Honduras are vertically integrated and market their product in international markets, artisanal producers sell to local processing plants, shrimp markets, or both. Saborío (2001) reported a total of 90 artisanal shrimp farming cooperatives operating in the Estero Real area of Nicaragua by the year 2000. These loosely defined community-based groups consisted of dozens of families holding a site concession (Jensen et al. 1997). Their extensive methods utilized tidal inflows to stock ponds, exchange water, and supply nutrients. The cooperative organization allowed artisanal shrimp farmers to supply markets more consistently.

In the United States, the average size of catfish farms has increased steadily over time (Figure 4.2). In merely 10 years, from 1992 to 2002, the average farm size has doubled, from 32 ha to 64 ha (80–160 acres). The increasing average farm size has been accompanied by a decrease in the number of catfish farms.

Unlike pond production industries that have tended to demonstrate a wide variety of sizes of individual farms across all the major species raised in ponds, net pen salmon farming is highly concentrated (Bjorndal et al. 2003). In Norway, the four largest net pen salmon farms controlled 28% of the country's production capacity whereas the 10 largest controlled 46%. The Chilean net pen salmon industry is even more highly concentrated, with the four largest firms accounting for 35% and the 10 largest 60% of exports in 2001. Moreover, there are growing Norwegian interests in the Chilean salmon industry.

Indoor recirculating systems may range from one 2-m (6 ft) diameter tank in a greenhouse that produces 100 kg (220 lb) of fish a year to large, industrial facilities with production capacity of several million kg a year. The smaller systems typically are targeted toward home consumption whereas the larger indoor systems frequently target the higher-priced live fish markets.

SUPPLY RESPONSE AND BIOLOGICAL LAGS

Given the variety of aquaculture species cultivated and their varying biological characteristics, the supply response varies a great deal from species to species and even with different production systems. Tilapia and shrimp raised in the tropics reach market size in six months whereas salmon and catfish may require 18–24 months for individual fish to reach market size.

Lengthy biological lags for animals to reach market size can cause supply to be inelastic (unresponsive to price changes). Inelastic supply makes it difficult for growers to respond to changing market conditions. Catfish, for example, require 18 months to reach market size. Kouka and Engle (1998) showed that response to price changes occurs at the hatchery/fry stage of production and that overall catfish farm supply is inelastic. Farmers must stock production units and begin feeding before knowing what the price will be when fish reach market size. Thus, it is difficult for farmers to adjust quickly to changing market conditions.

Prices of many aquaculture crops demonstrate seasonality effects across the year. For example, Figure 4.3 illustrates wholesale price seasonality by month for salmon, cultured yellowtail, and flounder in Japan (NMFS 2003). Salmon prices were lowest in September and October in 2003 in Japanese wholesale markets, prices of cultured yellowtail were highest in July through September, and flounder prices were highest in July and August. Price seasonality may vary by year. It is best to look at monthly prices over several years. Higher revenues may be generated if the production cycle or production combined

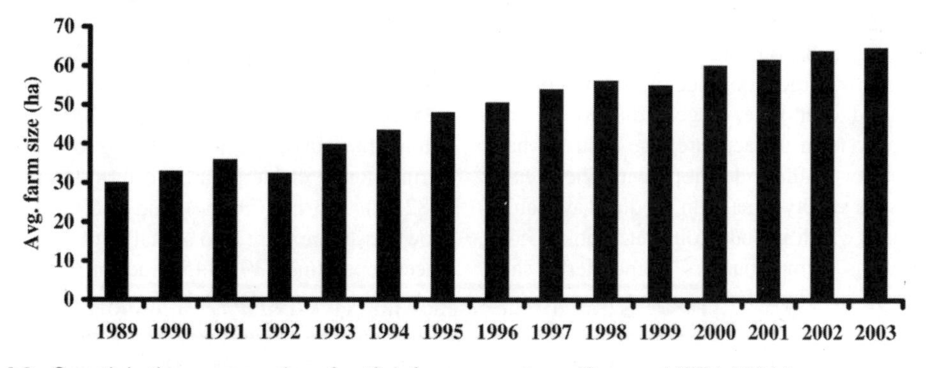

Figure 4.2. Growth in the average size of catfish farms over time. (Source: USDA 2004.)

Integrated Fish Farming Systems in the Mekong Delta of Vietnam

Aquaculture development in the Mekong Delta of Vietnam has traditionally been "resource use driven." Thousands of small farms utilized the available water bodies (canals between fruit trees, irrigation canals, ponds for water storage and irrigation, paddy fields, and so on) for fish culture, a complement to plant and fruit production, and animal husbandry. These practices are generally based on principles of nutrient recycling and efficient use of farmland and water resources. With the political and economic reform, or "doi moi," in the late 1980s in Vietnam, aquaculture development became more market driven as farmers had more freedom to make production and marketing decisions about which agricultural and aquacultural commodities would be produced. This led to agricultural intensification of the farms.

With increasing demand for agricultural and aquacultural products in the Ho Chi Minh City area and for export, and with the government providing farmers with the ability to make their own production and marketing decisions, a wide variety of integrated fish farming systems developed rapidly in the Mekong Delta region. There were as many different farming systems as there are agricultural commodities, including rice, nonrice crops, tree gardens, livestock, poultry, and aquaculture. Three main groupings of integrated fish farming systems include aquaculture in pond and garden canals, aquaculture integrated with animals and/or trees and other non-rice crops, and aquaculture integrated with rice cultivation. Buyers from Ho Chi Minh City provided farmers with domestic and international market information on prices and product demand and quantity and quality requirements. Small aquaculture hatcheries and nurseries developed to supply the farmers with fingerlings. Due to the flexible nature of their farming systems, farmers could easily and quickly adjust their production practices to meet changing market demands. The system was not without its problems. Farmers in the Mekong Delta region became heavily dependent on the buyers of their products, especially in terms of transportation of goods and price information. Farmers tended to have poor bargaining power with the buyers. Over time, farmers in the region have organized marketing associations and cooperatives to obtain inputs and credit, market their products, reduce risk, and obtain better prices.

Source: Eco-technological and socioeconomic analysis of fish farming systems in the freshwater area of the Mekong Delta. 1997. College of Agriculture, Cantho University, Cantho City, Vietnam. Contributed by Robert Pomeroy, University of Connecticut.

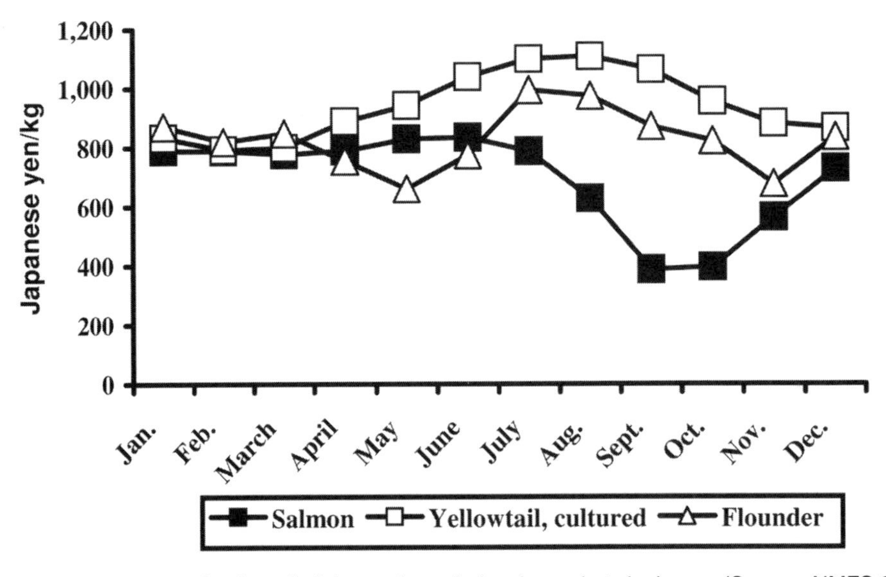

Figure 4.3. Price seasonality for three finfish species, wholesale markets in Japan. (Source: NMFS 2003.)

with a holding system can be managed to target a percentage of the crop toward marketing in months when price is high.

COMMODITY AND NICHE MARKETS

The species of aquatic organism to be raised, its biological requirements, the resulting supply characteristics, and the production system selected must all be appropriate for the specific markets to be targeted in the marketing plan. Commodity markets will be feasible only for larger businesses targeting high-volume markets in industries with processing capabilities and the ability to compete on price. A commodity is an economic good that can be produced legally and sold by almost anyone (Rhodes 1993). Niche markets for specialty, differentiated products may be the only feasible market outlets for small-scale growers. A differentiated product is an economic good that belongs to a single seller and that often may be patented, copyrighted, or trademarked to the exclusive use of that seller.

Large firms may also enter markets with differentiated products. Development of product brands can assist businesses to differentiate farmed and captured supply chains. Branding can occur at the species level, country-of-origin generic level, or as private brands within individual retailer chains (Burt 2000). Consumer perceptions related to origins of supply can be used as a basis for the development of brands for aquaculture product lines.

FARMERS' MARKETING ALTERNATIVES

SALES TO PROCESSORS

The majority of aquaculture products that are sold commercially are sold by farmers to processing plants. The larger seafood markets are those with higher-income consumers who have little interest in cleaning or dressing fish for consumption at home. Fillet products are the primary product sold, and the processing plant plays the role of transforming the form of the product into that preferred by the majority of consumers. Chapter 5 goes into more detail about seafood and aquaculture product processing.

Processing plants schedule delivery of loads of fish depending upon their current and anticipated orders. Farms that are able to regularly supply the volume and size of fish or shrimp desired by the processor or packing plant are those scheduled for regular deliveries to the plant. Thus, farmers need to have a clear idea of what the plant's specifications are for size tolerances, delivery volumes, timing of deliveries, quality control checks, flavor checks, and so on. Even firms that are vertically integrated may have separate cost centers such that the growout business will have to meet the processing center's delivery specifications.

Processing plants typically purchase the greatest overall volume of product, when compared to other potential market outlets, but also tend to pay the lowest price. This is because processing plants are frequently price takers in the market and there are many good substitutes for most fish and seafood products. Thus, many seafood products sold to processing plants are commodities that compete on price with other similar seafood products.

Some seafood processing plants are cooperatives to which the grower must belong to sell fish. Sales to cooperatives typically are in proportion to shares held by the member. New Generation Cooperatives often use delivery rights as a means to raise capital. If the cooperative is a successful business, the delivery rights acquire value through the ability to trade them with new members or with members who are seeking to expand their businesses.

SALES TO LIVEHAULERS

In the United States, EU countries, and other countries, fish farmers can sell fish to livehaulers, individuals who own large transport vehicles to haul fish to other distributors, or to retail outlets. Livehaulers typically purchase fish from farmers and resell them to either wholesalers or directly to retailers. Common carp in Europe are hauled extensively from fish farms in Hungary, Slovakia, and other Central European countries to markets in Germany and elsewhere. In the United States, Chinese carps are hauled from the southern part of the country to Asian ethnic grocery markets in New York City, Chicago, and San Francisco. Livehaulers frequently pay cash, but prices can be volatile.

Livehaulers may also haul fish to pay lakes and government-owned fishing lakes. Some states in the United States operate community and urban fishing programs. The scope of these programs frequently requires substantial quantities of fish to be purchased for stocking. Some of these programs prefer larger fish than do processing plants. Fish that are off-flavor may also be sold through these outlets.

SELLING DIRECTLY TO END CONSUMERS

Small-scale growers frequently have higher production costs due to the economies of scale common in many forms of aquaculture production. The higher production costs mean fewer years during which it is profitable to sell to a processing plant. Even when the price paid by processing plants is sufficiently high to allow for profit to be made by small-scale growers, the profit margin will be much lower than for larger-scale farms that can produce fish at a lower cost. Thus, it can be more feasible for the small-scale grower to develop markets based on direct sales to end consumers.

Sales to end consumers require the grower to do all the marketing by him- or herself. Thus, market-sized fish need to be either: (1) transported to customers in other locations (transportation marketing function), or (2) held on the farm in cages or tanks for farm-bank sales, or in ponds for fee fishing (storage marketing function). Which of these is more feasible would be determined in the market analysis and should be detailed in the marketing plan. Chapter 12 includes more details on how to develop and use the business' marketing plan.

Holding fish for sale allows small-scale producers to take advantage of higher prices obtainable through direct sales to the public. Adequate holding facilities and proper handling can make a big difference in a producer's profits. However, it is critical that fish be readily available and in good condition. Dead fish will turn customers away, as will lengthy waiting periods to catch fish.

Cages can be used for fish to be sold at irregular intervals, but tanks are best for supplying customers on a regular basis. Cages can be used to hold from 121–240 kg/m^3 (7.5 to 15 lb/ft^3) of cage, depending on the temperature (Rode and Stone 1994). For long-term holding, fish will need to be fed a maintenance diet (1% of fish weight), and supplemental aeration may be necessary. A pier often will be needed to retrieve fish quickly from cages to attend to customers. Cages anchored off shore without a pier will require a boat for both feeding and to retrieve fish, often with lengthy waits. Off-flavor problems in the pond where the fish cages are located will affect all fish and force sales to be curtailed until fish come back on flavor. Disease treatment can be difficult in cages. Investment costs in the U.S. for a small cage to hold about 455 kg (1,000 lb) of fish will be approximately $1,000.

Tanks can be used to hold fish and are preferable for shorter time periods for regular sales (Rode and Stone 1994). Fish can be held for several days at a rate of about 68 g/l (0.6 lb of fish/gallon) of water. A typical tank is 3–13 m long, 1–2 m wide, and about 1 m deep (10–40 ft long, 3–6 ft wide, and 2–3 ft deep). Tanks require a concrete slab with side walls of concrete block or poured concrete. Round or rectangular fiberglass tanks can be purchased. The tank facility is best covered with a roof so that sales can continue during inclement weather. An aeration or air blower system is needed to maintain adequate oxygen levels in the tank. Tanks can cost about $1,500 per vat for a tank with a capacity of 5,200 l (1,300 gallons).

Fish held in tanks are susceptible to theft. Because fish cannot be fed while in tanks, they can lose weight. More weight will be lost at higher temperatures. Channel catfish can lose as much as 4.5% of body weight in two days at a temperature of 22° C (71° F). Water in the tank will need to be exchanged at a rate of about 10% of the tank volume an hour to avoid buildup of ammonia. Well water is preferred because public waters have chlorine that is toxic to fish.

If the location of the farm is such that there is not a sufficient customer base to attract enough people to the farm for direct sales, fish will have to be transported to where the customers are. Fish can be hauled in hauling tanks to farmers' markets, street corners, parking lots, or directly to restaurants or grocery stores that purchase and sell live fish. The simplest type of hauling tank is a box made of marine plywood. Alternatively, aluminum or fiberglass tanks can be purchased in a range of sizes and dimensions to haul in the bed of a pickup truck, on the back of a bob truck, or for use with a flatbed or 18-wheeled tractor trailer rig (Rode and Stone 1994). Whichever type of tank is selected, internal dividers or baffles are necessary to reduce sloshing. Most fish species can be safely transported at a rate of 0.6 kg/l (5 lb of fish per gallon) of water (after fish added) for 16 hours. Thus, a tank with capacity of 800 l (200 gallons), loaded at 0.6 kg/l (5 lb/gal), can haul only 273 kg (600 lb) fish, with 480 l (120 gallons) of water remaining. Fish should be purged (held overnight to empty stomachs to avoid water-quality problems during hauling). Some ice added to reduce the temperature of the water will reduce stress on the fish during transport. The transport tank will need a source of air supplied either by an oxygen tank with diffuser bars, a blower, or small, 12-volt agitators. The investment cost for a 1,200 l (300-gallon) tank is about $1,200–$1,500. The oxygenation system cost is in addition to the cost of the tank.

The marketing plan for direct sales requiring transportation should indicate the dates and times of sales. Fish cannot be held indefinitely in a transport tank; customers need to be reliable and in condition to accept regularly scheduled deliveries of fish so that fish can arrive in good condition. The scale must be a certified scale.

Fee Fishing

Fee fishing, or pay lakes, are businesses that stock water bodies with fish and charge customers to fish in them. Some fee fishing operations are run by grow-out farmers whereas others are strictly in the business of buying fish to stock for resale by angling, an aquatic pick-your-own operation. Fish farmers can sell fish to a livehauler, who transports them to pay lakes elsewhere or can develop their own fee-fishing business. This section discusses the basics of what is needed to develop, own, and operate a fee-fishing business. Sales to livehaulers are discussed in another section of this chapter.

A fee-fishing business sells recreation. As with any other type of business, the location is essential. Successful operations are those located within 50–83 km (30–50 miles) of a population center with at least 50,000 or more people. Fee fishing will not work well in areas where good fishing in natural water bodies is readily available. However, a location near a major urban area with few opportunities to fish might be a prime spot for a fee-fishing business.

People go to fee fishing operations because they are looking for a family outing with good fishing, but also with amenities that ensure a fun but safe activity. Typical clientele for fee-fishing operations are families or grandparents with small children, elderly people, or physically handicapped people who find it difficult to get out on natural water bodies to fish. Adequate amenities for family activities are important, and the site must be aesthetically pleasing and comfortable. A natural setting that is screened from urban distractions, with easy access, good parking, and adequate security is best. Clean restrooms are essential. Concessions can generate important revenue to the business and contribute overall to the sense of a quality family activity. Sales or rental of bait and tackle are necessary to attract first-time anglers who may not own fishing gear. Chairs and umbrellas can be rented for the comfort of the customers. Snacks and drinks will help keep people at the fee-fishing site longer and will increase revenue. Sales of coolers with ice to take fish home can be supplemented with cookbooks, fish batter, and seasonings. Many customers will prefer to have their fish cleaned and will pay a fee to do so. Sunscreen and first aid supplies can be sold along with hats, and shirts with attractive designs can be sold and used as advertisement.

Word-of-mouth advertisement from patrons who have had a pleasant experience can be some of the best advertising. Roadside signs as well as ads in newspapers, on the television, radio, or in local shopper guides can be effective. Local fliers can be distributed at youth and community events.

Ponds should have banks with good grass cover and should be sodded if necessary. Small ponds are better than larger ones to ensure good fishing. With multiple ponds, patrons can be moved to those ponds with better fishing that day. Given that 30–50% of the fish in a pond may have learned to avoid hooks, ponds need to be seined regularly to remove fish that are not biting. An alternative market will need to be developed for those fish that will not bite a hook. Regular supplies of other fish must be added regularly to maintain good fishing.

Clearly placed signs should direct customers to parking and provide all necessary information. Prices, fishing regulations, rules related to various activities, and the times of operation need to be clearly visible on attractive signs in multiple locations. Activity rules must be posted clearly. Swimming, alcohol use, and abusive language should be prohibited. These activities are not conducive to the family atmosphere that is important to the majority of patrons of fee-fishing businesses. Any fishing gear restrictions need to be posted clearly.

The marketing plan for a fee fishing business may include offering group rates to youth groups. Boy and Girl Scout, church youth, 4-H, or school groups may be potential target market segments who may afterwards convince parents, grandparents, or other family members to return. Night fishing activities may offer an off-peak, exciting adventure to some organized youth groups. Stocking trophy-tagged fish provides an opportunity to generate excitement by advertising special prizes. Occasional additions of new species of fish will keep the experience new. Posting and selling instant pictures of customers with their catch can serve both to advertise the business and to generate revenue.

As with any other type of business, certain permits may be required. These vary with the state and country where located but may include permits related to starting a business, building ponds, having effluents from ponds, cleaning fish, and dealing with sales from the concessions.

There are a wide variety of pricing mechanisms used in fee-fishing businesses. Some charge an entrance fee. The advantage of having an entrance fee is that the revenue is generated immediately. However, costs may be high if the patrons catch many fish. A limit on the number or weight of fish caught can alleviate potential problems. Daily or seasonal entrance fees can be charged. An entrance fee will also discourage loitering and help to maintain the business's attractiveness to family-oriented groups. Other businesses charge by the unit weight of fish harvested.

HOME CONSUMPTION

Fish and other aquatic animals have been raised for home consumption for millennia. The earliest accounts of aquaculture chronicle farmers stocking and feeding fish in ponds for consumption by the family in Rome, Gaul, China, and Egypt (Bardach et al. 1972; Avault 1996). Around the world today, farmers and others raise fish in ponds, cages, and indoor systems to supply their homes with fresh fish and other aquatic products.

Fish production continues to be an important source of animal protein for rural populations (Engle et al. 1997). Surveys conducted in Rwanda, Thailand, Philippines, and Honduras showed that near-subsistence and small-scale farms with ponds may consume anywhere from nearly all of the crop produced to just a small percentage of the fish crop. Fish ponds can be a source of available and relatively inexpensive protein.

Early fish production for home consumption was based on the Chinese carp polyculture system and on oyster and mussel culture in tidal areas. Beginning in the 1950s, tilapia was stocked into ponds throughout Latin America, Africa, and Asia to supplement household supplies of animal protein (Engle et al. 1997; Engle and Skladany 1992; Edwards et al. 1991). Recent documented examples of tilapia grown for home consumption include Mexico (Fitzsimmons 2000; World Bank 1997), Rwanda (Engle et al. 1993) and Jamaica (Chakalall and Noriega-Curtis 1988), but the practice has continued throughout Asia and Africa. Some hobbyists in the United States and Europe raise fish to supply their households. Fish is also used for barter in some areas.

FORMING A MARKETING COOPERATIVE

One option that is often suggested for small-scale aquaculture growers is to form a marketing cooperative. Pooling production from different farms creates the larger volumes of product required to fulfill larger, longer-term contracts. More stable prices may be obtained with longer-term contracts. However, many cooperatives fail. Successful cooperatives are those that have strong, well-respected management that is viewed as being fair to all and insists that all members follow the policies established by the cooperative. Allocating sales to members is especially difficult when production and prices are seasonal.

FUTURES MARKETS FOR AQUACULTURE PRODUCTS?

Futures markets have been used to hedge against price fluctuations by farmers for many years and, hence, can be used to reduce market risk for both buyers and sellers of shrimp. Futures contracts are standardized, legally binding agreements to either deliver or receive a certain quantity and grade of a specific commodity during a designated delivery period (the contract month). The contract includes information on where the product would be delivered and any adjustments on price from substituting a different species or size. Commodities need to be standardized in futures markets so that exchanges can occur. This makes it easier for anyone to enter into a futures contract and know exactly what is being purchased.

Futures markets augment cash markets because contracts can be traded without delivery or receipt of the product. Feeder and fed (live) cattle, hogs, pork bellies, cotton, canola, and wheat are all traded at exchanges such as the Chicago Board of Trade, the New York Cotton Exchange, the Winnipeg Commodity Exchange, and the Minneapolis Grain Exchange.

The Minneapolis Grain Exchange began trading futures contracts for farm-raised and wild white shrimp in 1993 and added a contract for farm-raised giant tiger shrimp in 1994. The two main shrimp contracts offered were the following: (1) 2,273 kg (5,000 lb) of raw, frozen, headless, shell-on, 41–50 count white shrimp (*Penaeus vannamei, P. occidentalis, P. schmitti, P. merguiensis, and P. setiferus*); and (2) 2,273 kg (5,000 lb) of raw, frozen, shell-on, 21-25 count farm-raised giant tiger shrimp (*P. monodon*). These shrimp futures contracts in the United States were discontinued after 2000.

AQUACULTURE MARKET SYNOPSIS: BAITFISH

Baitfish are small minnows that are sold to recreational fishermen, who use them as bait to catch sport fish. Fish and crustaceans are raised and sold as bait all over the world. However, baitfish production in most countries is either small scale, incidental, or simply serves to sell fish and crustaceans that are too small to meet foodfish market requirements. However, the U.S. baitfish industry provides an example of baitfish production that has been developed into a large and important industry.

The recreational fishing industry in the United States is a multibillion dollar industry. In the 1950s and 1960s, when many new reservoirs were built, demand for live bait grew rapidly. The baitfish industry developed and grew to meet this demand. Nevertheless, competition continues from the sale of artificial lures for fishing and from wild-caught fish that are sold as bait in the United States.

Baitfish farming is a unique type of aquaculture in many respects. The baitfish industry produces and sells vast numbers of various sizes of small fish. Arkansas alone produces more than 6 billion baitfish annually (Stone et al. 1997a). Litvak and Mandrak (1993) estimated the retail value of baitfish sold in North America (including both farm-raised and wild-caught) to be $1 billion annually.

The majority of farm-raised baitfish sold in the United States is produced in ponds in Arkansas. Overall, $37.5 million of baitfish were sold in the United States in 1998 according to the USDA Census of Aquaculture (USDA 1998). Of these, 61% were sold from Arkansas. Ninety-three percent of baitfish farms are small businesses. Most baitfish farms are primarily family farms and partnerships. A few farms have diversified into distribution and wholesale functions and serve as market outlets for the smaller operations (Stone et al., in press).

Baitfish are sold as a live product and differentiated by size. Different sizes of minnows are selected by recreational fishermen, depending on the type and size of sport fish they would like to catch.

Baitfish farmers have developed extensive marketing and distribution networks over time (Stone et al. 1997b). Although some farms sell directly to fishermen, most sell through networks of wholesalers and distributors. Some large farms function as wholesale distributors for smaller farms (Fig. 4.4). Other distributors own their own holding facilities and have developed retail networks in a given sales area. Fish are then distributed from these warehouses to retail bait shops or other wholesale operators, who then resell to bait shops (Fig. 4.5). Baitfish are hauled by transport trucks long distances across the United States and handled several times en route. Thus, the fish must be vigorous and hardy enough to withstand the travel and handling and still be hardy and vigorous when bought by the consumers. The industry standard for customer service is to replace any fish losses incurred by the distributors and wholesalers regardless of the cause of the mortalities. A strong commitment to their customers is one of the characteristics of successful baitfish farms.

The greatest challenge to baitfish producers is that the demand for their product varies with the amount

Figure 4.4. Grading and loading shed for baitfish. Photo courtesy of Dr. Nathan M. Stone.

Figure 4.5. a. Retail bait shop, with hauling tank on back of pickup truck, Jimmy's Bait Shop, Pine Bluff, Arkansas. Photo courtesy of Nathan M. Stone, with permission of Jimmy's Bait Shop. **b.** Retail bait shops, River City Marine, Pine Bluff, Arkansas. Photo courtesy of Dr. Nathan M. Stone, with permission of River City Marine.

of recreational fishing, which is highly dependent on the weather. Moreover, the demand for different sizes of fish will depend on the weather conditions at different times of the year in different parts of the country. This requires baitfish farmers to maintain stocks of all sizes of fish on hand at all times to have the supply ready to cover whatever the market will want that particular year.

Golden shiners (*Notemigonus chrysoleucas*) are the major baitfish species raised. Nearly half of all baitfish raised in the United States are golden shiners (Fig. 4.6). Feeder goldfish (*Carassius auratus*) are the second most commonly cultured baitfish. Goldfish are popular fish to keep as pets in either aquaria or pools in water gardens. Their value as ornamental fish is included in the synopsis in Chapter 13. However, goldfish are also raised on farms to sell as feeder fish (for customers, to feed pet carnivorous fish) or as trotline or other bait. Trotlines are a type of fishing tackle used in rivers and lakes. It is a fishing line that has a series of hooks that are baited with live fish overnight and checked in the morning. Fathead minnows (*Pimephales promelas*) are the other major species of baitfish raised. Fathead minnows tend to be raised more extensively and are sold most commonly as fishing bait, similarly to sales of golden shiners.

SUMMARY

Aquaculture growers produce widely diverse types of aquatic plants and animals. Demand for seafood and aquaculture products tends to vary with species that traditionally have been captured in local waters. Markets and marketing systems for aquaculture products reflect this diversity.

The biology of the species raised and the production system used play major roles in the volume and seasonality of supply with implications for prices received by farmers. The chapter summarized a number of widely different examples of aquaculture production and how these relate to marketing.

Marketing alternatives for aquaculture growers may include sales to processors, to livehaulers, or directly to end consumers. Some of the market requirements unique to each of these market outlets were discussed. Forming cooperatives to compete for larger contracts may be a viable option, but marketing cooperatives can be difficult to manage and the failure rate is high.

STUDY AND DISCUSSION QUESTIONS

1. Explain, using examples, how the choice of species and production system can affect the marketing alternatives available to an aquaculture grower.

2. What are important questions to ask when considering selling to a processing plant?

3. What are the major difficulties associated with forming a marketing cooperative?

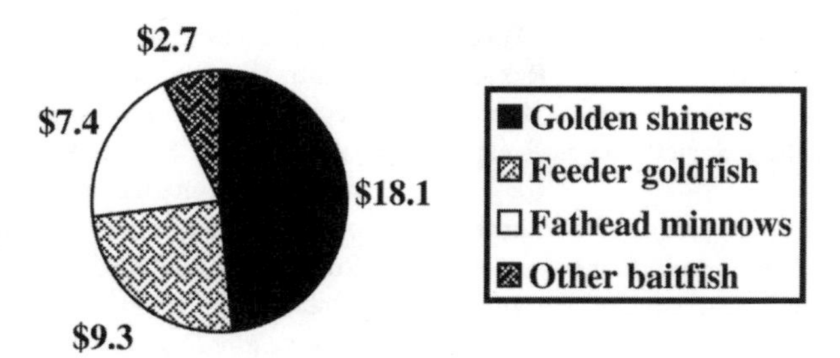

Figure 4.6. Major baitfish species, United States ($ million). (Source: USDA 1998.)

4. What advantage is it to a small-scale producer to hold fish that are market size?

5. What are the advantages and disadvantages of holding fish in cages?

6. What are the advantages and disadvantages of holding fish in tanks?

7. What are the keys to success for fee fishing operations?

REFERENCES

ANDAH. 1997. Boletín InformativoTécnico. Asociación Nacional de Acuicultores de Honduras. ANDAH, Choluteca, Honduras.

Avault, J.W. 1996. *Fundamentals of Aquaculture*. AVA Publishing Company, Inc., Baton Rouge, Louisiana.

Bardach, J.E., J.H. Ryther, and W.O. McLarney. 1972. *Aquaculture*. Wiley-Interscience, a division of John Wiley & Sons, Inc., New York, New York.

Bjorndal, T., G.A. Knapp, and A. Lem. 2003. Salmon—a study of global supply and demand. *FAO/Globefish* Volume 73, Food and Agricultural Organization, United Nations, Rome, Italy.

Boyd, C.E. and H. Egna (eds.). *Dynamics of Pond Aquaculture*. CRC Press, New York.

Burt, S. 2000. The strategic role of retail brands in British grocery retailing. *European Journal of Marketing* 34:875–890.

Chakalall, B. and P. Noriega-Curtis. 1988. Tilapia farming in Jamaica. Gulf and Caribbean Fisheries Institute, Charleston, South Carolina.

Csavas, I. 1994. Important factors in the success of shrimp farming. *World Aquaculture* 25(1):34-55.

Dunning, R.D. 1989. Economic optimization of shrimp culture in Ecuador. Master's thesis. Auburn University, Auburn, Alabama, USA.

Edwards, P., H. Demaine, and S. Komolmarl. 1991. Toward the improvement of fish culture by small-scale farmers in Northeast Thailand. *Journal of Asian Farm Association* 1:287-302.

Engle, C.R. and N.M. Stone. 2005. Aquaculture: Production, processing, marketing. *Marcel Dekker Encyclopedia of Animal Science*. Marcel Dekker, Inc., New York, NY.

Engle, C.R. In press. Marketing and economics. In: Webster, C. and C. Lim. *Tilapia Nutrition, Feeding, and Culture*. The Haworth Press, Binghamton, NY.

Engle, C.R. 1997. Economics of tilapia (*Oreochromis* sp.) aquaculture. In: Costa-Pierce, B. and Rakocy, J. (eds.), *Tilapia Aquaculture in the Americas, Volume 1*. World Aquaculture Society, Baton Rouge, Louisiana.

Engle, C.R. and G. L. Pounds. 1993. Trade-offs between single- and multiple-batch production of channel catfish: An economics perspective. *Journal of Applied Aquaculture* 3:311–332.

Engle, C.R. and M. Skladany. 1992. The economic benefit of chicken manure utilization in fish production in Thailand. Pond Dynamics/Aquaculture CRSP Research Report 92-45, Office of International Research and Development, Oregon State University, Corvallis, Oregon.

Engle, C.R., R. Balakrishnan, T.R. Hanson, and J.J. Molnar. 1997. Economic Considerations. Pages 377–395 in Egna, H.S. and C.E. Boyd. *Dynamics of Pond Aquaculture*. CRC Press, New York, New York.

Engle, C.R., M. Brewster, and F. Hitayezu. 1993. An economic analysis of fish production in a subsistence agricultural economy: the case of Rwanda. *Journal of Aquaculture in the Tropics* 8:151–165.

FAO. 2002. The state of world fisheries and aquaculture 2002, Food and Agriculture Organization, United Nations, Rome, Italy.

Fitzsimmons, K. 2000. Tilapia aquaculture in Mexico. Pages 171–183 in Costa-Pierce, B.A. and J.E. Rakocy (eds.), *Tilapia Aquaculture in the Americas, Volume Two*. The World Aquaculture Society, Baton Rouge, Louisiana.

Green, B.W. and C.R. Engle. 2000. Commercial tilapia aquaculture in Honduras. Pages 151–170 in Costa-Pierce, B.A. and J.E. Rakocy (eds.), *Tilapia Aquaculture in the Americas, Volume Two*. The World Aquaculture Society, Baton Rouge, Louisiana.

Hanley, F. 2000. Tilapia aquaculture in Jamaica. Pages 204–214 in Costa-Pierce, B.A. and J.E. Rakocy (eds.), *Tilapia Aquaculture in the Americas, Volume Two*. The World Aquaculture Society, Baton Rouge, Louisiana.

Jensen, G.L., G. Treece, and J. Wyban. 1997. Shrimp farming in Nicaragua: market opportunities and financing options for new investments. *World Aquaculture* 28(2):45–52.

Kouka, P.J. and C.R. Engle. 1998. An estimation of supply in the catfish industry. *Journal of Applied Aquaculture* 8(3):1–15.

Lin, K.W. 1995. Progression of intensive marine shrimp culture in Thailand. Pages 13–23 in C.L. Browdy and J.S. Hopkins (eds.), Swimming through troubled water. Proceedings of the special session on shrimp farming. Aquaculture '95. World Aquaculture Society, Baton Rouge, Louisiana.

Little, D. 1995. The development of small-scale poultry-fish integration in Northeast Thailand: potential and constraints. Pages 25–276 in J. Symoens and J. Micha (eds.), The Management of Integrated Fresh Water Agro-Piscicultural Ecosystems in Tropical Areas: Proceedings of Seminar. Brussels, Belgium.

Litvak, M.K. and N.E. Mandrak. 1993. Ecology of freshwater baitfish use in Canada and the United States. *Fisheries* 18(12):6–13.

Lovshin, L.L., B.B. Schwarz, V.G. de Castillo, C.R. Engle, and L.U. Hatch. 1986. Cooperatively managed rural Panamanian fish ponds: the integrated approach. Research and Development Series No. 33. Alabama Agricultural Experiment Station, Auburn University, Alabama.

NMFS. 2003. Average wholesale prices and sales volume of selected fishery products at 10 major central wholesale markets in Japan. http://swr.nmfs.noaa.gov/fmd/sunee/salesvol. Southwest Regional Office, National Marine Fisheries Service, Long Beach, California.

Pillay, T.V.R. 1990. *Aquaculture: Principles and Practices*. Surrey, England: Fishing News (Books) Inc. 575 p.

Popma, T.J. and F. Rodriquez, B. 2000. Tilapia aquaculture in Colombia. Pages 141–150 in Costa-Pierce, B.A. and J.E. Rakocy (eds.), *Tilapia Aquaculture in the Americas, Volume Two*. The World Aquaculture Society, Baton Rouge, Louisiana.

Pullin, R.S.V. and Z.H.Shehadeh. 1980. Integrated agriculture-aquaculture farming systems. ICLARM Conference Proceedings 4, International Center for Living Aquatic Resources Management, Manila, The Philippines.

Pullin, R.S.V., T. Bhukaswan, K. Tonguthai and J.L. Maclean (eds). 1987. The second International Symposium on Tilapia in Aquaculture. Manila, The Philippines. ICLARM Contribution No. 530. 623 p.

Rhodes, V.J. 1993. *The Agricultural Marketing System*, Fourth Edition. Gorsuch Scarisbrick Publishers, Scottsdale, Arizona. 484 p.

Rode, R. and N. Stone. 1994. Small scale catfish production: holding fish for sale. FSA 9075. University of Arkansas Cooperative Extension Service, Little Rock, Arkansas.

Rosenberry, R. 1999. World Shrimp Farming, 1999. *Aquaculture Digest,* San Diego, CA.

Saborío, A. 2001. La Camaronicultura en Nicaragua—Año 2000. Centro de Investigación del Camarón. Universidad Centroamericana, Managua, Nicaragua.

Setboonsarng, S. and P. Edwards. 1998. An assessment of alternative strategies for the integration of pond aquaculture into small-scale farming system of Northeast Thailand. *Aquaculture Economics and Management* 2(3):151-162.

Smith, L.J. and S. Peterson. 1982. *Aquaculture Development in Less Developed Countries: Social, Economic, and Political Problems*. Westview Press, Boulder, Colorado.

Stanley, D.L. 1993. Optimización economica y social de la maricultura hondureña. Pages 94–118 in Memorias del Segundo Simposio Centroamericano Sobre Camarón Cultivado. Federación de Productores y Agroexportadores de Honduras and Asociación Nacional de Acuicultores de Honduras, Tegucigalpa, Honduras.

Stone, N., A. Goodwin, R. Lochmann, H. Phillips, C.Engle, and H. Thomforde. In press. Baitfish Culture. In A. Kelly (ed.), *Aquatic Species Cultured Since 1897*. AFS Special Publication.

Stone, N., E. Park, L. Dorman, and H. Thomforde. 1997a. Baitfish culture in Arkansas. *World Aquaculture* 28(4):5–13.

Stone, N., E. Park, L. Dorman, and H. Thomforde. 1997b. Baitfish culture in Arkansas: golden shiners, goldfish and fathead minnows. MP 386, Cooperative Extension Program, University of Arkansas at Pine Bluff, Pine Bluff, Arkansas.

Tucker, C.S., J.L. Avery, and D. Heikes. 2004. Culture methods. Pages 166–195 in Tucker, C.S. and J.A. Hargreaves (eds.), *Biology and Culture of Channel Catfish*. Elsevier, Amsterdam, The Netherlands.

United States Department of Agriculture. 1998. Census of Aquaculture (1998). 1997 Census of Agriculture Volume 3, Special Studies, Part 3. United States Department of Agriculture, Washington, D.C.

Valderrama, D. and C.R. Engle. 2001. Risk analysis of shrimp farming in Honduras. *Aquaculture Economics and Management* 5(1/2):49–68.

World Bank. 1997. Mexico aquaculture development project. Report 16476—M.E. World Bank, Washington, D.C.

5
Seafood and Aquaculture Product Processing

Fish processing is quite varied in terms of types of operation, scales of production and outputs, and species of fish. Marine fish account for more than 90% of fish production, with the remainder being freshwater fish and fish produced from aquaculture.

Processing in the seafood and aquaculture industry encompasses all the steps that the fish goes through, from the time of harvest to the time that the seafood and aquaculture products reach the consumer. As with other sectors of the food industry, processing of seafood and aquaculture products is designed to provide products that are safe and meet consumers' demand requirements for quality and convenience. Thus, processing seafood and aquaculture products must aim at increasing shelf life, reducing microbial content, preserving fish/seafood products, and providing convenience. The trend in processing is generally driven by consumer demands and technological advances. In 1999, several manufacturers of retail seafood products reported a significant shift by consumers from buying higher priced premium items, such as grilled fish and specialty items, to lower-priced items such as basic breaded fish products and minced fish products. More recently, the trend in food processing has moved toward the production of ready-to-eat and ready-to-serve products that need only to be heated in the oven or microwave to be served.

Processing of seafood and aquaculture products is very diverse and depends on species and products. Processing of products ranges from simple cleaning, dressing, and icing to elaborate grading and processing schemes. A simple dressing process typically entails removing viscera and gills of fish, leaving the head on. Dressing produces semi-products in rudimentary condition that usually undergo further processing into ready-to-cook and ready-to-serve products. Secondary processing treatments include heading and gutting, cutting products into chunks, deboning, filleting, buttering, breading, stuffing, canning and packing.

Generally, the processing operations involve receiving raw seafood/fish, washing, deheading, peeling/skinning, grading, washing, blanching, cooking, cooling, freezing (IQF), glazing, glazing freezing, packaging, and cold storage at about $-20°$ C. Specific examples will help to illuminate the processes involved.

The largest seafood processing plant in the United States, owned by Alaska Seafood International, is based in Anchorage, Alaska. The plant is 202,000 sq ft and has a capacity of processing 57 million kg (125 million lb) of seafood per year. Value-added products produced from the plant include glazed and coated portions of salmon, halibut, and cod, salmon burgers, and fish tenders products. For glazed and coated fish portions, the processing operation involves removal of the fish collar and skin, filleting, and injection with a marinade. Fish products are then cut into portions that include the loin and the center cut or tail. These are then weighed and sorted automatically into 113–170 g (4–6 ounce) portions. The portions are then frozen and either packaged for shipment or further processed by the addition of glazes or transformed into other value-added products (Mermelstein, 2002). For further processing of fish portions, a press stamps out the portions that feed into the breading machine or into the steaking machine to produce steaks of salmon, halibut, swordfish, and tuna. A glazing line applies flavored glazes to the fish portions. Some of the flavors used are garlic, butter, and honey-sesame ginger to enhance the flavor. Flavored marinade are either injected directly into the fillets or incorporated by vacuum tumbling.

The processes to make salmon burgers at the Alaska plant involve chopping frozen salmon fillets in a bowl chopper with other ingredients and then forming products into a 28–57 g (1–2 ounce) burger shape that is battered, breaded, and then par-fried. The burgers simply require heating in an oven at 191–204° C (375–400° F) for 10–15 min to be ready to eat. The bowl chopper has the capability of vacuum infusing the product with marinade. Liquid carbon dioxide can also be added as a fog to lower the temperature to below freezing to facilitate forming (Mermelstein 2002). There are various packaging lines, including vacuum packaging for retail sales and bulk packaging. Natural portions are packaged in Styrofoam trays with a clear over-wrap or are vacuum packed and frozen for retail sale. Some products, such as the Cheese Salmon Tenders, are frozen and packaged in 1-kg (2.5-lb) standup polybags specifically for Sam's Club.

The basic first stage in processing channel catfish, the species raised by the largest segment of aquaculture in the United States, is to produce a form referred to as a whole-dressed fish (headed, gutted/ eviscerated, and skinned, or HGS). Whole fish may sometimes be headed and gutted (or H & G) with tail and fins intact. Whole-dressed catfish was the major product form sold in the early years of the industry, but changes in fish and seafood markets have led to decreasing demand for whole-dressed catfish. This whole-dressed product typically undergoes further processing in which it is cut into a variety of forms that includes fillets (with belly flap), shank fillets (boneless fillet with belly flap or nugget removed), strips/fingers or fillet strips (boneless finger-size pieces cut from shank fillets), nuggets (belly flap section removed from fillet), and steaks (Fig. 5.1). Secondary processing also includes the production of breaded fillets and nuggets, heat-set, breaded fillets, portions and nuggets, marinated fillets, and smoked fillets and dressed fish (Silva and Dean 2001).

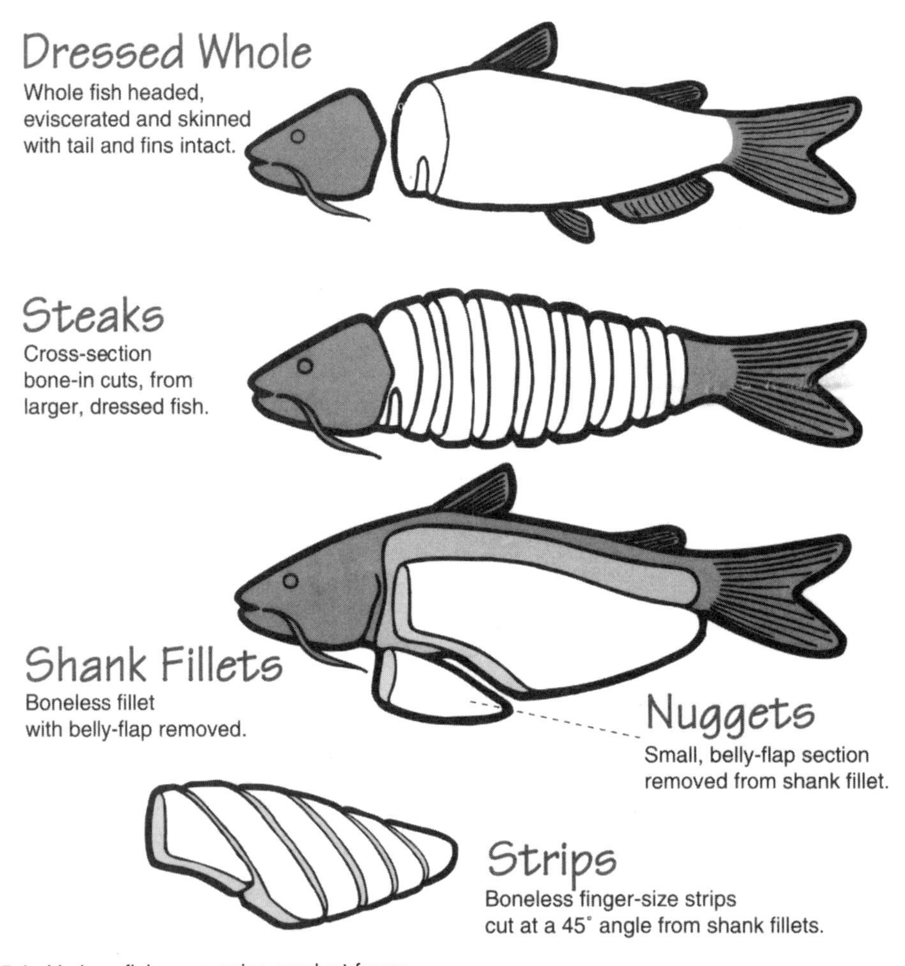

Figure 5.1. Various fish processing product forms.

Figure 5.2 is a flow chart of the production technology for processing trout and carp fillets in Poland, and Figure 5.3 is a flow chart for processing marine white fish in the United States. From Figure 5.3, the fish are first gutted, cleaned, and sometimes deheaded on board the fishing vessel before landing. The fish are kept on ice in boxes until delivered to the processing plant. At the plant, pretreatment of the fish involves the removal of ice, washing, grading, and deheading, if not done previously. Large fish may also be scaled before further processing. Filleting is the next step in the process, typically done by mechanical filleting machines. The filleting machines comprise pairs of mechanically operated knives, which cut the fillets from the backbone and remove the collarbone. Some fish fillets may also be skinned at this stage. Trimming involves removal of pin bones and any remaining irregularities.

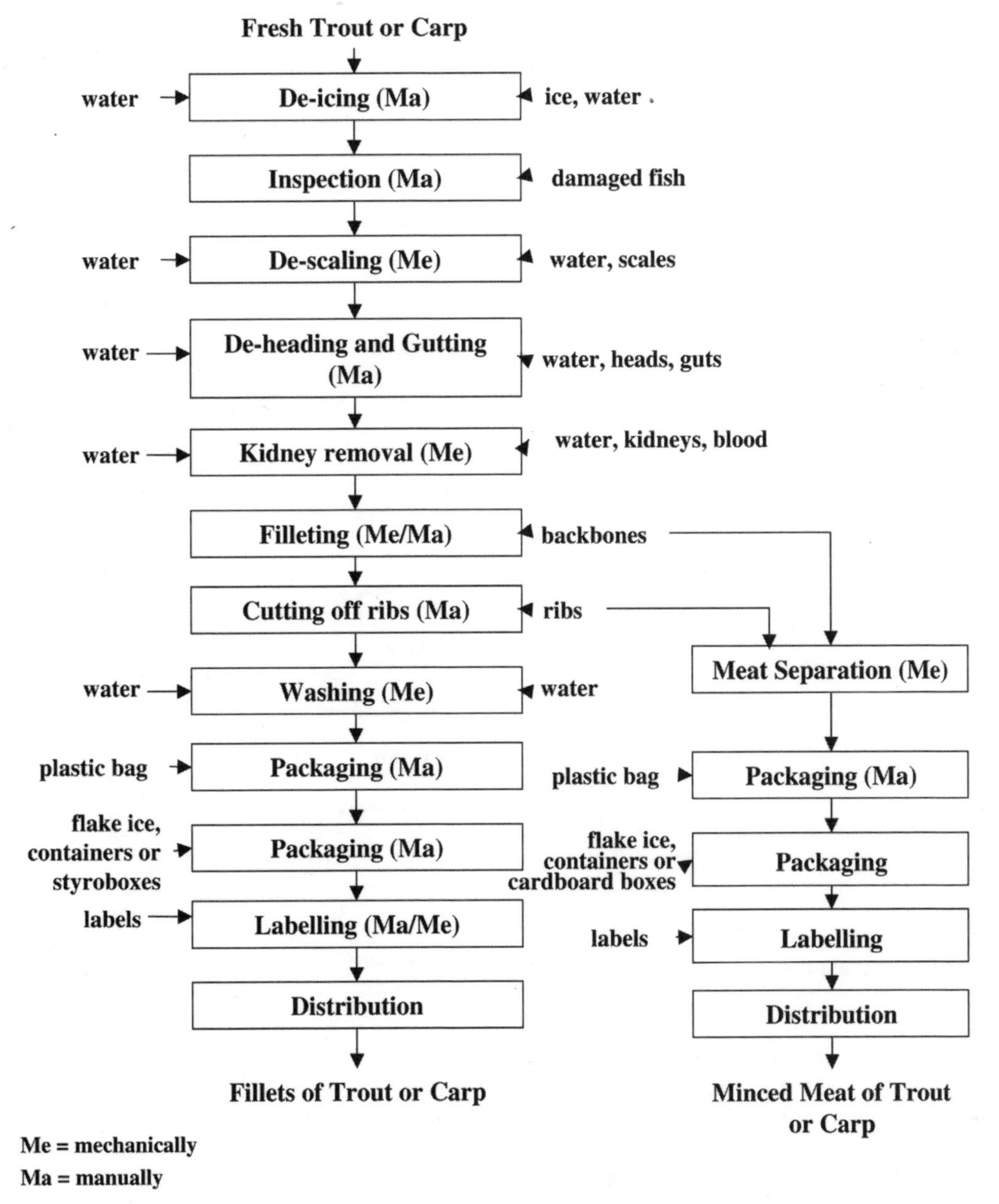

Figure 5.2. Production technology of trout and carp fillets in Poland (Bykowski and Dutkiewicz 1996).

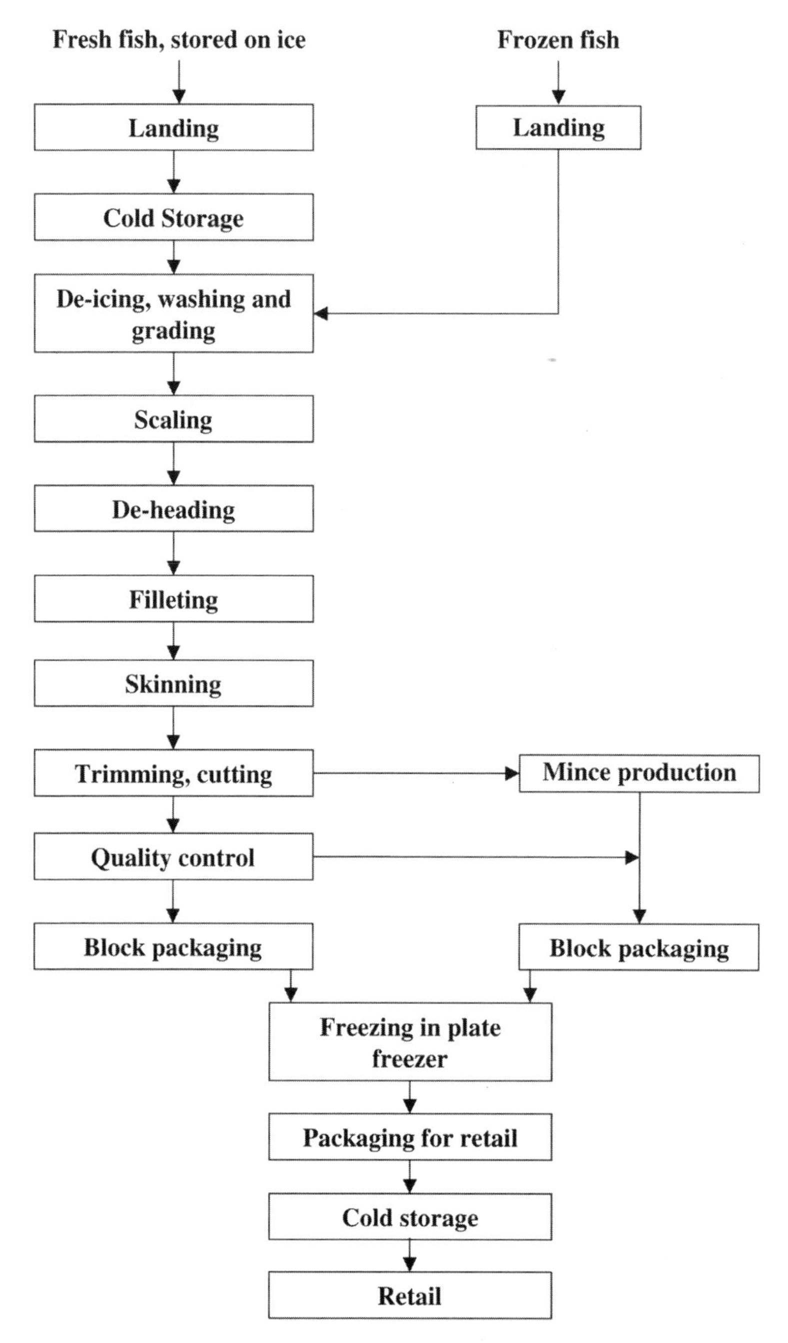

Figure 5.3. Process flow diagram for the filleting of white fish.

Fillets are then inspected, defects and any parts that are of inferior quality are removed, and off-cuts or mis-cuts are collected and minced. The fillets then are cut into portions according to weight depending on the final product or are divided into parts such as the loin, tail, and belly flap. The final step is inspection of the fillets to ensure that they meet product standards before being packed for shipment. Fresh products are packaged in boxes with ice, the ice being separated from the products by a layer of plastic. Frozen products can be packed in a number of ways. Fillets or pieces can be individually frozen and wrapped in plastic, but the most common method is packing them as frozen fillets in 6- to 11-kg blocks in waxed cartons. The blocks are typically frozen and then kept in cold storage.

Processed seafood products may be sold fresh, frozen, smoked, seasoned, canned, dried, or dehydrated. Inedible and substandard portions of processed products are usually used to produce fishmeal products used for animal feed.

Processing may also involve an extremely complex set of techniques and ingredients that transforms raw products into food products that are tasty, nutritious, and ready-to-eat food products requiring minimal preparation and cooking, or formulated food products such as surimi. Surimi is an important fish product, with the majority of catches for some fish species used solely for its production. Complex processing results in production of what is generally known as value-added aquaculture and seafood products. Adding value in seafood processing generally implies a degree of processing that makes the seafood product more desirable to consumers, which may relate to better appearance, taste, texture, flavor, or more convenience. Value adding may also relate to processing products to improve shelf stability and functionality.

Freezing and storage are important to maintaining the quality of processed products. High-quality storage provides the processor with a means of controlling products to ensure consistency in supply, quality, and shipments to distributors or retailers. Supply of raw fish materials either from aquaculture facilities or natural catches from the oceans can be seasonal. Freezing and storage provide a means of stabilizing product temperatures and accumulating complete lots or loads for direct shipment to buyers with minimal repacking, transfer, and temperature fluctuation (Kolbe and Kramer 1997). The recommended temperature, relative humidity, and approximate storage life for selected seafood products are presented in Table 5.1.

Table 5.1. Recommended Temperature and Relative Humidity, and Approximate Transit and Storage Life for Seafood.

Product	Temperature		Relative humidity (%)	Approximate storage life
	°C	°F		
Haddock, Cod, Perch	−1 to 1	31 to 34	95 to 100	12 days
Hake, Whiting	0 to 1	32 to 34	95 to 100	10 days
Halibut	−1 to 4	31 to 34	95 to 100	18 days
Herring, kippered, smoked	0 to 2	32 to 36	80 to 90	10 days
Mackerel	0 to 1	32 to 34	95 to 100	6 to 8 days
Menhaden	1 to 5	34 to 41	95 to 100	4 to 5 days
Salmon	−1 to 1	31 to 34	95 to 100	18 days
Tuna	0 to 2	32 to 36	95 to 100	14 days
Frozen fish	−29 to −23	−20 to −10	90 to 95	6 to 12 mo
Clams (shucked meats)	−1.7	29	85 to 90	5 days
Crabmeat, pasteurized	0 to 1.1	32 to 34		6 mo
Crabs, King, Snow, cooked, frozen	−18	0		12 mo
Crabs, Dungeness, cooked, frozen	−18	0		3 to 6 mo
Scallop meat	0 to 1	32 to 34	95 to 100	12 days
Shrimp	−1 to 1	31 to 34	95 to 100	12 to 14 days
Lobster, American, live	5 to 10	41 to 50	in water	indefinite
Lobster, American fresh meat	−1.1 to 0	30 to 32	90 to 95	3 to 5 days
Lobster, American, frozen, shell	0	−18		3 to 6 mo
Lobster, meat, cooked, frozen	0	−18		6 to 9 mo
Lobster, Spiny, frozen, shell	0	−18		10 to 12 mo
Oysters, meat	0 to 2	32 to 36	100	5 to 8 days
Oysters, clams, in shell	5 to 10	41 to 50	95 to 100	5 days
Frozen shellfish	−29 to −20	−20 to −4	90 to 95	3 to 8 mo

Source: The Refrigeration Research and Education Foundation, 1996; American Society of Heating, Refrigeration, and Air Conditioning Engineers, Inc., 1994.

STRUCTURE OF THE SEAFOOD AND AQUACULTURE PRODUCT PROCESSING INDUSTRY

Processing of seafood and aquaculture products primarily takes place in processing establishments, but some large fishing vessels that operate in deep waters have facilities on board where seafood and fish are processed. Some fishing vessels both catch and process seafood and fish while other vessels serve mainly as processing ships.

The National Marine Fisheries Service (NMFS) periodically reports the results from annual surveys of all seafood processors that operate in the United States. In 1994, the United States was home to 1,504 seafood-processing plants, which employed nearly 55,744 workers (National Marine Fisheries Service 1995). However, by 2002 there were only about 935 processing plants engaged in processing fresh and frozen seafood and aquaculture products, employing about 44,489 workers (National Marine Fisheries Service 2004). These establishments are primarily engaged in one or more of the following: (1) eviscerating fresh fish by removing heads, fins, scales, bones, and entrails; (2) shucking and packing fresh shellfish; (3) manufacturing frozen seafood; and (4) processing fresh and frozen marine fats and oils. Processed fresh and frozen products include fish fillets and steaks, fish sticks and portions, and breaded shrimps. In 1995 the U.S. production of raw (uncooked) fish fillets and steaks, including blocks, was 175 million kg (385.3 million lb) valued at $840.9 million. Alaska pollock fillets and blocks led all species with 61.6 million kg (135.5 million lb). Production of groundfish fillets and steaks was 98.5 million kg (216.7 million lb). The combined production of fish sticks and portions was 147.9 million kg (325.4 million lb) valued at $429.7 million in 1995. The production of breaded shrimp in 1995 was 45.6 million kg (100.4 million lb) valued at $298.7 million. In 2003, however, the U.S. production of raw (uncooked) fish fillets and steaks, including blocks, was 277.9 million kg (611.4 million lb). These fillets and steaks were valued at $1.1 billion. The combined production of fish sticks and portions was 88 million kg (193.6 million lb), valued at $261.7 million. The production of breaded shrimp in 2003 was 69 million kg (152.0 million lb), valued at $465.3 million (National Marine Fisheries Service 2004).

The National Marine Fisheries Service (NMFS) reports cited above indicate that between 1994 and 2003, the number of processing plants decreased by about 38% and employment by 20%, but production of fresh and frozen products increased by about 18%. The trend is a reflection of the expansion through mergers and acquisitions in the food processing industry in the United States during the past two decades.

Economists and policy makers are often interested in the structure of these companies because the structure could have implications for market performance. The structure of the seafood and aquaculture products industry in the United States relates to the concentration of the industry, the degree of vertical integration, product characteristics, and freedom of entry and exit (Figure 5.4). Each of these industry features is discussed below.

The number and quality of firms competing in an industry is sometimes thought to determine the nature of competition in the industry, depending on the industry concentration. When the number of firms operating in the industry is sufficiently large, and the product handled by the industry is homogenous or standardized, the industry is said to be perfectly (pure) competitive. Pure monopoly applies to the industry when there is only one firm in the industry with a unique product that has no close substitute. In between pure competition and pure monopoly are monopolistic competition and oligopolistic competition. Monopolistic competition is a blend between monopoly and perfect competition. The oligopolistic structure has few firms, and products may be differentiated (Table 5.2). The structure of the seafood and aquaculture products industry is typical of a monopolistic competitive industry, with a relatively large number of firms operating competitively in the

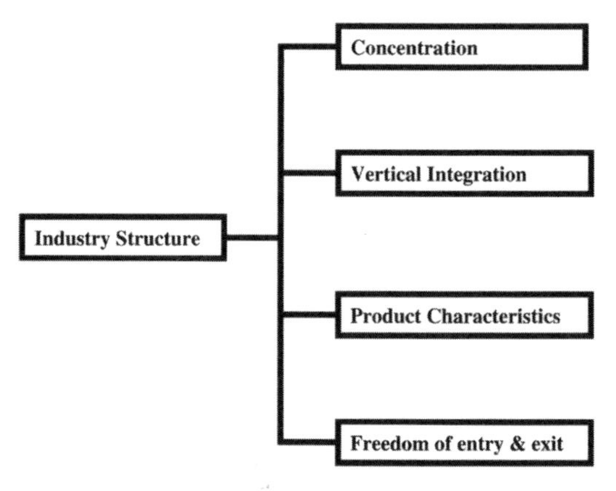

Figure 5.4. Theoretical characteristics of industry structure.

production of differentiated products (U.S. Department of Commerce—National Oceanic and Atmospheric Administration 2004). However, Dillard (1985) and Kouka (1995) concluded that the catfish-processing sector is somewhat oligopolistic because the industry is dominated by a few relatively large companies, with frequent price wars that have tended to keep prices received by processors below cost. The price wars resulted in a high turnover in the industry. Wiese (2004) reported that 85 catfish processing plants entered the industry whereas 72 plants exited the industry since 1981. The average length of time in the catfish processing business was five years. However, there are some "successful" processing plants that have been in business for more than 20 years. Although some seafood processing industries may be relatively concentrated, easy entry makes the industry quite competitive and less concentrated because of relative prices, abundance of raw materials, and government policies.

The market structure of the processing industry includes both the market for processed fish products and raw fish material, in which processors serve as buyers. The logical counterpart of a monopolist market is the monopsonistic market, in which there is only one buyer. Similarly, an oligopsonistic market implies few buyers of homogeneous or differentiated products, and there is costly entry or exit of buyers. The market structure in the market for raw seafood products may be somewhat different. Hackett and Krachey (2004) suggest that various markets for landed fish ranged between being moderately concentrated and concentrated on the buyer side. Radtke and Davis (2000), in their study of U.S. West Coast processors, reported that in 1997, California processors could be characterized as oligopsonists in the market for fish because the 15 largest processing companies or parent groups processed 65% of the fish by volume and 46% of the total fish by value. The authors report that the processing industry in California has experienced additional consolidation since 1997.

Concentration

Industry concentration is the percentage of business (share of total value of shipments) accounted for by a number of businesses in the industry. However, an industry with a large number of firms may not always be competitive. The U.S. Census Bureau (USCB) reports the concentration ratios for the 4, 8, 20, and 50 largest fresh and frozen seafood and aquaculture products processing companies in 1997 as 14%, 23%, 42%, and 66%, respectively (U.S. Department of Commerce—U.S. Census Bureau 2001). The seafood and aquaculture product processing industry is much less concentrated than other meat processing industries such as the beef and pork packing industries, in which the concentration ratios for the 4, 8, 20, and 50 largest companies are 57%, 71%, 82%, and 90%, respectively (U.S. Department of Commerce—U.S. Census Bureau 2001). The four-firm concentration ratios increased from 14–41% in poultry, 33–43% in hogs, and 26–71% in the cattle industry from 1963 to 1992 (Figure 5.5).

Concentration ratios have implications for competition and economic performance in the industry. Although a concentrated industry may be viewed by some as less economically efficient, high operational and financial performance of firms have often led to expansions, mergers, and acquisitions that result in a concentrated industry. Food processors often specialize by product line, but the trend has been to diversify and add additional product lines. Brands allow processors to differentiate product and certify product quality.

Larger firms may have the advantage of economies of scale where more can be produced at a

Table 5.2. Types of Product Market Structures.

	Characteristics			
Type of structure	Number of firms	Type of product	Control over price	Freedom of entry and exit
Perfect competition	Numerous	Homogenous	None	Very easy
Monopolistic competition	Many	Differentiated	Some	Relatively easy
Oligopoly	Few	Homogenous or differentiated	Some	Partially restricted
Monopoly	One	Unique	Considerable	Absolutely restricted

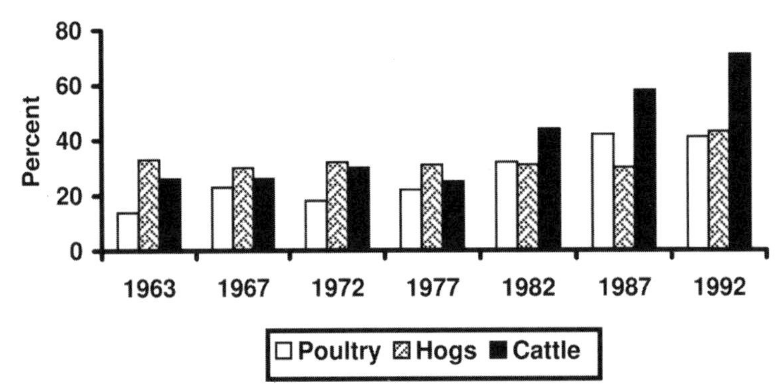

Figure 5.5. Four-firm concentration ratios for poultry, hog, and cattle industries, 1963–1992. (Source: USDA-ERS.)

lower per unit cost, but mergers and acquisitions are usually made with the intention of increasing market share. In 2000, Trident Seafoods Corporation acquired the seafood division of Tyson Foods, while Bumble Bee Seafood Incorporated acquired Tyson's surimi seafood business. Bumblebee Seafood Incorporated and Trident Seafoods Corporation were ranked second and fourth, respectively, among North America's top 25 seafood suppliers by the Seafood Business Magazine in 2000. Similarly, the merger in 1999 between Stolt SeaFarms and International Aqua Foods made Stolt SeaFarms the largest farmed salmon producer in North America.

Antitrust laws exist, however, to promote competition. The federal government usually challenges any cooperation and merger between firms that will result in a monopoly or near monopoly. Exceptions occur when one of the merging firms is on the verge of bankruptcy, in which case the federal government may allow such a merger. In some cases, however, action is taken toward existing concentrations if there is evidence that the firm that has more than 60% of the market used deliberate conduct to achieve dominance (for example, the federal government's antitrust actions against Microsoft in 1998 alleging violations of the Sherman Act[1]). The merger of Stolt SeaFarms and International Aqua Foods was the result of the Fisheries Act, which required a minimum of 75 percent ownership by Americans (see Appendix 5.A, at the end of this chapter, for details of this legislation).

Mergers often result in closure of some processing plants and reductions in the labor to streamline the production base and improve overall operating efficiencies and competitiveness. Between 1992 and 1997, the number of establishments engaged in processing fresh and frozen seafood and aquaculture products declined by about 2.3% (U.S. Department of Commerce—U.S. Census Bureau 2001). The decline is largely the results of mergers and acquisitions.

Vertical Integration

Some seafood and aquaculture products companies are vertically integrated because they operate fish farms as well as fish processing plants. When firms operate at more than one level of a series of levels leading from raw materials to the final consumer, they are considered to be vertically integrated. In some cases, firms become vertically integrated as a result of merging firms at different stages of the production process (vertical merger). Firms integrate vertically for several reasons, including: (1) cost reduction; (2) economies of scope and diversification; and (3) greater business strength. Firms could also integrate vertically to monitor and maintain quality along the production process.

With vertical integration, the whole production process of seafood and aquaculture products from harvest to the final consumer is divided and undertaken by separate firms. The Red Chamber (Group) Company is a family-owned business that was ranked third in the 2000 Seafood Business Magazine's ranking of the top 25 seafood companies. Their seafood operating companies include a processing company, Neptune Foods, and the fishing companies, Avalon Bay and Tampa Bay Fisheries. Carolina Classics Catfish, an aquaculture company, is a vertically integrated company that produces feed (operated under Carolina Fish Feeds), grows fish, and processes and delivers catfish products. Idaho Trout Processors Company operates the trout farms

Rim View Trout, Rainbow Trout Farms, and Clear Lakes Trout Farm as well as a processing company. Some of these vertically integrated companies involve a group of producers who collectively own the processing company. For example, Delta Pride Catfish, Inc. is a catfish processing company owned by about 115 catfish producers in the Mississippi Delta area.

Product Characteristics

A monopolistic competitive industry often consists of firms producing a differentiated product, such that each firm's output is distinguishable from any other firm's output. Products may be differentiated through physical attributes, functional features, material make-up, packaging, advertising, and branding. In the seafood and aquaculture products industry, there are diverse products including shellfish, finfish, scaled fish, and other unclassified fish. Final products in the form of buttered, breaded, stuffed, dried, marinated, and canned products and the different brands of these products make seafood and aquaculture products different from one another. The U.S. Department of Commerce lists as many as 120 different seafood and aquaculture products and more than 1,600 brands of products that are produced by fish processing establishments (Table 5.3).

Table 5.3. Inspected Fishery Products Produced in USDC-Approved Establishments.

Raw portions, sticks, nuggets, etc.	Marinated fillets
Raw steaks	Breaded raw portions, sticks, nuggets, etc.
Raw fillets	Breaded raw strips
Raw whole	Breaded raw fillets
Raw dressed head off/on	Breaded precooked portions, sticks, nuggets, etc.
Raw dressed and boned	Breaded precooked strips
Breaded raw portions, sticks, nuggets, etc.	Breaded fully cooked minced cakes, patties, or burgers
Breaded raw fillets	Seafood frozen
Breaded precooked fillets	Breaded precooked minced cakes, patties, or burgers
Fish frozen	Crab fresh/refrigerated
Raw portions, sticks, nuggets, etc.	Breaded raw cakes, patties, or burgers
Raw steaks	Crab frozen
Raw stuffed	Raw cakes, patties, or burgers
Raw cakes, patties, or burgers	Raw soft shell
Raw fillets	Cooked soups
Raw blocks	Breaded precooked cakes, patties, or burgers
Raw minced	Crab canned
Raw dressed and boned	Dips and spreads
Breaded raw portions, sticks, nuggets, etc.	Lobster fresh/refrigerated
Breaded raw steaks	Live
Breaded raw strips	Shrimp frozen
Breaded raw stuffed	Raw headless
Breaded raw fillets	Raw whole
Breaded raw meat	Marinated meats
Breaded precooked portions, sticks, nuggets, etc.	Breaded raw imitation
Breaded precooked croquettes	Breaded raw meats
Breaded precooked portions, sticks, nuggets, etc.	Breaded raw whole
Breaded precooked strips	Breaded precooked crisps
Breaded precooked cakes, patties or burgers	Breaded precooked dinners
Breaded precooked fillets	Breaded precooked meats
Breaded fully cooked portions, sticks, nuggets, etc.	Batter coated precooked meats
Breaded fully cooked fillets	Breaded precooked minced
Batter-coated precooked portions, sticks, nuggets, etc.	Peeled raw meats
Batter-coated precooked strips	Peeled raw deveined
Batter-coated precooked fillets	Peeled raw whole

(*continued*)

Table 5.3. Inspected Fishery Products Produced in USDC-Approved Establishments (*continued*).

Breaded raw minced portions, sticks, nuggets, etc	Raw dressed and skinned
Breaded precooked minced portions, sticks, nuggets, etc.	Peeled cooked deveined
Breaded precooked minced cakes, patties or burgers	Oyster fresh/refrigerated
Breaded fully cooked minced portions, sticks, nuggets, etc.	Live
Batter coated precooked minced portions, sticks, nuggets, etc.	Raw shucked
Farm-raised catfish fresh/refrigerated	Oyster frozen
Raw bellies	Breaded raw meats
Raw portions, sticks, nuggets, etc.	Breaded raw whole
Raw steaks	Scallop fresh/refrigerated
Raw strips	Raw shucked
Raw fillets	Breaded raw whole
Raw dressed and skinned	Scallop frozen
Marinated bellies	Raw shucked
Farm-raised catfish fresh/refrigerated (continued)	Breaded raw whole
Marinated steaks	Breaded precooked whole
Marinated fillets	Batter coated precooked whole
Marinated dressed and skinned	Squid fresh/refrigerated
Farm-raised catfish frozen	Raw whole
Raw bellies	Squid frozen
Raw portions, sticks, nuggets, etc.	Breaded raw tubes and/or rings
Raw steaks	Surimi fresh/refrigerated
Raw strips	Cooked analog
Raw fillets	Surimi frozen
	Cooked analog
	Breaded precooked minced analog

Source: U.S. Department of Commerce—NOAA (2004).

Entry into the Industry

New firms enter an industry if they expect to make a profit. However, in a monopolistic competitive industry, firms make nominal profits. Normal profit is the amount of profit necessary to induce the firm to stay in business without either excessive or minimal levels of profit. Natural barriers that could restrict free entry into the seafood and aquaculture products processing industry include economies of scale, the large capital outlays, ownership of essential raw materials, advertising and product differentiation, sunk costs, and government policies. Knapp et al. (2001) suggest seven important "reality checks" for anyone planning a fish processing plant at the village or local level: (1) availability of fish; (2) current and future competition from other processors; (3) availability of good plant management; (4) availability of skilled production workers; (5) availability of water and power, and waste disposal; (6) marketing of products; and (7) availability of reliable transportation to take products to market. The authors conclude that it is only after considering these issues that one can move on to the planning phase. Planning involves addressing issues such as products to produce, markets for the products, kinds of building and equipment needed, financing, and, most important, whether the plant can earn enough money to stay in business.

Economy of size relates to the efficiency of large firms; thus large-scale processing operations with the associated large capital outlay could be a hindrance to the entry of new firms. Vertical integration allows firms to control the raw materials of catch or production of seafood and fish needed for the processing establishment. Product branding is a major cue to consumer behavior, and processing companies have different lines of branding to differentiate their products from similar products produced by competing companies. Nonrecognition of new brands by consumers could be a barrier to potential new entrants to the industry. All sorts of government policies including ownership requirements, licensing, trademark protection, and regulations can become barriers to entry. A key objective of the American Fisheries Act is the 75% minimum U.S. ownership of fishing vessels operating in U.S. waters. Prior to the Act, some major fishing companies had majority foreign ownership. The seafood and aquaculture products industry has seen more mergers and acquisitions during the past decade than new entrants into the industry.

PLANT LOCATION

Proximity to inputs, availability of services, and the type of marketing system needed by a company greatly determine the location and the size of processing plants. Generally, firms would expect the costs associated with obtaining raw materials and essential services including technology, labor, communication and transportation, and access to the markets for their outputs to be low. There is a web of linkages among industries because the output of some firms and industries constitutes the inputs of other firms and industries. This linkage allows firms to realize substantial cost advantages due to proximity. Thus, economies of location play a significant role in the choice of location for processing plants. Bykowski and Dutkiewicz (1996) suggest that the most important factor that should be considered for locating a processing plant is adequate size for both present needs and future development. A well-designed building should comprise sufficient space for work to be conducted under adequate hygienic conditions, an area for machinery, equipment and storage, separation of operations that might contaminate food, adequate natural or artificial lighting, ventilation, and protection against pests (Bykowski and Dutkiewicz, 1996).

Bykowski and Dutkiewicz (1996) suggest a location close to public transport such as rail or road, access to electricity, water, and steam, and adequate waste disposal. The local authorities should be actively involved in the process in order to avoid problems in the future. In the United States, seafood processing plants are commonly located along the seaboards, whereas processing plants for aquaculture products are located in major aquaculture production regions. The Quonset-Davisville Port and Commerce Park in Rhode Island is a location with extensive infrastructure and facilities including deep-water access and airport, rail, and highway connectors. Bridges on the rail lines accommodate double-stack containers. Seafood companies such as Seafreeze Ltd., American Mussel Harvesters, and others are located at the port. The infrastructure and facilities at the port continue to attract new seafood processing companies. Major catfish processing plants are located within a 50-mile radius in the Mississippi delta region, where more than 80% of catfish production takes place. The region also has access to state and interstate highways, railways, and regional industrial parks.

Though fish processing operations are located close to commercial fishing areas, catches may be transported long distances or exported for processing in some cases. The *EUROFISH Magazine* (January/February 2001 issue) reports that Danish exports of unprocessed fish have almost doubled since 1983 but there has been a decline in exports of processed fish product within the same period. This is because more of the fish landed by the national fleet is exported unprocessed, resulting in the processing sector's becoming increasingly dependent on imported raw materials. Also, the Polish market is one of the fastest developing fish markets in Europe. Poland has a strong fish processing sector but a weak fishing sector. Therefore, Poland imports fish for processing, including herring, mackerel, Alaska pollock, hake, salmon, and cod (Publishing House MPR Ltd. 2004).

LAW OF MARKET AREAS

The location decision for any business involves a consideration of the sales potential that exists in the area or region. The marketing areas of processors involve sales to wholesalers and retailers. Combinations of geographic, demographic, economic, and competitive factors will determine the market area that a processing company will service. In particular, human resources and costs of operation are important. Market areas have different demographic characteristics, competitive factors, and sales potentials. The size of the market area offers opportunities for or constraints to potential sales and expansion and business development in general. Mountain ranges, rivers, and road patterns can define market areas and influence the nature of transportation systems that are important to the distribution of processed products. Highway speed limits, the nature of roads, highway access, bridges across rivers, and general topographic features determine trading patterns within and between market areas. Wholesale distribution of products would depend on the number and proximity of potential outlets to serve and the quality of transportation that would enable delivery to clients in near and distant market areas. Transportation characteristics affect transportation costs, delivery policies, and delivery structure of products.

Businesses usually target markets in urban areas and areas of larger population because of high demand. Apparently, these large market areas are where more direct competition among suppliers exists. Each major market area has its own unique characteristics in terms of age distribution, number and types of households, income levels, work patterns, shopping patterns, retail sales levels, economic health, and so on. Therefore, demographic interest considered in assessing a market area would focus on personal income, education level, age, and lifestyles of potential workers as well as customers within the market area.

CAPACITY UTILIZATION

Economic theory suggests that low-capacity utilization or excess capacity is one of the characteristics of a monopolistic competitive industry. Individual companies do not disclose their processing capacity for confidentiality reasons, but the U.S. Census Bureau reports the rate of production capacity utilization on an industry-wide basis (U.S. Department of Commerce—U.S. Census Bureau 2003). The rate of production capacity utilization is the ratio of total capacity utilized relative to the total processing capacity available. Industries with full utilization of processing capacity are characterized by two and three daily shifts of production workers, but this is usually not the case with the seafood and aquaculture processing products industry. The rate of capacity utilization in seafood and aquaculture processing has averaged 60% over the last 10 years as compared to 85% for the meat (beef, pork, and poultry) processing industry (U.S. Department of Commerce—U.S.Census Bureau 2003).

In Canada, optimistic projections for fish stocks in the 1970s led to a significant increase in fish processing capacity in the Atlantic provinces, particularly with the anticipated extension of Canada's economic zone to 200 miles. Excess processing capacity was also built to meet the peak landings from the seasonal inshore fishery (Fishing Industry Renewal Board 1997).

Product differentiation is another characteristic of a monopolistic competitive industry. The more products are differentiated, the less elastic the demand curve for the products. With inelastic demand, production does not take place at the minimum of the average cost curve. This leads to low-capacity utilization or excess capacity. The seafood and aquaculture industry produces differentiated products that provide a number of varieties of seafood products to meet the diverse tastes of seafood consumers. Consumers have a wider range of seafood and aquaculture product choices. The industry offers a variety of marine and aquaculture products in terms of fish species, cuts and portions, preparations, and cooking forms. For example, American Pride Seafoods produces a variety of seafood products including fresh and frozen Atlantic salmon fillets, batter-dipped cod, Alaska pollock and whiting, baked, broiled, and breaded cod and pollock portions, fried, natural-shaped pollock, skinless boned cod loins and cod, pollock and haddock fillets, raw, breaded cod, haddock and flounder portions, minced cod, fresh and frozen catfish fillets, breaded and marinated catfish, whiting, sea scallops, and frozen whole sea scallops.

INNOVATION AND BRANDING

The art of marketing is recognizing the factors underlying the seafood preferences of buyers and producing to meet those preferences. Meeting the diverse preferences of buyers of seafood and aquaculture products requires innovation in the development of new products, modification of old products, and presentation of new products in better ways to buyers. Companies that usually want an edge over the competition find innovative ways of staying ahead through the development of new products, new processing operations or formulation of new ideas for marketing products. Associated with this is branding of products for company identification and product differentiation.

At the 2004 Boston Seafood Show, a variety of new seafood products were showcased. Some of the highlights include products such as presliced cold smoked salmon (1-1.4 kg; 2-3 lb) with mild distinctive flavor; a hot smoked salmon variety pack of four portions of three types of Cajun and cracked pepper; frozen salmon portions with Chef Paul Prudohomme's Magic Seasoning in a 1-kg (2-lb) retail resealable bag; and 142–170 g (5–6 oz) salmon portion individually vacuum packed and with deep skin off. Aquafarms International exhibited these products. Claw Island Foods of Maine exhibited 85-g (3-oz) hand-made lobster cakes, and Clear Spring Foods displayed boneless Maplewood smoked Idaho rainbow trout fillets, plain or smoked with cracked black pepper, or with lemon or roasted garlic. Other products exhibited at the show were top-crusted seafood fillets including Mediterranean Style Crusted Salmon, Summer Herb Crusted Cod, Coconut Crusted Tilapia, Potato Crusted Cod, Tortilla Crusted Tilapia, and Sesame Crusted Salmon; Tilapia with Butter Crumb Pecan Topping; Gourmet Cuisine "Low Carb" Skillet Meals; Fettuccine Primavera with Shrimp and Broccoli; wild salmon portions that were boned, trimmed, seasoned, and preseared with grill marks; and marinated bay scallops with Asian, Polynesian, roasted garlic, garlic, and rosemary flavors.

Branding of products is achieved through the use of a brand name, brand mark, logo, registered brand (®) or a trademark (™). Registered brands and trademarks are protected by law and meant for the exclusive use of the registered owner. Many consumers use brand names as cues for purchasing products because it provides some sense of satisfaction and security. Thus, product branding requires that companies ensure consumer familiarity with their brands. In the seafood and aquaculture industry, brand names are associated with company reputation that may relate to specific products, quality, price, packaging, or organic qualities. Some companies and brand names are associated with one line of products. An example is StarKist® which offers different tuna products including Flavor Fresh Pouch™, Naturally Low Sodium-Low Fat Tuna, Chunk Light Tuna, Lunch To-Go, Solid White Albacore Tuna, Gourmet's Choice Tuna Fillets, Select Prime Light Fillets, and Tuna Creations™. Contessa® is associated with shrimp, which is available cooked or uncooked and tail-on or tail-off. Other seafood companies such as Bumble Bee® offer several product lines including albacore tuna, salmon, shrimp, crab, oysters, clams, and ready-to-eat tuna salad. Bumble Bee prides itself on quality of products. Wild Oats Market, Inc. boasts of being a leader in the natural and organic food industry.

Traditionally, many manufacturing companies have adopted a new-product development cycle of four to five years. However, the dynamic nature of consumer preferences for food products has necessitated a rather accelerated process of new product development. Many new products introduced to the market have focused on attributes such as convenience and health. New products and ideas stimulate excitement and curiosity in consumers that can translate into purchase. Some of the new ideas and products by the seafood industry in the past decade include ready-to-eat products, resealable retail packs, reduced fat products, and fish products with no preservatives.

CHALLENGES IN AQUACULTURE PRODUCT PROCESSING

One of the objectives of the U.S. Department of Commerce is to increase the value of domestic aquaculture production from the present $900 million annually to $5 billion, which will help offset the $6 billion annual U.S. trade deficit in seafood by the year 2025 (U.S. Department of Commerce 2004). The major farmed foodfish in the United States include salmon, trout, channel catfish, striped bass, tilapia, clams, crawfish, mussels, oysters, and shrimps.

The production of aquaculture is expanding and intensifying in the United States, and the total supply of fish available for consumption depends on future trends in the aquaculture industry. One of the challenges confronting the aquaculture processing industry is to realize the potential growth in the market for aquaculture products and pursuing a stable, sustainable, and competitive processing sector. The United States imports about 60% of its seafood needs, and with free trade and other bilateral and multilateral trade agreements in place, the domestic industry faces an increasing level of import competition. The major competition to processed foodfish has come from imported salmon and trout from Chile and Canada, basa/tra from Vietnam, crawfish from China, and oysters from South Korea. In 2002, the U.S catfish industry filed an antidumping suit against Vietnam because of depressed catfish prices (see details in Chapter 10 on this action).

Aquaculture products have traditionally been fresh and frozen raw products. However, consumer demand requirements for health, quality, and convenience necessitate a production process that is consumer oriented in order to take advantage of the market. Consumer demand for fish and seafood products continues to be strong due to the many nutritional and health benefits of consuming fish and seafood products. Scientific reports and government food guides continually cite fish and seafood products as being low in fat, easily digestible, and a good source of protein, important minerals, and vitamins. Besides seeking the safety and quality of these products, consumers with their busy schedules are looking for fish and seafood products and meals that can be cooked and served fresh in a matter of minutes. The challenge to the processing industry is to produce these types of products cost effectively to ensure that the domestic industry competes effectively.

The development of chain formation in the distribution and marketing system is another challenge that confronts the processing sector. Although some processing firms have developed their own sales organizations and marketing arms that are responsible for selling processed products, others have established strategic alliances and partnerships with retail and restaurant outlets in order to give better guarantees in terms of continuity and quality. For example, Idaho Trout Processors Company supplies fresh and frozen dressed whole trout products specially produced for the grocery giant Albertson's; similarly, the Alaska Seafood International produces frozen Cheese Salmon Tenders packaged in 1-kg (2.5-lb) standup polybags specifically for Sam's Club. Such partnerships help to achieve better coordination between links in the marketing chain while strengthening the distribution function of products. The challenge is to systematically utilize the information on the dynamics of the markets for the purpose of coordinating supply and demand, which could translate into the development of new products and product concepts.

The aquaculture industry is confronted with a series of environmental and health concerns relating to fish feed and pollution. Environmental groups and advocates of wild-caught fisheries have raised concerns relating to effluents from aquaculture production facilities and the quality of aquaculture products. The challenge for the processing industry is to address these issues and provide consumers with guarantees relating to the general quality, the functional quality, and the healthy image of aquaculture products as well as the eco-sustainability of aquaculture production practices. The key environmental issues associated with fish processing are the high consumption of water, the generation of effluent streams, the consumption of energy, and the generation of by-products. For some plants, noise and odor may also be concerns.

A few seafood processing plants concentrate on a single species, such as tuna, salmon or shrimp, but most plants process several different species to take advantage of the different fisheries in their region. This is not the case for aquaculture processing plants, which process mainly a single species of fish. Diversification into multispecies processing would afford the aquaculture processing sector the opportunity to better utilize processing capacity, become cost effective, and reduce marketing risks.

AQUACULTURE MARKET SYNOPSIS: U.S. CHANNEL CATFISH

Channel catfish production in the United States takes about 18 months from fry to a foodfish size of approximately 0.57 kg (1.25 lb). Ponds are partially harvested when there are 4,500 to 18,000 kg (10,000–40,000 lb) of market-sized fish. Fish are placed in aerated tank trucks and shipped live to the processing plant. Samples of catfish are first checked for flavor through a taste-test two weeks, one week, and the day before the fish are harvested. Fish are flavor checked again at the processing plant before the fish are unloaded. At the processing plant, catfish are kept alive, shocked, and processed within minutes before being packed on ice or frozen to temperatures of $-40°$ C ($-40°$ F) using a quick-freeze method. Quick freezing allows the flavor, taste, and quality to remain in the fish for longer periods of time.

In the United States, Mississippi, Alabama, Arkansas, and Louisiana account for 95% of total catfish production, with about 75% of catfish processing occurring in the state of Mississippi (Mississippi State University 2003). About 95 to 98% of all catfish production goes to processing plants, with a small percentage sold through other channels. Other channels include live-haul, fee and recreational fishing facilities, government agencies, and directly to consumers, retailers, and restaurants.

The size of catfish processors ranges from very small enterprises to relatively large businesses that produce various fresh, frozen, and value-added catfish products for wholesale and retail sales. Figure 5.6 presents trends in the number of catfish processors versus the quantity of catfish processed and sold. In Mississippi, there are 14 catfish processing plants that process a total of between 2.7 to 3.2 million kg (5–7 million lb) of catfish per week, with the capacity to produce up to 4 million kg (8.8 million lb). The largest catfish processor is Delta Pride in Mississippi, a producer cooperative company that processes an average of 0.7 million kg (1.5 million lb) a week. Heartland Catfish Company in Mississippi and Harvest Select Catfish in Alabama each has a processing capacity of about 0.35 million kg (0.77 million lb) per week. Haring's Pride Catfish in Louisiana processes about 0.25 million kg (0.55 million lb) per week in two plants, and Texas Aquaculture Cooperative has the capacity to process about 0.06 million kg (0.13 million lb) of catfish a week. There are a number of other catfish processing plants in Florida, North Carolina, Texas, Louisiana, Arkansas, and Kentucky.

Despite a fairly stable number of firms in the sector, processing capacity has expanded since 1981 to accommodate the increasing farm production capacity. Processing capacity represented by the quantity of catfish processed increased by about 200% between 1986 and 2002 (Fig. 5.6). Processed catfish products are differentiated in the form of dressed whole fish, fillets, nuggets, steaks, or value-added products (Fig. 5.7). Processed products are also sold fresh, frozen, breaded, marinated, or in some other value-added forms such as patties, smoked, and precooked frozen dinners or entrees prepared as heat-and-serve items to provide catfish buyers with a variety of catfish products. Generic advertising and

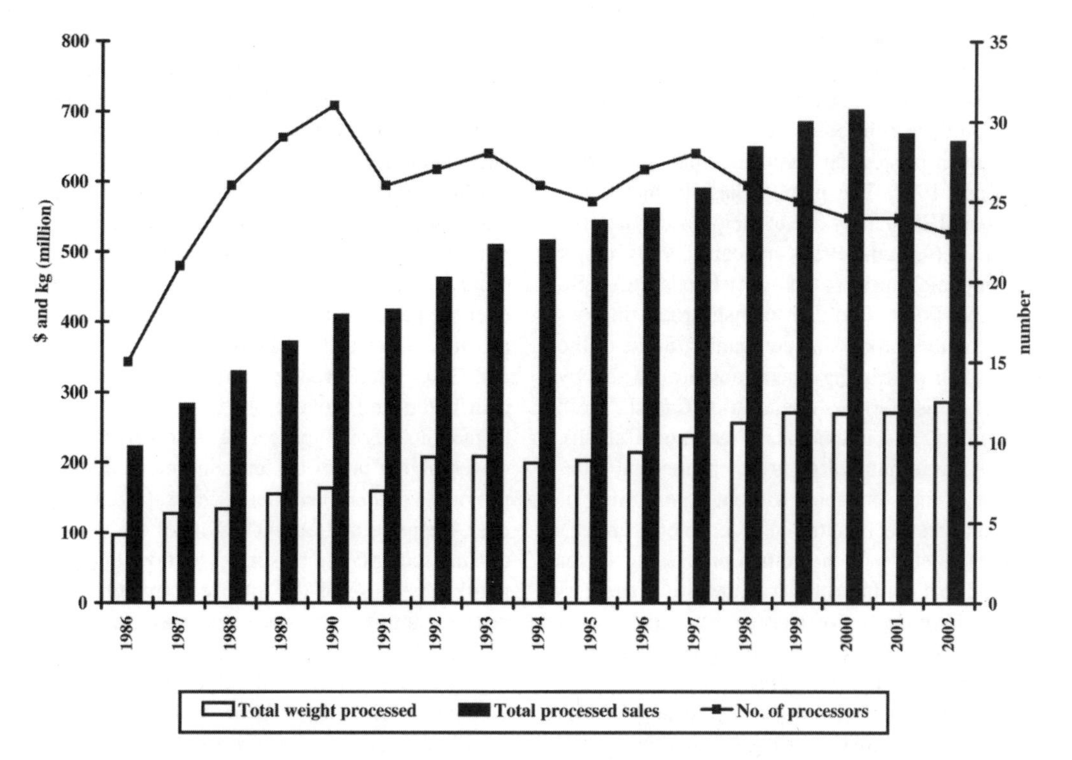

Figure 5.6. Number of catfish processors versus quantity of catfish processed and sold. (Source: USDA 2004.)

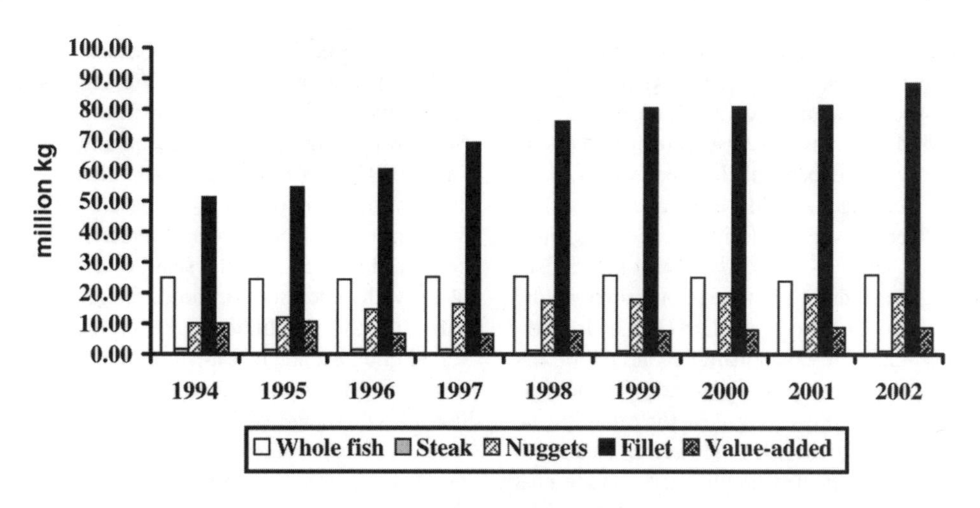

Figure 5.7. Processor sales volume of various catfish products (million kg). (Source: USDA 2004.)

promotional activities for catfish are the responsibilities of The Catfish Institute, which is funded through a feed check-off program. However, individual processing companies also try to differentiate their products through advertising, packaging, services associated with sales, and use of trademarks and brand names.

Theoretically, the catfish-processing sector conforms to a monopolistic competitive industry because the industry comprises a fairly large number of processing companies that compete with each other to produce a differentiated product. The number of catfish processing companies has not varied much since 1981. The number has fluctuated between 19 and 28, mainly due to mergers and acquisitions (Fig. 5.6). In the 1980s and early 1990s, major food companies such as ConAgra Foods®, Cargill, and Hormel Foods sold their catfish processing operations to other processing companies. Some of the major catfish processing operations are owned by groups of catfish farmers (Delta Pride Catfish, Inc.), farm families, and individuals (Heartland Catfish). The fairly large number of firms ensures the independence of the companies without a possibility of collusion to restrict quantity in order to boost price.

In 1981, a study of the catfish processing sector concluded that the industry structure was characterized by a high degree of market concentration. At that time, five of the 14 processing firms reporting to USDA were handling 98% of the total live weight of fish processed (Miller et al. 1981). Miller et al. (1981) also observed that there is a high degree of mutual interdependence among the processing companies in terms of pricing and other business policies. Excess processing capacity was also found in the sector.

Dillard (1985) suggested that the structure of the catfish processing sector falls somewhere between oligopoly and monopolistic competition, but perhaps more toward oligopoly because catfish processors are mutually interdependent. In other words, each processor recognizes that its output and pricing decisions influence its rival's decisions, and vice versa. However, Dillard was quick to add that the catfish-processing sector does not strictly conform to all the characteristics of oligopoly. Dillard (1985) also suggested that there are no short-run economic profits accruing to the catfish processing industry and estimated an average processing cost for catfish to be \$5.31/kg (\$2.41/lb) in 1994. However, Dean and Hanson (2003) estimated preliminary average cost of catfish processing in Mississippi to be \$1.84/kg (\$0.84/lb) in 2002. The difference in cost is probably due to improved processing technology, but the cost of production depends largely on the product mix in the production process. Figure 5.8 is an illustration of a breakdown of approximate yields and product mix of various catfish product forms based on the conversion of 4,545 kg (10,000 lb) of live catfish to processed products (Silva and Dean 2001). Further processing of catfish from the whole fish product results in lower yields and more waste and increased cost per kg of marketable product. The product mix varies with processors and depends largely on the processor's marketing strategy and customer demands.

Catfish processors cannot be considered to be mutually interdependent, because each processor's output and pricing decisions are independent of the other processors. Processors have limited control over the prices of their products given that products are not sold directly to consumers but to individual and chain retail grocery store outlets, food service distributors and brokers, and to individual and chain restaurants. Some processors, however, have received higher prices by servicing niche markets and providing special customer services. In general, however, the price and output results of the catfish processing sector could be similar to those of pure competition because of the intense competition among processors to supply processed products, mainly fillets, to the food service sector. Furthermore, the highly elastic nature of the demand curve for individual catfish processing companies suggests that pricing and quantity results are near pure competition. Catfish processors commonly lose customers to one another.

The increasing trends in processing capacity and the accompanying increase in the sales of catfish fillets over the past decade have been interrupted by trade competition. Specifically, Vietnam became the largest exporter of frozen basa/tra fillets to the United States beginning in 1998. Consequently, the market share of U.S. farm-raised catfish fillets peaked in 1997, with the introduction of basa/tra fillets from Vietnam to the U.S. market. Thereafter, the market continued to decline in the share of domestic fillets along with processor and producer prices until the early 2003. The declining market share and the associated price decline of catfish prompted various actions from the industry that resulted in the imposition of an average of 39% tariff on basa/tra from Vietnam. The average wholesale price of U.S. farm-raised frozen catfish declined from an average of \$6.22/kg (\$2.83/lb) in 2000 to an average of \$5.31/kg (\$2.41/lb) in 2003 (Figure 5.9).

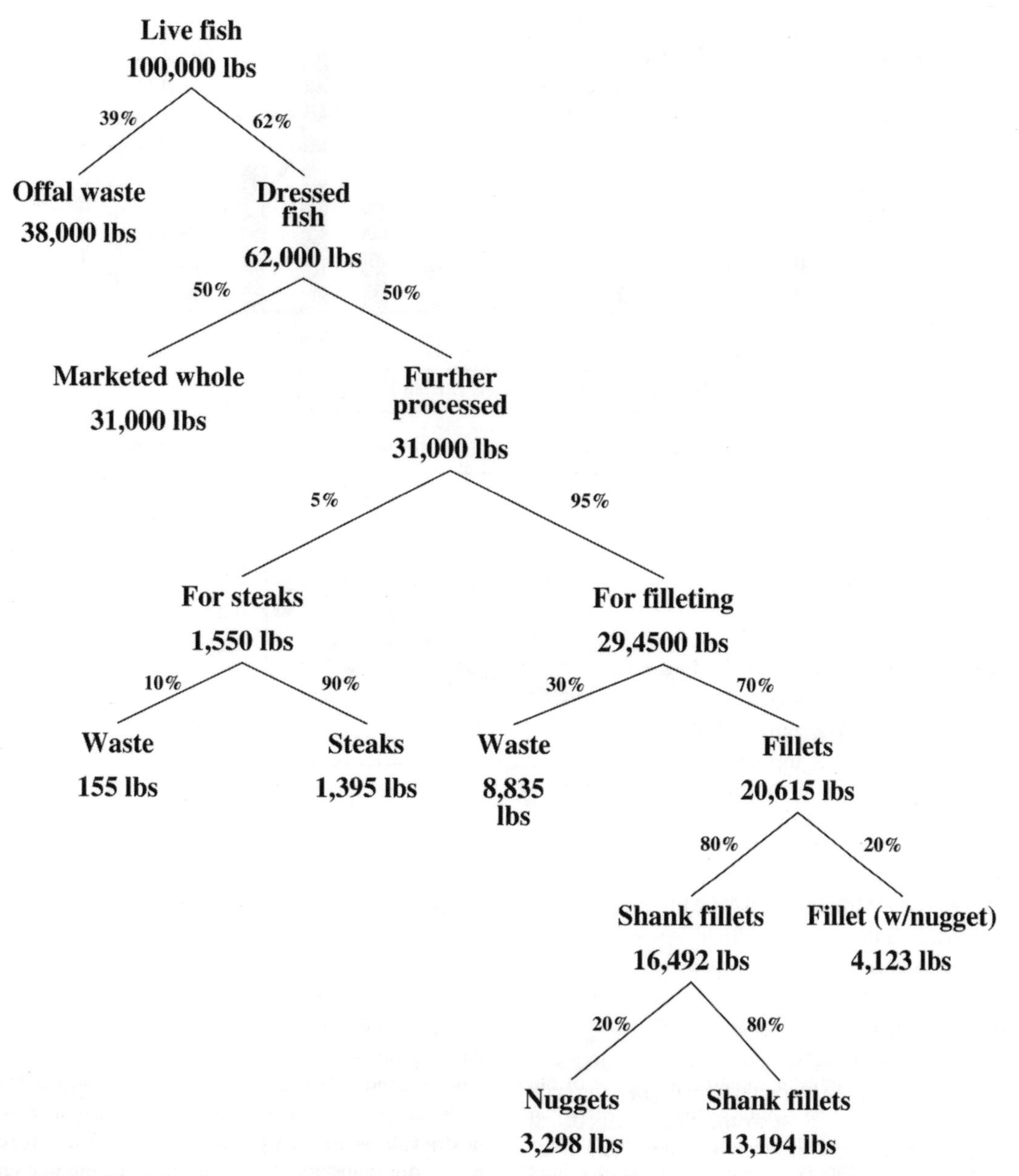

Figure 5.8. Catfish processing input-output chart illustrating a hypothetical product mix (Silva and Dean 2001).

On the consumption side, however, per capita demand for catfish increased from about 0.32 kg (0.7 lb) in 1990 to about 0.52 kg (1.14 lb) in 2003, an increase of 64% and an annual rate of increase of 5% (Figure 5.10). The increase in catfish consumption could be attributed partly to a general increase in fish consumption, intensive marketing efforts within the industry, and changes in consumption patterns with respect to demand for new fish products. The southern region of the United States is the traditional market area for catfish, which includes Oklahoma, Texas, Arkansas, Kentucky, Tennessee, Mississippi, and Alabama. Some market expansion has been accomplished in other parts of the nation, particularly California, Illinois, the midwest, and the northeast.

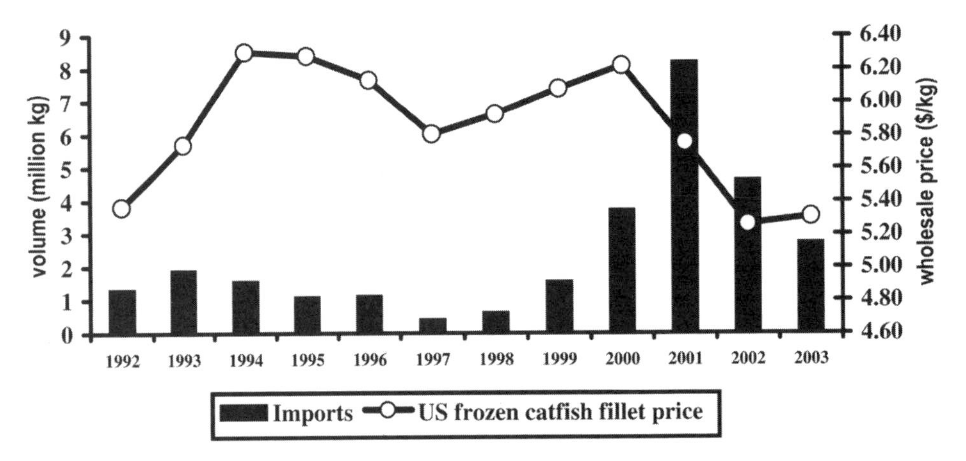

Figure 5.9. Processor sales volume of various catfish products (million kg). (Source: USDA 2004.)

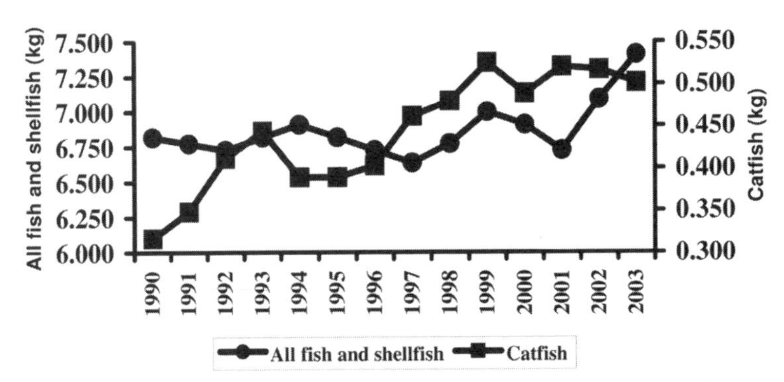

Figure 5.10. Per-capita consumption of catfish and all fish and shellfish in the United States. (Source: NFI 2004.)

SUMMARY

Processing in the seafood and aquaculture industry encompasses all the steps that food goes through, from the time of harvest to the time seafood and aquaculture products reach the consumer. The structure of the seafood and aquaculture products industry relates to the concentration of the industry, the degree of vertical integration, product characteristics, and freedom of entry and exit. The industry is typical of a monopolistic competitive industry, with a relatively large number of firms operating competitively in the production of differentiated products. The U.S. Census Bureau reports the concentration ratios for the 4, 8, 20, and 50 largest fresh and frozen seafood and aquaculture products processing companies in 1997 as 14%, 23%, 42%, and 66%, respectively. The seafood and aquaculture product processing industry is much less concentrated than other meat processing industries such as the beef and pork packing industry. Also, the rate of capacity utilization has averaged 60% over the last 10 years compared to 85% for the beef, pork, and poultry processing industries. This indicates relative underutilization of processing capacity. Some of the challenges confronting the aquaculture processing industry include realizing the potential growth in the market. Because the U.S. imports about 60% of its seafood needs, producing healthful, quality, and convenient seafood products in a cost-effective way that also meets environmental standards is essential for domestic processors to capture an increased share of their domestic market.

STUDY AND DISCUSSION QUESTIONS

1. What is the importance of seafood processing in the food system?

2. List two factors that drive seafood and aquaculture product processing.

3. What is value-added processing?

4. What factors characterize the structure of an industry? What is the importance of each of the factors?

5. Explain the concept of concentration in an industry.

6. What is the most concentrated food industry in the United States?

7. Distinguish between a competitive market and a monopolistic market.

8. What are anti-trust laws? Give five examples.

9. What is the difference between vertical integration and horizontal integration?

10. Give two advantages of vertical integration.

11. What is production capacity utilization rate?

12. What are some of the challenges facing the U.S. seafood and aquaculture processing industry? Suggest various ways by which the industry can address these challenges.

APPENDIX

SEC. 202. STANDARD FOR FISHERY ENDORSEMENTS.

(a) STANDARD.–Section 12102(c) of title 46, United States Code, is amended to read as follows–

"(c)(1) A vessel owned by a corporation, partnership, association, trust, joint venture, limited liability company, limited liability partnership, or any other entity is not eligible for a fishery endorsement under section 12108 of this title unless at least 75 per centum of the interest in such entity, at each tier of ownership of such entity and in the aggregate, is owned and controlled by citizens of the United States.

"(2) The Secretary shall apply section 2(c) of the Shipping Act, 1916 (46 App. U.S.C. 802(c)) in determining under this subsection whether at least 75 per centum of the interest in a corporation, partnership, association, trust, joint venture, limited liability company, limited liability partnership, or any other entity is owned and controlled by citizens of the United States. For the purposes of this subsection and of applying the restrictions on controlling interest in section 2(c) of such Act, the terms 'control' or 'controlled'–

"(A) shall include–

"(i) the right to direct the business of the entity which owns the vessel;

"(ii) the right to limit the actions of or replace the chief executive officer, a majority of the board of directors, any general partner, or any person serving in a management capacity of the entity which owns the vessel; or

"(iii) the right to direct the transfer, operation or manning of a vessel with a fishery endorsement; and

"(B) shall not include the right to simply participate in the activities under subparagraph (A), or the use by a mortgagee under paragraph (4) of loan covenants approved by the Secretary.

NOTES

1. Sherman Act, 15 U.S.C. § 1 (Trusts, etc., in restraint of trade illegal; penalty): Every contract, combination in the form of trust or otherwise, or conspiracy, in restraint of trade or commerce among the several States, or with foreign nations, is declared to be illegal. Every person who shall make any contract or engage in any combination or conspiracy hereby declared to be illegal shall be deemed guilty of a felony, and, on conviction thereof, shall be punished by fine not exceeding $10,000,000 if a corporation, or, if any other person, $350,000, or by imprisonment not exceeding three years, or by both said punishments, in the discretion of the court.

Sherman Act, 15 U.S.C. § 2 (Monopolizing trade a felony; penalty). Every person who shall monopolize, or attempt to monopolize, or combine or conspire with any other person or persons, to monopolize any part of the trade or commerce among the several States, or with foreign nations, shall be deemed guilty of a felony, and, on conviction thereof, shall be punished by fine not exceeding $10,000,000 if a corporation, or, if any other person, $350,000, or by imprisonment not exceeding three years, or by both said punishments, in the discretion of the court.

REFERENCES

Bykowski, P. and D. Dutkiewicz. 1996. Freshwater fish processing and equipment in small plants. Food and Agriculture Organization Fisheries Circular No. 905, Rome, FAO.

Dean, S. and T. Hanson. 2003. Economic impact of the Mississippi farm-raised catfish industry at the year 2003. Mississippi State University Extension Service, Mississippi Agricultural and Forestry Experimental Station, Publication 2317, Mississippi State University, Mississippi.

Dillard, J.D. 1985. Organization of the catfish industry: a comment on market structure-conduct-performance. Paper presented at the Catfish Farmers of America Annual Meeting, Memphis, Tennessee.

EUROFISH Magazine. 2001. The fish processing industry in Denmark—Europe's major fish exporter imports raw materials. January/February 2001 Accessed October 22, 2004, at http://www.eurofish.dk/indexSub.php?id=570&easysitestatid=-1801202820.

Fishing Industry Renewal Board (FIRB). 1997. A policy framework for fish processing. Fishing Industry Renewal Board Final Report, Government of Canada.

Hackett, S.C. and M.J Krachey. 2004. California's wetfish industry complex: an economic overview. Humboldt State University Working Paper. Humboldt State University. Accessed October 18, 2004, at http://www.humboldt.edu/~sh2/wetfishpaper.htm

Knapp, G., C. Wiese, J. Henzler, and P. Redmayne. 2001. A village fish plant: yes or no?—a planning handbook. Institute of Social and Economic Research, University of Alaska, Anchorage, Alaska.

Kolbe, E. and D. Kramer. 1997. Planning seafood cold storage. Marine Advisory Bulletin No. 46, Alaska Sea Grant College Program, University of Alaska, Fairbanks, Alaska.

Kouka, P.J. 1995. An empirical model of pricing in the catfish industry. *Marine Resource Economics* 10: 161–169.

Mermelstein, N.H. 2002. Processing in the largest U.S. seafood plant. *Food Technology* 56(2).

Miller, J.S., J.R. Coner, and J.E. Waldrop. 1981. Survey of commercial catfish processors—structural and operational characteristics and procurement and marketing practices. Agricultural Economics Research Report No. 130. Mississippi State University, Mississippi.

Mississippi State University. 2003. Aquaculture: catfish current situation. Mississippi State University Extension Service, Mississippi Agricultural and Forestry Experimental Station, Mississippi State University, Mississippi.

National Marine Fisheries Service. 2004. Fisheries of the United States—2003. Office of Science and Technology, Fisheries Statistics Division, David Van Voorhees, Chief, Elizabeth S. Pritchard, Editor, Silver Spring, Maryland.

National Marine Fisheries Service. 1995. Preliminary report for 1995 on commercial and recreational fisheries of the United States. Office of Science and Technology, Fisheries Statistics Division, Silver Spring, Maryland.

Publishing House MPR Ltd. Accessed October 22, 2004, at http://www.mpr.nets.pl/anglia/start.php

Radtke, H. and S. Davis. 2000. Description of the U.S. west coast commercial fishing fleet and seafood processors. Pacific States Marine Fisheries Commission.

Silva, J.L. and S. Dean. 2001. Processed catfish—product forms, packaging, yields, and product mix. Southern Regional Aquaculture Center Fact Sheet No. 184, Mississippi State University, Mississippi.

U.S. Department of Commerce (USDC)—National Oceanic and Atmospheric Administration (NOAA). 2004. USDC participants list for firms, facilities and products, USDC-NOAA, 19(2).

U.S. Department of Commerce—US Census Bureau. 2001. Concentration ratios in manufacturing. 1997 Economic Census, Subject Series, June 2001 EC97M31S-CR.

U.S. Department of Commerce—US Census Bureau. 2003. Survey of plant capacity: 2001. Current Industrial Reports, February 2003 MQ-CI (01).

Wiese, N. J. 2004. Market characteristics of U.S. farm-raised catfish. Master's thesis, Aquaculture/Fisheries Center, University of Arkansas at Pine Bluff, Pine Bluff, Arkansas

6
Participation and Leadership in Marketing Channels

Chapter 3 introduced some of the terms and concepts related to marketing channels. This chapter goes into more depth and reviews the dynamics of channel organization, ownership, and control in aquaculture marketing. Contrasts will be made with trends in agribusiness marketing. The market synopsis on trout highlights a mature industry that has moved into value-added products to develop new markets.

MARKETING CHANNELS FOR PRIMARY SEAFOOD PRODUCTS

A marketing channel (also called channel of distribution) is a combination of interrelated intermediaries (individuals and organizations) who direct the physical flow of products from producers to the ultimate consumers. Producers make a narrow range of products in large quantities, but consumers want a broad range of goods in small quantities. The flow of products is accomplished through the marketplace where there is buying and selling or some contractual arrangements. Market channels can be very simple and direct, as with direct sales from farms to consumers, or can be very complex and comprise an array of brokers, sales agents, traders, distributors, wholesalers, foodservice operators, and importers. In the United States, the major customers of seafood and aquaculture food products are businesses in the retail and food service sectors. Other customers are government organizations and exporters.

Seafood Distribution in Subsistence Economies

Although some market channels can be very complex, others can be simple. The complexity often depends on the type of seafood and whether the nation's economy is well developed or subsistence in nature. In Honduras, for example, fish marketing involves middlemen and fish traders who fulfill the functions of brokers, wholesalers, wholesaler-retailers, and retailers (Fig. 6.1). The fish traders buy and sell all kinds of freshwater, brackish water, and marine fish. Commercial marketing channels for farm-raised tilapia in Honduras are not complex (Leyva 2004). However, in the rural areas where the economy is based more on near-subsistence and small-scale production, fish farmers maintain some of their produce in small ponds to use for home consumption. Medium- and large-scale fish farmers tend to sell all of their harvest. Molnar et al. (1996) reported that the percentage of farmers keeping tilapia for home consumption decreased as the pond area increased, indicating that increased pond area was associated with increased entry into the cash market economy.

Seafood Distribution in Developed Economies

In most western economies, seafood-marketing channels consist of a wide variety and a large number of actors that include importers, agents, traders, wholesalers, processors, retailers, restaurants, and others. Large retail chains are very much involved in the distribution of seafood products to their outlets. In Germany, the retail sector is highly concentrated such that the top five retailers account for 63% of the total retail market value and put a squeeze on food distribution (Lahidji et al. 1998). The German seafood market is heavily dependent on imports of seafood products to meet domestic demand. For example, the salmon supply in Germany is almost solely dependent on imports. Figure 6.2 suggests that the retail and foodservice sectors are heavily involved in the flow of salmon in the seafood distribution system in Germany (Johnsen and Nilssen 2001).

In Northern Ireland, marketing of Dublin Bay Prawns (also called Norwegian Lobster or simply Nephrops) follows two main channels, and the route

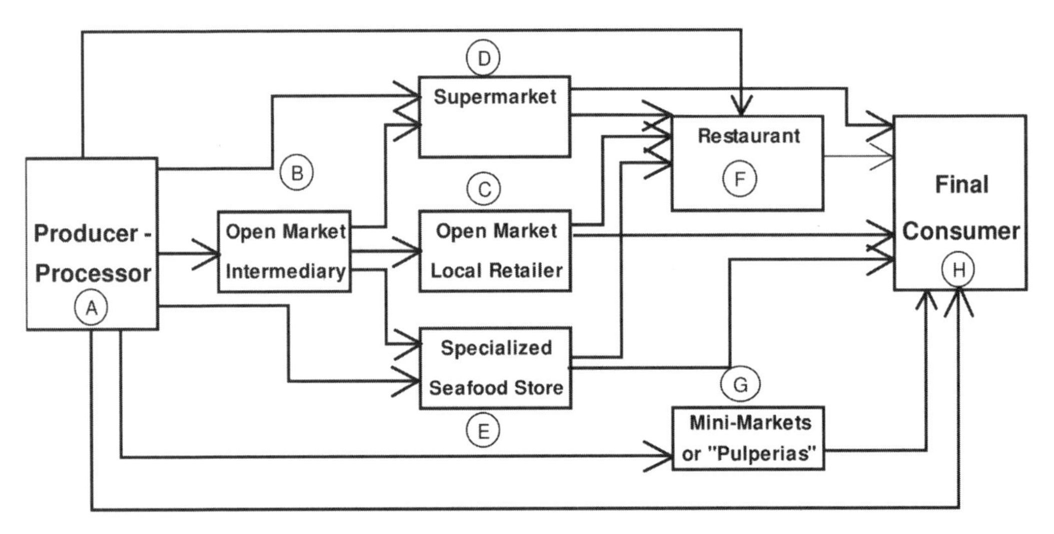

Figure 6.1. Market channels for tilapia in Honduras (Leyva 2004).

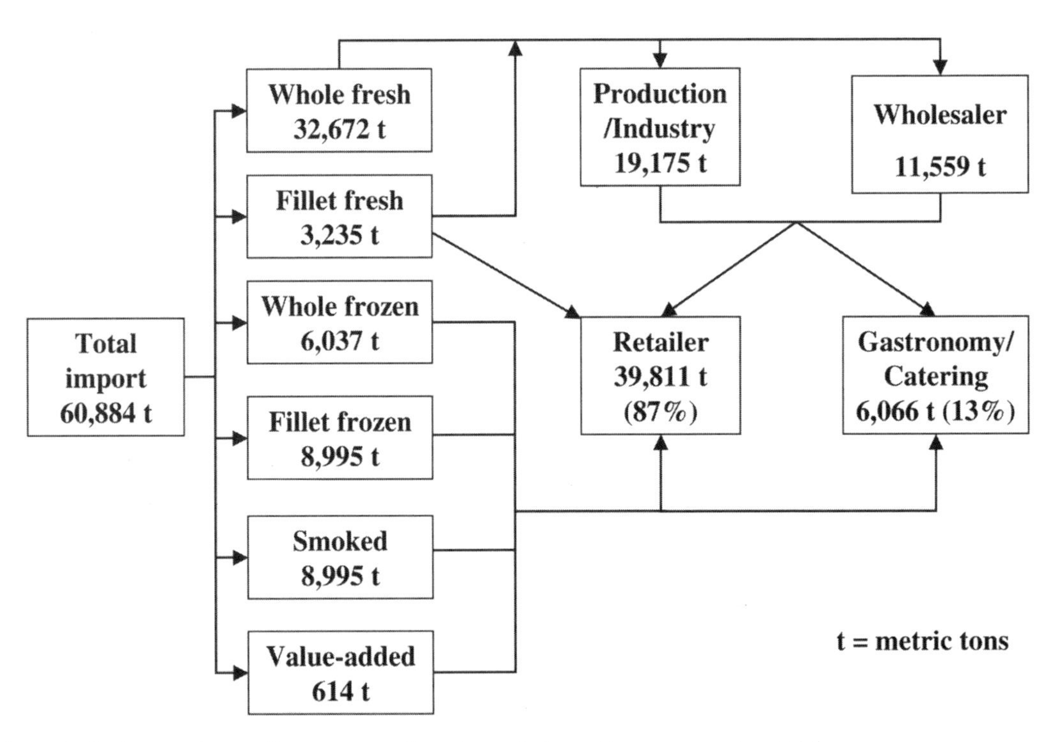

Figure 6.2. Salmon flow in German marketing channels (quantities in product weight).

depends on whether the prawns are sold whole or just the tails are sold. The prawns are the most important seafood species of Northern Ireland's fishing industry (Rogers 2000). The tails are bought mainly by the local prawn scampi processing sector, which processes them into a range of breaded and peeled scampi products mainly for supermarkets and catering outlets in England. Figure 6.3 presents a schematic flow of the marketing channels for Dublin Bay prawns that was developed from a sample of 44 seafood businesses surveyed in March 2000 (Rogers 2000). Compared to prawns, the channels for whitefish, including cod, haddock, hake, dogfish and whiting, are very different from those in Northern Ireland (Fig. 6.4) and involve hawkers (small businesses employing fewer than 10 full-time employees whose

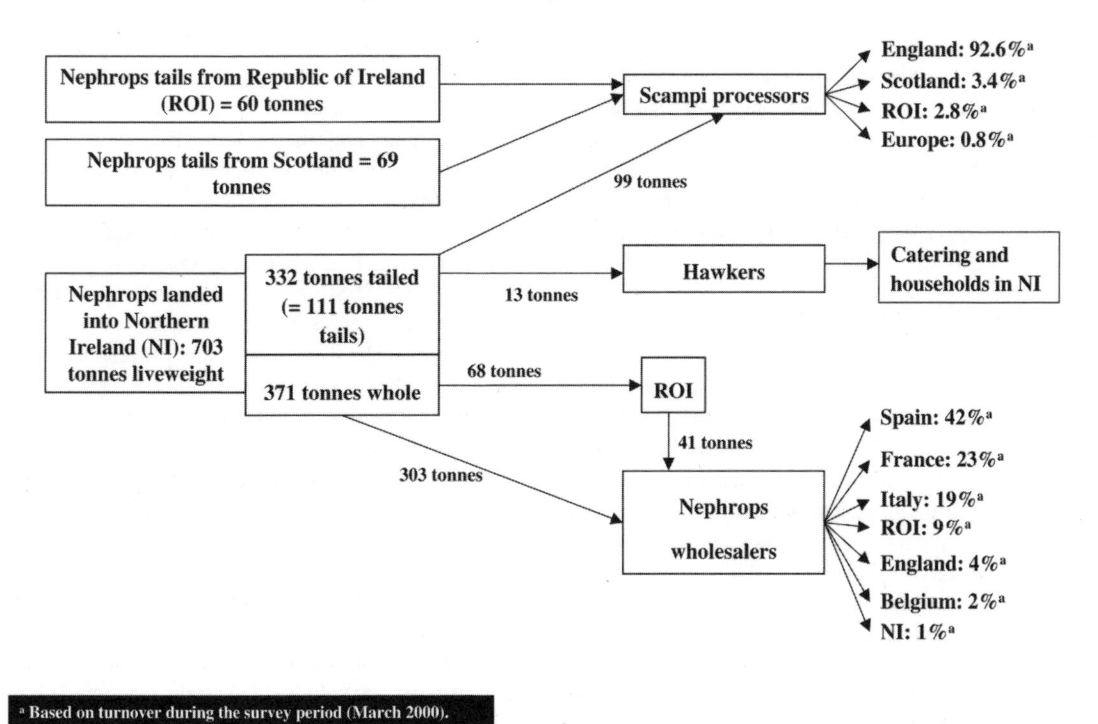

Figure 6.3. Nephrops marketing channels, March 2000 (Rogers 2000).

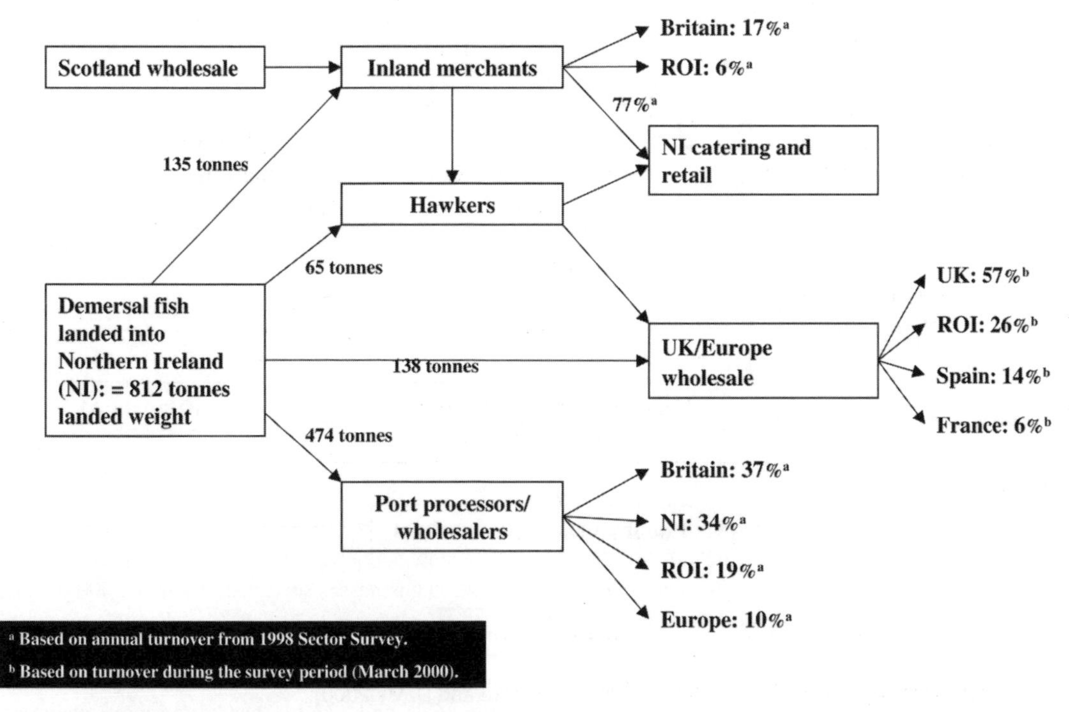

Figure 6.4. Whitefish marketing channels, March 2000 (Rogers 2000).

primary activity is filleting fish for catering and retail markets in Northern Ireland), inland merchants (businesses that process and wholesale marine fish situated more than 17 km [10 miles] from a major fishing port) and port processors/wholesalers (businesses that process and wholesale marine fish, located within 17 km [10 miles] of a major fishing port) (Rogers 2000).

Seafood Distribution in the United States

The seafood distribution business in the United States is highly competitive and fragmented, with several examples of flows in the distribution channel because of the wide variety of actors involved in seafood distribution. Distribution of seafood is complex, but a typical flow of commercial fisheries products is depicted in Figure 6.5. Although some seafood products flow directly to the consumer, others flow to the consumer through processors, brokers, distributors, and retailers, and value may be added to the product at any stage in the channel (Radtke and Davis 2000). The top five seafood products in 2003 were shrimp, canned tuna, salmon, pollock, and catfish. The demand for seafood far exceeds production from commercial fishing and aquaculture, and the shortfall in domestic supply is most severe for shrimp, tuna, and the fish blocks. Consequently, more than 50% of the seafood consumed in the United States is imported.

For shrimp, about 85% of the total supply is imported into the United States, primarily from Southeast Asia (NMFS 2004). Domestic farmed shrimp production accounts for less than 5% of the total U.S. supply. The flow of domestic processed shrimp begins with fishermen bringing harvested fresh shrimp to the dock. Some may be deheaded, sorted, and frozen. Processors buy the fresh shrimp at the dock and may sell processed shrimp to distributors/wholesalers, brokers, or directly to retailer customers such as chain grocery and restaurant companies (U.S. International Trade Commission 2004). The market is similar for imported shrimp. Some importers process shrimp into other value-added products such as marinated, sauced, or breaded shrimp and then sell the value-added products to retail chain grocery and restaurant companies and other customers. Shrimp importers are part of seafood com-

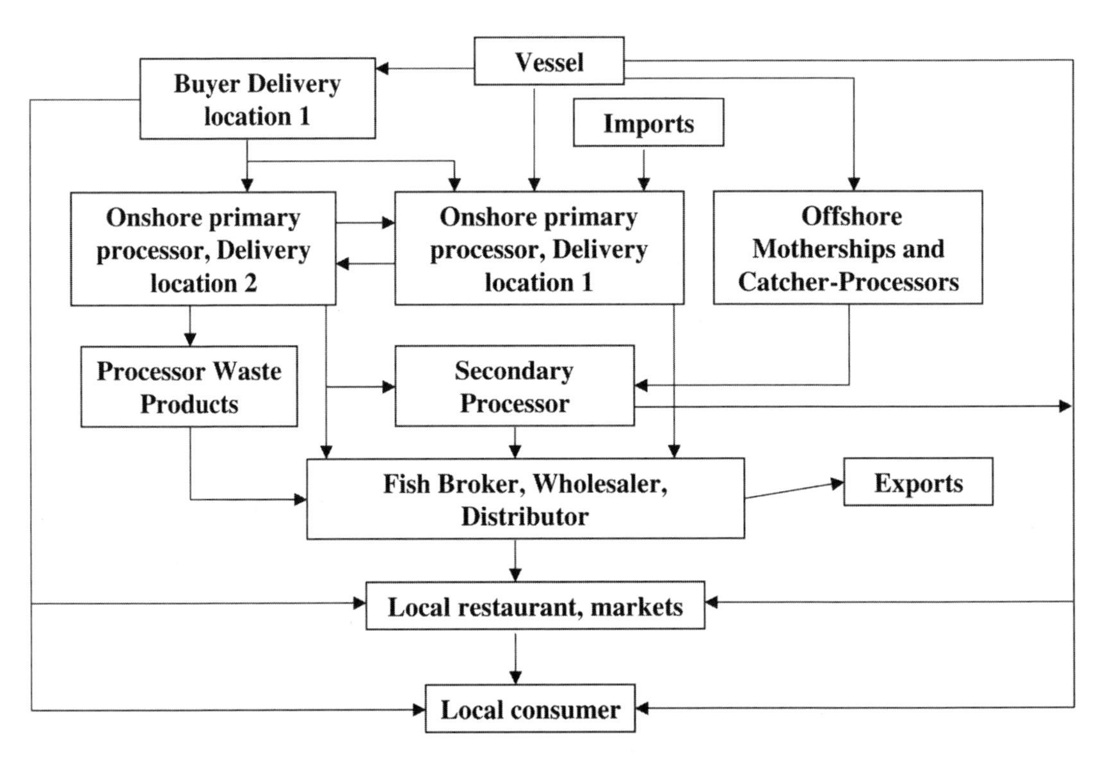

Figure 6.5. U.S. seafood product distribution chain (Radtke and Davis 2000).

panies engaged in the import, distribution, and marketing of general seafood, most of which may be considered to be independent and family-owned. In 2003, the National Oceanic and Atmospheric Administration of the U.S. Department of Commerce (USDC-NOAA) estimated that there were more than 3,500 seafood dealers operating in the United States, and approximately 1,000 were in the business of importing fish and shellfish. The major shrimp importers include companies such as Slate Gorton, Ocean Garden Products, Empress International, and Darden Restaurants. Both shrimp processors and importers serve national, regional, and multiple-market areas.

The Alaska fisheries sector is the largest fisheries sector in the United States and leads all states in the United States in landings of fish including Pacific salmon, Pacific cod, flounder, hake, Pacific ocean perch, Alaska pollock, and rockfishes. Marketing of Alaska seafood is by the Alaska Seafood Marketing Institute (ASMI), which is responsible for promotional materials, trade show exhibitions, buyer's guides, recipes, and other information. The institute arranges sales with major U.S. broad-line foodservice distributors, major supermarket chains, and importers of Alaska seafood. ASMI has overseas marketing representatives in Japan, China, and the European Union responsible for promoting and selling Alaska seafood.

MARKETING CHANNELS FOR MAJOR PRIMARY COMMODITIES

The physical flow of agricultural commodities in the United States through the marketing channel varies with commodity groups. The major agricultural commodities in the United States are grains (for example, corn, soybean, and wheat) and livestock (cattle, hogs, and poultry). Understanding marketing channels for other types of agricultural products may shed light on opportunities for aquaculture growers. The following sections describe marketing channels for grain, livestock, and fresh produce crops.

MARKETING CHANNELS FOR GRAINS

A schematic representation of the physical flows through the marketing channel for grains is provided in Figure 6.6. When grains are harvested from farms, they are first stored either on farm for later sale or shipped to local country (rural) or terminal elevators. Primary processing functions such as cleaning, sorting, and grading are performed at the elevators. There are three types of local country elevators:

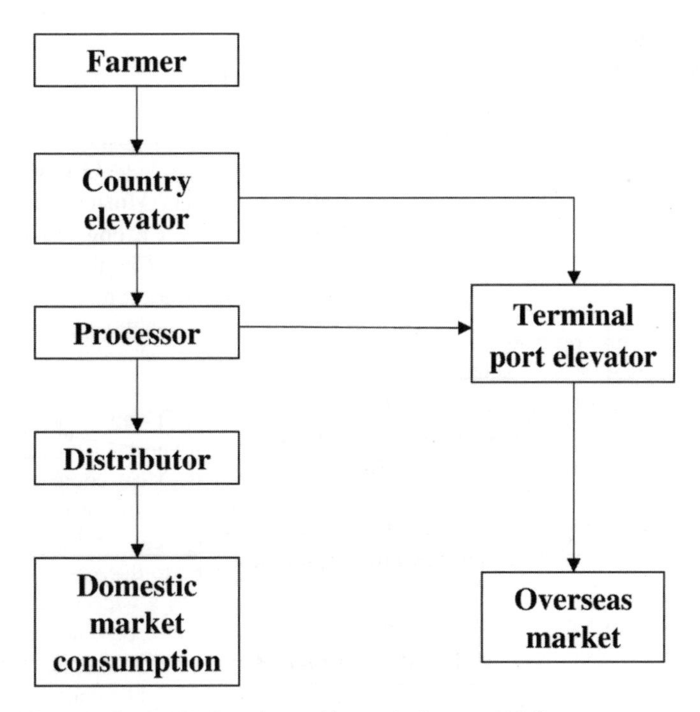

Figure 6.6. Physical flows in the marketing channel for grain (Lence 2000).

independent, cooperative, and line elevators (Lence 2000). Independent elevators are operated and controlled by the individual owners. Cooperative elevators are owned by farmer groups and organizations, whereas line elevators mainly consist of groups of elevators that are owned and operated as a chain by a major corporation. Such a corporation may be involved in other steps of the marketing channel such as milling, oil extraction, exporting, or some other marketing function. Local country elevator companies may sell grain products to food processing companies or to terminal elevator companies that usually handle larger volumes of grains than do local country elevator companies. Terminal elevator companies sell grains to food processing companies as well as to the export market. Harvested grains are mostly transported by road from the farm to local country elevators and, to a smaller extent, processing plants. Traditionally, however, the rail network is the major source of transportation used to move grains from country elevators or terminal elevators to major processors or export ports. Some U.S. grain is transported from the terminal elevators to the processor or export port locations by barges on rivers.

Although some of the actors in the marketing channels for grains may be distinct and separate entities of independent companies, some channels are vertically integrated (that is, have common ownership) at various levels, with companies owning more than one level of the channels. For example, major agricultural companies such as Cargill and Archer Daniels Midland (ADM) are vertically integrated. Cargill operates as an international marketer, processor, and distributor of agricultural and food products. It also operates chains of elevator facilities, exports grains, and processes food. Cargill's processing businesses produce and sell sweeteners, food and industrial starches, various starch derivatives, wheat flour, wheat proteins, malt, and other products. Archer Daniels Midland procures, transports, stores, processes and markets a wide range of agricultural products that includes soybean, corn, and wheat. ADM's processing activities include processing corn into corn oils, high fructose corn syrup or lecithin, corn flours, nutraceuticals such as natural-source vitamin E, renewable fuel sources, and numerous other products. Some of the products from the processing operations of these agricultural companies are used in further food processing in the manufacture of bakery products, breakfast cereals, brewery products, soft drinks, snack foods, and other grain products. Distribution of final consumer products are addressed in Chapter 8.

Marketing Channels for Livestock

For livestock, animals ready for the market are shipped to a slaughterhouse where they are slaughtered and processed into various products (Fig. 6.7). In the livestock sector, there are few distinct and separate entities of independent companies in the channel of distributions. Most of the channels are vertically integrated at various levels and are dominated by a few companies such as Tyson Foods, Inc., Excel Foods (owned by Cargill), Swift and Company, National Beef, and Smithfield Foods.

Tyson Foods, Inc. is the world's largest supplier of poultry products. Tyson also handles beef and pork products. For poultry, Tyson has production contracts with independent poultry growers who raise chickens for the company using inputs supplied by Tyson. It takes four to -six weeks to grow chicks to market size (others take seven to eight weeks for an average sized four-pound bird). Birds are transported by fleets of Tyson trucks to one of the 83 Tyson processing facilities in 20 states across the United States. Processed poultry products include IQF (individually quick frozen) chicken, oven-roasted chicken, wings and drumsticks, cornish game hens, boneless chicken, skinless breasts, chicken patties, nuggets, tenders, fillets, meal kits or entrees, and full dinners. Processed products are distributed through Tyson's own warehousing and distribution division. The division serves more than 1,200 food-service operators, grocers, and caterers on a weekly basis and hauls more than 350 loads per week for other wholesalers, distributors, and brokers.

Beef and pork packing and processing operations also depend on independent livestock producers. Most meat packing plants are located strategically near large cattle and hog growout operations. Excel Foods (owned by Cargill) dominates the beef packing industry, while Smithfield dominates the pork packing industry, although Tyson also operates beef and pork packing plants. There are hundreds of fresh beef and pork products produced from meat packing and processing plants. These include beef loins, chuck, and ground beef in addition to a variety of other products. Meat companies produce and sell fresh, chilled, and frozen meat products to grocery retail establishments, food processors for further processing, and to the export market. Tyson for example, produces a wide variety of retail processed meats, including bacon, hot dogs, luncheon meats, hams, and sausage sold under various brand names. Similarly, Smithfield's major products include ham, loins, bacon, hot dogs, luncheon meats, smoked and

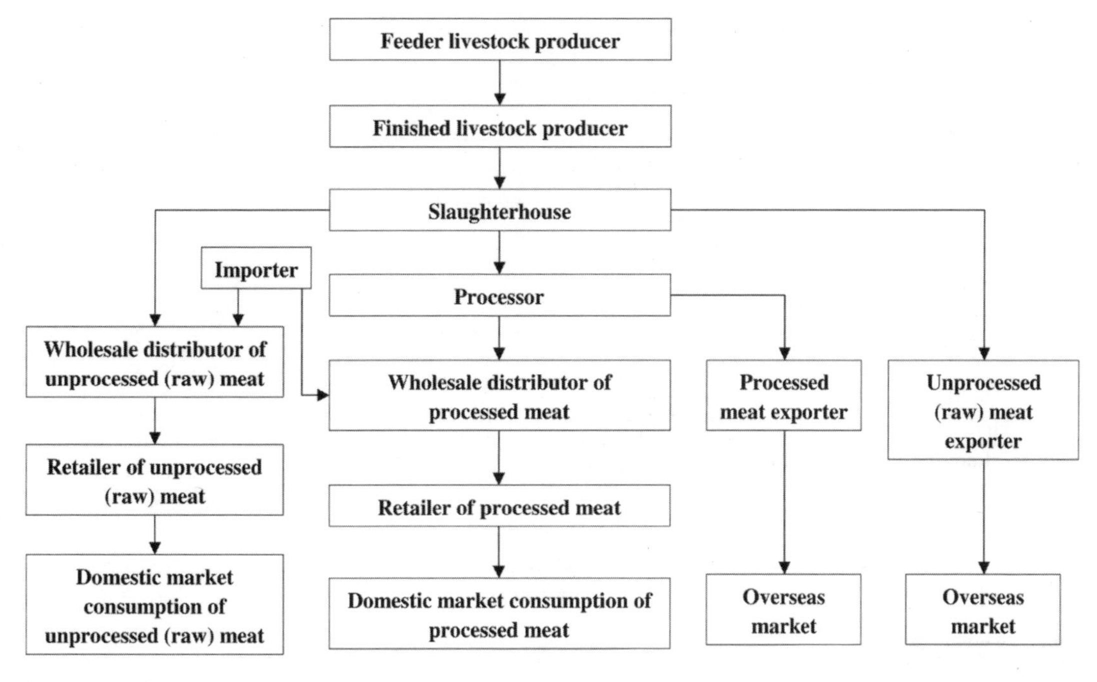

Figure 6.7. Physical flows in the marketing channel for livestock (Lence 2000).

canned hams, packaged lard, dry salt meats, sausage products, and others. Smithfield also produces boxed, processed, and ground beef.

Processed meat products from beef and pork packing operations go through the distribution chain to the final consumer. These are shipped in refrigerated trucks to retail and foodservice customers, where they are typically processed further or prepared before being sold to the consumer.

Marketing Channels for Fresh Produce

The distribution channel for fresh produce is depicted in Figure 6.8. Fresh fruit and vegetable products flow quickly through the marketing system to avoid spoilage. After harvest, shippers or grower-shippers handle and pack fresh produce, which may be exported or sold directly to consumers, retail stores, or foodservice establishments. Sometimes, sales from grower-shippers to retailers and foodservice establishments pass through wholesalers or brokers, but the growth in value-added and company-branded products has resulted in the integration of wholesaling, retailing, and importing operations that handle produce year round (Dimitri et al. 2003). Some grower-shippers and shippers sell to the full range of market channels, but others specialize in supplying particular types of customers (Glaser et al. 2001). Large retail stores have increased volumes of produce purchased directly, bypassing produce wholesalers. In 1997, $1.1 billion worth of produce was sold directly to the consumer, $34.3 billion through retail stores, and $35.4 billion through foodservice establishments (Dimitri et al. 2003). From 1987 to 1999, the share of sales of produce from independent wholesalers to retailers declined from 38.1% to 34.6%, but the share of sales of produce from wholesalers to the foodservice sector increased from 8.4% to 21.2% (Dimitri et al. 2003).

PRICE DISCOVERY FOR PRIMARY COMMODITIES

Evidence from the food distribution system in the United States indicates differences in the relative importance of specific commodity flows, how channel agents facilitate commerce, and price-discovery mechanisms. Price discovery is the process of buyers and sellers arriving at a transaction price for a given quality and quantity of a product at a given time and place. Price discovery involves several interrelated concepts such as the market structure, market behavior, market information, and price reporting, as well as futures markets and risk-management alternatives. Price discovery begins with the market price level.

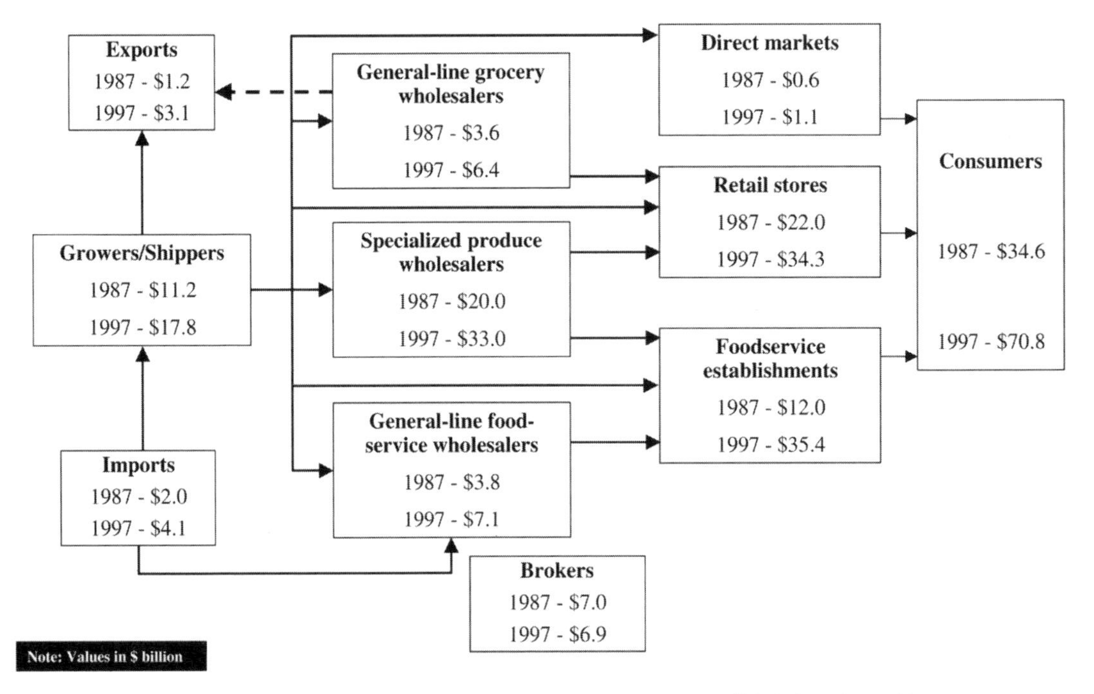

Figure 6.8. Fresh fruit and vegetable marketing channels, 1987 and 1997 (Dimitri et al. 2003).

Contracting and Vertical Integration in U.S. Agribusiness

Developments in agricultural markets indicate a trend of reduced dependence on open market transactions or trading on a spot market basis and more dependence on efficient and longer-term trading relationships, such as forward contracts, production and marketing contracts, marketing agreements, and vertical integration. For example, in the U.S. poultry industry, contracts and vertical integration now account for more than 90% of the production of broilers, turkeys, and eggs (Martinez 2002). Poultry and turkey contracts usually involve a negotiated delivery schedule, pricing method, and product characteristics. The contract covers the provision of services such as management, labor, utilities, housing, and equipment rather than the sale of goods, and the birds remain the property of the company. The company supplies the feed and medication, and provides supervisory field personnel. The contract includes incentives to encourage the production of quality birds with minimal feed because the grower is compensated on the basis of a formula based on the farmer's production efficiency (the number and weight of birds harvested compared to the number of chicks and pounds of feed delivered). In most situations, the farmer's compensation is adjusted based on a comparative ranking with a group of other poultry growers in the same geographic area whose birds were also processed by the company during the same time period (Martinez 2002).

In the 1980s, most fed cattle were priced on a live-weight, spot market basis, with prices affected by factors such as boxed beef cutout values, live cattle futures market prices, cattle quality (including sex, weight, quality grade, and yield grade), number of days between purchase and delivery of cattle, individual packing plants or firms, packing plant utilization, day of the week, time of year, and extent and type of captive supplies (cattle acquired by packers through forward basis contracts) (Ward and Schroeder 2001). Today, a larger percentage of hogs and cattle are traded through marketing contracts, and carcass weight and grid pricing have increased in importance. With grid pricing, a price is discovered for each animal whereby higher-quality cattle receive higher prices and lower-quality cattle receive lower prices. Most grids consist of a base price with specified premiums and discounts for carcasses above and below the base or standard quality specifications. Individual packers have their own grids with alternative base prices and varying premiums and discounts (Ward and Schroeder 2001).

Formula pricing in contracts is also becoming a popular pricing mechanism in the livestock industry

(Martinez 2002). As does grid pricing, formula pricing establishes a transaction price for a particular quality and quantity of livestock based upon an external reference price; additionally, formula pricing specifies a time and place. Formula pricing is commonly used in grain marketing whereby local cash prices are often tied to a terminal price or nearby futures price, less a transportation and local grain elevator handling margin. For hogs, nearly all formula-priced transactions are based on the Iowa-Southern Minnesota spot market price quote (Martinez 2002; McBride and Key 2003). The contract requires a producer to deliver a certain quantity of hogs to the processor at a certain place and time. The producer may receive a formula-based price, with premiums or discounts based on the size and quality of the hogs. Formula pricing is also used in cattle markets.

Production contracts are common in the cattle and hog industries, where especially large companies provide management services, feeder animals, veterinary services, and other inputs. The grower provides land, facilities, and labor to feed the animals. Most cattle and hog contracts are in the four- to five-year range (Martinez 2002). The increasing percentage of livestock held in captive supplies by packers through contracts is resulting in fewer incentives for meat packers to bid aggressively on livestock offered through spot markets (McBride and Key 2003). Smithfield Foods, the largest pork processor, combines hog production and processing in a vertical integration strategy. Through its hog raising and pork processing subsidiaries, the company exercises complete control over its products, from its genetic lines and nutritional regimen to processing, packaging, and delivery to customers.

For fresh produce, retailer-shipper marketing agreements are based on quantity, quality, brand, and price considerations. The market is characterized by daily and weekly cash sales, short- and medium-term contractual arrangements, and some form of other marketing arrangements. Quality and brand reputation play a major role in pricing, allowing some firms to receive a premium over spot-market price (Glaser et al. 2001). Glaser et al. (2001) reported that the majority of lettuce shippers were price takers and that sales to wholesalers and grocery retailers were daily on a spot market basis, whereas sales to foodservice customers were in the form of advance pricing as well as daily spot market basis. A small percentage of lettuce sales to grocery retailers and foodservice customers was produced on annual contracts. However, all lettuce sales to mass merchandisers such as Kmart, Wal-Mart, Sam's Club, and Costco were based on long-term contracts (Glaser et al. 2001). For bagged salad, Glaser et al. (2001) reported negotiated price and multiyear contracts. Generally, transactions include off-invoice marketing and trade practices that cover both fees (such as volume discounts and slotting fees) and services such as automatic inventory replenishment, special packaging, and requirements for third-party food safety certification (Dimitri et al. 2003). Dimitri et al. (2003) reported that, in 1999, short-term contracts accounted for 11% of the total sales of grapes, oranges, grapefruit, and tomatoes, whereas long-term (annual or multiyear) contracts accounted for 7% of total sales. Sales of lettuce and bagged salad were also on annual or multiyear contracts to retailers. A typical produce sale takes place between a shipper and a large supermarket retailer under a standing agreement or contract specifying various conditions and terms, including marketing services provided by the shipper, volume discounts, and other price adjustments and quality specifications (Dimitri et al. 2003).

Contracting and Vertical Integration in U.S. Seafood Business

Various pricing mechanisms are used in U.S. seafood marketing, including negotiation on a boat-by-boat basis at the time of landing, short-term marketing agreements, and sale on consignment (Anderson 2003). The specific provisions of transactions between buyers and sellers of seafood are generally proprietary with little available public information. However, some form of contracting and marketing arrangements exists in the seafood business. For example, some fish processing companies contract with fishermen by providing them with inputs such as nets, boats, motors and gear. Fishermen, in turn, supply their catch to processors at some negotiated price (University of Alaska 2001). A study by the U.S. International Trade Commission (USITC) revealed that some shrimp contracts involved fixed price or quantities and covered periods ranging from two months to a year. Long contracts for up to two years usually had fixed prices and quantities and involved volume discounts (USITC 2004). A fair amount of catfish transactions between processors and farmers involve delivery rights. In Mississippi, some catfish processors sell delivery rights to fish farmers, which requires delivery of a certain quantity and quality of catfish for a specified period at some negotiated price. Marketing arrangements between major seafood buyers such as large wholesalers, mass merchandisers, brokers, restaurant chains, and grocery chains would involve quantity and price considerations, off-invoice marketing, and some form of trade practices involving services, special packaging, and requirements for third-party food safety certification. Contracting helps these large-scale buyers to guarantee supply and provide some price stability.

Vertical integration in the distribution chain helps seafood companies to reduce distribution costs and enhance control over product supply and price. Unlike the grain and livestock sectors, the seafood sector is much less vertically integrated. However, there are trends toward integration in the commercial fisheries sector because of the uncertain nature of commercial fisheries, sizes of spawning runs, management, fishing regulations, subsistence fishing regulations, and quality standards. Production of high-quality processed seafood products begins with high-quality fish, which means changes in the ways fish are handled and shorter time periods between harvesting and delivery for processing. The competitive nature and international scope of the seafood business have resulted in vertical integration of several processing companies in Alaska and the U.S. Pacific northwest. Companies have purchased vessels and processing plants and have developed distribution networks that allow them to offer customers a wide variety of seafood products to, and from around, the world.

The aquaculture sector is becoming increasingly integrated with ownership from hatchery operations through processing to distribution to retail and foodservice customers. A classic example is Clear Springs Foods. It is a vertically integrated company involved in trout farming, fish feed manufacturing, trout processing, and distribution. The farm operations include a broodstock facility that produces about 80 million rainbow trout eggs a year, and farms that raise rainbow trout foodfish. The company also owns a feed mill that produces feed formulations for the farming operations. The research and development center produces vaccines, monitors water quality on the farms, and provides an array of fish health services to the farms. The center is also engaged in research projects on nutrition, waste management, genetics, and fish culture. The research division provides a complement of quality assurance services to the other divisions of the company. Rainbow trout harvested at the farms are shipped live to the company's processing facility and then packaged under an HACCP quality assurance program. In terms of distribution, Clear Springs operates its own fleet of refrigerated trucks to deliver products to customers across the United States. Clear Springs supports the sale of its products through a national network of regional sales managers and broker sales representatives.

Other Transaction Types in U.S. Seafood Business

Pricing of landings from small commercial fisheries is on an individual basis. Spot market price and prices offered by processors depend on processors' marketing arrangements for processed fish, such that ex-vessel prices offered by processors depend on their price spread or marketing margin (University of Alaska 2001). The Fulton Street Market in New York City is the largest open spot market in the United States where many food retailers and restaurateurs come to buy seafood. Other commercial fisheries are sold through fish marketing cooperatives that negotiate prices with large processors and buyers. In the domestic shrimp market, shrimp are usually sold in the spot market and pricing may be handled as a transaction-by-transaction negotiation, a spot market price, or the price reported in Urner Barry's industry price reports (U.S. International Trade Commission 2004).

There are also some fish auctions in the United States. These include the Portland Fish Exchange, New England fish exchange auction, New Bedford whaling city seafood display auction, Gloucester seafood display auction, and fish auctions in Honolulu and Hilo where buyers engage in competitive bidding for seafood products.

Futures markets have been used in the grain and livestock industries as a price discovery mechanism for a long time, but have not been successful long term in the seafood industry. In the broadest terms, futures markets are about anticipated future prices of basic commodities based on current market and industry information. Futures markets are concerned with such questions as what the price of the commodity will be the next December. Futures contracts are contractual agreements made between two parties through a regulated exchange in which the parties agree to buy or sell an asset (such as livestock) at a certain time in the future at a mutually agreed upon price. Each futures contract specifies the quantity and quality of the item, expiration month, the time of delivery, and virtually all the details of the transaction except price, which the two parties negotiate based on current market conditions. Some futures contracts call for the actual, physical delivery of the underlying commodity at contract termination, but others simply call for a cash settlement at contract termination. Two futures contracts for shrimp introduced in the 1960s on the Chicago Mercantile Exchange were terminated after a brief period because of low trading volume. In 1994, the Minneapolis Grain Exchange began a frozen white shrimp futures, but that was also discontinued due to lack of interest. Japan has a futures contract for frozen black tiger shrimp on the Kansai Commodities Exchange in Osaka.

FISH AUCTIONS

Fish auctions play an important role in fish and seafood marketing in a number of countries. In Malaysia, a fish auctioning system was introduced in 1986 that functioned primarily at the local level (Muhaiyadin 1994). More than 600 auction markets were located at fishing ports in Indonesia in 1994; these function as the primary starting point for exchanges (Karundeng and Pawiro 1994). Fishermen are price takers and must accept the price that results during the auction market. In India, fishermen turn their catch over to an agency authorized to unload, sort, grade, weigh, and display it (Chauhan 1994). Fishermen pay a commission for the auction-handling services. In Bangladesh, most fish auctions are run with open, incremental bidding, but there are some syndicate-controlled fixed prices of fish at the port (fish are iced, packed, and sent to other, larger markets to be auctioned off), limited bidding, and tender systems (Hussain 1994). With limited bidding, the auctioneer fixes the price through negotiation or partial bidding. Price-tendering systems are typically used by owners of trawler catches. In Thailand, auctions are held at wholesale markets by commission agents who act as brokers (Piamsomban 1994). For the most part, only high-valued species are sold by auction. Bids are written on a piece of paper and handed to the auctioneer, with the highest-priced bidder purchasing the fish.

PARTICIPATION IN MARKETING CHANNELS

SALES AGENTS AND BROKERS

Independent sales agents and brokers of seafood products perform similar functions by locating buyers and negotiating a sale. Sales can be negotiated between seafood processors and buyers such as wholesalers, retailers, exporters, or foodservice establishments. As pointed out in Chapter 3, the services of independent sales agents and brokers are compensated with commission fees when sale is completed. The National Oceanic and Atmospheric Administration (NOAA) of the U.S. Department of Commerce lists about 571 major seafood brokers in 2003 (U.S. Department of Commerce—National Oceanic and Atmospheric Administration). However, some seafood and aquaculture product processing companies also maintain a staff of sales personnel who promote and sell only the company's products.

Homziak and Posadas (1992) interviewed 72 U.S. and Canadian tilapia brokers. The brokers that handled tilapia had mean annual gross sales significantly greater than the average for all seafood companies, and controlled nearly 10% of the seafood market and provided a diversity of seafood products. Approximately 52% bought fresh tilapia and 43% handled frozen fish, but these companies primarily purchased lower-priced, whole tilapia products (48%).

Brokers play an important role in seafood marketing channels in many developing countries. Legaspi (1994) points out that in the Philippines, brokers have access to information on prices and supply of other types of fish and seafood and financial transactions that growers do not have. This information, along with access to boat facilities and supplies of ice can put brokers in a monopsonistic (one buyer) position of control in the market.

DISTRIBUTORS

A typical food distributor operates warehousing facilities and transportation services. The main function of a distributor is to receive, warehouse or store, invoice, and deliver goods. Distributors usually handle a wide range of food products that includes aquaculture and seafood products, but there are distributors that handle only seafood and aquaculture products. Examples of distributors that specialize in seafood include H & M Bay, Inc. of Maryland and Preferred Freezer Services of New Jersey. There are several others.

Distributors, in general, do not have the responsibility of selling products to the delivery points. They usually cover a multistate region and are contracted by seafood and aquaculture processing companies to deliver to their customers. For example, the Idaho Trout Processors Company contracts with distributors to deliver fresh and frozen dressed whole trout products to the warehouses or distribution centers of their major customers, which are grocery wholesalers. The wholesaler then distributes the products further through the food system. Many other companies look for logistics providers that can meet their needs including warehousing, data management, shipping, and invoicing. Consequently, major trucking companies such as JB Hunt, TNT, and US Freightways operate warehouse and distribution services and handle dry, frozen, and refrigerated food products. Ocean Spray, a large agricultural cooperative of cranberry and citrus growers and processors in North America, uses Schneider Logistics, a major trucking company in North America for freight-related services for its processing plants, warehouses, and distribution centers in Canada and the United States.

WHOLESALERS

In contrast to distributors, food wholesalers own products that they handle. The food wholesale and distribution system is generally separated into five components

Integrated Grocery Wholesalers

These wholesalers serve the grocery and retail industry consisting of supermarkets, warehouse clubs, convenience stores, and others. These wholesalers own retail store chains and deliver most of the products they sell in their stores. They operate their own transportation and warehouses or distribution centers from where distribution is made to their retail stores. Seafood and aquaculture processors and food manufacturers usually deliver their products to the warehouses or distribution centers of integrated grocery wholesalers using their own transportation network or contracting the services of a distributor. Examples of integrated grocery wholesalers include Kroger, Albertson's, Safeway, Wal-Mart, and Publix.

Nonintegrated Grocery Wholesalers

Nonintegrated grocery wholesalers are general-line grocery wholesalers that normally do not own the retail or grocery stores that they serve. They normally procure grocery products, both food and nonfood, for independent grocery and retail stores and smaller retail chains that do not own and operate buying offices, warehouses, trucking fleet, and store delivery services. The primary function of nonintegrated grocery wholesalers is to serve independent grocery outlets although some wholesalers, such as Supervalu, own a percentage of the retail grocery outlets they serve. As of February 2003, Supervalu served as primary supplier to approximately 2,460 stores, 29 Cub Foods franchised locations, and Supervalu's own regional banner store network of 267 stores, while serving as secondary supplier to approximately 1,500 stores.

Integrated Foodservice Wholesalers

Integrated foodservice wholesalers own self-distributing retail foodservice operations. They deliver a greater percentage of products to consumers at their restaurant outlets. They operate their own transportation and warehouses from where distribution is made to their foodservice establishments. Major restaurant chains such as McDonald's and Shoney's are examples of integrated foodservice wholesalers.

Nonintegrated Foodservice Wholesalers

These wholesalers serve hotels, restaurants, commercial cafeterias, hospitals, schools, and hotels and do not own any of the foodservice establishments that they serve. Examples of such wholesalers are Sysco, U.S. Foodservice, and Alliant.

Specialized Wholesalers

These are companies primarily engaged in the wholesale distribution of a particular line of product such as seafood, meats, or fresh produce (fruits and vegetables). There is a wide variety of different types of wholesalers. Merchant wholesalers are independent businesses that own the merchandise they purchase. Jobbers are specialized versions of merchant wholesalers that have been important historically in delivering seafood from fishermen to restaurants or retail grocery stores. With the growing influence of large food service distributors such as Sysco and U.S. Foodservice, the role of jobbers in seafood marketing has diminished.

RETAILERS

Most aquaculture growers will not be in a position to supply product directly to retailers. Most retailers prefer processed product forms and will purchase from brokers, processors, or other distributors. Chapter 4 includes a section on selling aquaculture products directly from the farm to the end consumer. Thus, this chapter only summarizes a few trends in retailing without entering into a great deal of detail.

The top 11 supermarket chains in the United States include Wal-Mart, Kroger Company, Albertsons, Safeway, Ahold USA, Publix, Delhaize America, Winn-Dixie, H.E. Butt Grocery, Supervalu, and Great A & P Tea Company (Johnson 2004). Of these, only Wal-Mart and Kroger are nationwide chains. The others are all regional chains. Although Wal-Mart is the undisputed leader in retail sales, Kroger, Albertsons, and Safeway all have more stores than does Wal-Mart.

Retail sales of frozen seafood increased by 5% overall between 2003 and 2004 (Johnson 2004). However, individual vendor performance varied from a decrease in sales of 20.9% to an 88.7% increase in sales. Refrigerated seafood sales decreased overall by 4.7% from 2003-2004. The refrigerated seafood category was dominated by Louis Kemp crab and lobster delights. Private labels were also important in this category.

Regional differences can be significant in retail sales of seafood to consumers. For example, Johnson (2004) showed that relatively more consumers in the Northeast consume seafood at home, whereas those in the Midwest are more likely to eat seafood away from home. Moreover, Hispanic consumers are far more likely to consume seafood at home than are the national average or other ethnic groups.

Foodservice sales in the United States increased in 2003 (Johnson 2004) and are forecasted to continue to increase in 2004. Red Lobster is by far the leading restaurant chain in terms of seafood sales. Landry's/Joe's Crab Shack/Chart House is second in terms of sales. Long John Silver's ranks third in sales but has more restaurants than the other two. Captain D's Seafood, McCormick & Schmick's, Legal Seafood, Bubba Gump Shrimp Company, Bonefish Grill, McGrath's Fish House, Rockfish Seafood Grill, and Shells Restaurants complete the top 11 seafood restaurants in the United States.

CHANNEL OWNERSHIP AND CONTROL FOR SECONDARY PRODUCTS

Most of the production of farmed fish products is customer oriented, enabling processors to take advantage of the trends in the marketplace. The products must be the right products, available in the right quantity, at the right place, and at the right time for the target customers. Every seafood and aquaculture processor strives to achieve efficiency and reliability in terms of product supply because an efficient channel system for the processors is important for customer loyalty and could greatly improve one's market share. Consequently, the process of distribution of a company's product requires careful planning and execution to help determine the overall success of the marketing effort.

One of the fundamental issues that processors consider is the choice of intermediary to adopt for the distribution of their products. The choice depends on a number of factors, including type of customers that they have, performance capabilities of the intermediary, and the costs associated with the distribution of the product. The decision also depends on a company's overall management and sales strategy, how seafood consumers purchase seafood and fish products, and the extent to which processors want to perform any of the many levels of channel functions in a cost-effective manner. Thus, depending on the needs of a processor, a choice is made of the intermediary to perform the desired functions at various levels in the channel system. Alternatively, a processing company may decide to perform all the necessary functions itself. Whatever the choice of marketing channel, the seafood and processing company should tailor the choice to support the overall marketing strategy of the company. In certain instances, processing companies form distribution alliances with other processing companies. Such partnerships help to expand product distribution and allow more customers to have access to diverse products. The alliance also offers an opportunity for companies to grow through the distribution relationship.

Seafood and aquaculture processing companies usually rely on a particular set of channels for the distribution of their products. However, a new breed of retail establishment that involves an increased level of mergers and acquisitions, expansion into food retailing by retail discount stores, and the use of information technology has raised some concerns among food processors about their ability to adapt to the changing needs of large volume buyers. Many consumer-products companies and retailers have been in power struggles over merchandising standards, marketing control, pricing, and markdown management.

Most processor sales reside with a set of distribution partners, so it is not surprising that many companies prefer to avoid jeopardizing these key relationships out of fear that they will be excluded from doing business with high-volume buyers. For example, if a relatively large percentage of a processing company's products is sold through a distributor/wholesaler such as Sysco, threatening the relationship would be disastrous for such a processor. Bargaining power appears to have shifted to buyers as suppliers strive to do business with only one large-volume buyer. Such a large buyer could potentially dictate the terms of transactions. The relationship is key to the survival of many processors because, with many processors, any one could be replaceable. However, it would be difficult for a processor to find another buyer of comparable size. Thus, it is worthwhile for processing companies to carefully analyze the nature of the distribution network to understand where the power truly resides.

MARKET POWER AND CHANNEL CONTROL

The issue of control of channels often lies with the bargaining power. Various stages of the marketing chain exhibit some level of buyer and/or seller concentration that creates the potential for market power. Farmers generally face commodity markets in which they can sell commodities at the market price, but have no individual bargaining power to negotiate prices or transaction terms. However, farmer organizations and cooperatives allow farmers to pool sales and input purchases, providing them some level of control over their commodities and bargaining power with farm commodity buyers. In addition, farmers are increasingly engaged in contracts and vertical integration in some agricultural sectors to ensure steady market and stable prices for their commodities.

At the food processing sector, continued consolidation, acquisitions, and industrialization are being pursued to gain economies of size, economies of scope, specialized production, more capital-intensive technology, higher productivity, and efficiencies from vertical coordination (Harris et al. 2002). However, gains in economies of size can also be viewed in the light of the share of the market controlled by food processing firms, which in turn translates into some level of bargaining power. Therefore, one commonly cited reason by companies for mergers and acquisitions is to maintain bargaining power with other stages of the supply chain such as food wholesaling and retailing. For example, in red meat packing, market share of the four largest firms increased from 47% in 1987 to about 61–63% in 1993. Particularly in steer and heifer slaughter, the four largest firms controlled about 81% in 1999 compared to 70% in 1989, and in hog slaughter, the four largest companies controlled 57% of the industry in 1999 compared to 70% in 1989. In the pasta industry, the four largest processors had a 78% market share in 1992; in malt beverages, the four largest firms controlled 90% in 1992. In 1998, companies with $800 million or more in sales accounted for 69% of U.S. dairy sales (Harris et al. 2002).

At the retail sector, rising concentration and consolidation of sales among large supermarket chains and supercenters have made retailer market power in the food industry a topical issue. As more and more products compete for space in supermarkets, retailers have gained increasing power to determine what should be displayed on store shelves. There is a significant trend toward store brands that compete with national brands. Some food wholesalers also have brands, thus food manufacturers, wholesalers, and retailers are aggressively competing to achieve product differentiation.

However, market power can be difficult to measure. Jaffry et al. (2003) used nonlinear three-stage least squares to estimate a dynamic error correction model to test for market power in the salmon retail sector in the United Kingdom. The analysis showed that the market was competitive in both the short and long run and that supermarket retailers are behaving in a competitive manner. Asche et al. (2002) found that, because the Law of One Price holds in salmon value chains in the United Kingdom, lower salmon prices are being passed on to consumers.

Competition among diverse products has resulted in retailers demanding slotting fees as a means for signaling and screening new products and as a basis for achieving efficient cost sharing and risk shifting among manufacturers and retailers. Slotting fees are also thought to lead to more efficient shelf-space allocation and demand/supply apportionment. In contrast, opponents of slotting fees see the fees as an abuse of power by large retailers who use the fees to gain competitive advantage over smaller rivals as well as to discriminate among food manufacturers.

CHANNEL COORDINATION AND LEADERSHIP FOR SECONDARY PRODUCTS

Coordination of the distribution channel is critical for the effective management of the distribution channel. This is because the distribution system involves several members pursuing different objectives and possibly conflicting with each other. Coordination mechanisms are therefore necessary to formally cater to transactions along the system. Various mechanisms have been suggested to coordinate the conflicting interests of channel members for mutual profit maximization. These include: (1) a market-based mechanism that coordinates channels through short-term exchanges; (2) administered channel coordination through nonmarket incentives such as promotions; (3) contractual channel coordination through long-term contracts including franchising; and (4) vertical integration that coordinates the channels through ownership and authority of members at various levels of the system.

In the food marketing system, coordination mechanisms take the form of specialized contracts between a food processor and a wholesaler or retailer. Such contracts often involve profit sharing or quantity discount arrangements that allow risks and revenues to be shared by all members. Because of concentration, revenue sharing may not always be equitable. A contract would normally involve periodic or stochastic orders from the wholesaler, retailer, or restaurateur for specified quantities at some agreed-upon prices, with provisions to order additional quantities of products within the period. When a greater proportion of processor sales is concentrated in a few distributors, wholesalers, retailers, or restaurateurs, processors can become dependent on wholesalers, distributors, retailers, or restaurateurs. Major retailers such as Wal-Mart, Albertsons, and Kroger capture more value through practices such as levying slotting fees on food processors for placement of products in prime shelf areas. Slotting fees are lump-sum fees that suppliers pay to retailers for introducing new products to the supermarket shelves or for securing prime shelf areas. The fees have long been used in the supermarket industry for dry grocery items and have recently entered the fresh produce and other departments in supermarkets. Undoubtedly, large wholesale and retail giants have introduced purchasing might, logistics expertise, and category management skills to the seafood marketing business.

Different models of contracts are practiced depending on the product. These include quantity flexibility contracts, backup agreements, buy-back or return policies, incentive mechanisms, revenue-sharing contracts, allocation rules, and quantity discounts. Contracts would usually specify the rights, responsibilities, rewards, and sanctions for nonconformity for each member of the channel in the system. Food processors may use different marketing channels to reach diverse target markets, each channel involving a different set of intermediaries.

Although many marketing channels are organized by consensus among the members, some are organized and controlled by a single leader, also called the channel leader. The channel leader may be a processor, wholesaler, or retailer. The channel leader normally possesses channel power, the ability to influence another channel member's goal achievement. Nevertheless, channel cooperation is vital if each member is to gain from the system. There are several ways to improve channel cooperation. If a marketing channel is viewed as a unified supply chain competing with other systems, then individual members will be less likely to take actions that create disadvantages

Contract Farming in the Thailand Shrimp Industry

To support the development of the shrimp aquaculture industry in Thailand in the early 1980s, a unique arrangement was established between large feed and processing firms and farmers. At the time, farmers were reluctant to build ponds and invest the necessary capital into the new industry without some technical assistance and guarantee of a market. The feed and processing firms wanted to supply the industry with their products and help it to develop. To attract private farmers to grow shrimp, agreements were established between the feed and processing firms and the farmers in which selected supply inputs, such as feed and seed, were supplied and for which technical assistance was provided. In some cases, the firms provided loans to build the ponds or contracted a farmer to grow the shrimp in a firm-owned pond. The farmer and the buyer agreed to terms and conditions for the future sale and purchase of the shrimp. This "contract farming" arrangement proved to be a very successful model, and the shrimp industry in Thailand has flourished as a result.

Contract farming can be defined as an agreement between farmers and processing or marketing firms (or both) for the production and supply of aquacultural products under forward agreements, frequently at predetermined prices. As in the case of Thailand, these arrangements often involve the purchasing firm's providing support through the supply of inputs and the provision of technical assistance. The contractual arrangement between the farmer and the purchaser involves a commitment on the part of the farmer to provide a specific quantity and quality of the aquaculture product as determined by the purchaser, and for the purchaser to support the farmer's production and to purchase the product at a specified price. Depending upon the specifics of the agreement, the farmer agrees to follow recommended production methods and practices.

Contract farming can be carried out in a variety of different ways depending upon the depth and complexity of the market and the resource and management provisions in the contractual agreement. Contract farming is not without its problems, however, because conflicts can arise over rigid production requirements and prices. In aquaculture, contract farming can be a means to develop markets and to bring about the transfer of technical skills that can be profitable to both the farmer and the firm.

Contributed by: Robert Pomeroy, University of Connecticut

for other members. Channel members should agree to direct efforts toward common objectives so that channel roles can be structured for maximum marketing effectiveness, which in turn can help members achieve individual objectives.

One of the mechanisms of coordination in the channel system is the Electronic Data Interchange (EDI) that is being utilized by various members of the channel. EDI is a computer-to-computer exchange of formatted business transactions in a standard format. The system allows a company to send information over communications links. The system can read information from an incoming invoice, such as net total or vendor name or address, and send it directly to the company's accounting application for payment preparation. The EDI system is used for inventory control, stock replenishment programs, warehouse management, customer management, pricing, and financial reporting. Some related software could be used to rank customers, products, and services by profitability, to optimize inventory and customer service levels and to assist with business-to-business management.

Various channel stages may be combined under the management of a channel leader either horizontally or vertically. Vertical channel integration involves a combination of two or more stages of the channel under one management. An example is when one member of a marketing channel purchases the operations of another member, or one member simply performs the functions of another member, eliminating the need for that intermediary as a separate entity. Normally, members of a channel system work independently, but in vertical channel integration, members coordinate efforts to reach a desired target market. The integration allows a single channel member to coordinate or manage channel activities to achieve an efficient, low-cost distribution system. The vertical marketing systems can be summarized as taking one of three forms: (1) a corporate system in which all stages of the marketing channel, from processor to consumers, is under a single owner; (2) an administered system in which channel members are independent but a high level of interorganizational management is achieved by information coordination; and (3) a contractual system in which channel members are linked by legal agreements spelling out each member's rights and obligations. This is the most popular type of vertical marketing system.

Combining channels at the same level of operation under one management constitutes horizontal channel integration, that is, a merger between companies at the same level in a marketing channel. Although horizontal integration allows efficiencies and economies of scale in purchasing, marketing research, advertising, and specialized personnel, it is not always the most effective method of improving distribution.

CHANNEL AGREEMENTS

Tying Agreements

A tying agreement is when a processor or a supplier supplies a product to a channel member with the stipulation that the channel member must purchase other products as well. For example, the feed-for-fish program of Southern Farm Services provided feed on credit in return for guaranteed delivery of market-sized fish to the processing plant. Related to this type of agreement is what is commonly known as "full-line forcing." Here, a supplier requires that channel members purchase the supplier's entire line of products to obtain any of the supplier's products. Tying agreements are legal provided that the supplier alone can provide a line of products of a certain quality, when the intermediary is free to carry competing products as well, and when a company has just entered the market. Most other tying agreements are considered illegal.

Exclusive Dealing

Exclusive dealing occurs when a processor or supplier forbids an intermediary to carry products of competing suppliers or processors. This type of agreement is illegal if the agreement blocks competing suppliers from as much as 10% of the market, if the sales revenue involved in the transaction is larger, and if the supplier is much larger and thus more intimidating than the intermediary. Exclusive dealing is legal if intermediaries have access to similar products from competitors or if the exclusive dealing contract strengthens an otherwise weak competitor.

CHANNEL CONFLICT

Conflicts arise among channels members due to various issues such as self-interest, misunderstandings, disappointments, false expectations, communication difficulties, and disagreements. There appears to be no single method for resolving conflict. Nevertheless, partnerships can be maintained where there is a clear understanding of the role of each channel member and when measures are in place for channel coordination that may require leadership and benevolent exercise of control. An important element in maintaining good relationships among channel members is ensuring that each member meets agreed-upon contract guidelines. Potential conflict areas include processor rebates, product promotion, billing payments, resellers with different brands, territorial issues, and direct sales.

AQUACULTURE MARKET SYNOPSIS: TROUT

Rainbow trout (*Oncorhynchus mykiss*) have been cultured for more than a century and have been introduced into countries over the entire world. Recreational trout fishing has become preferred angling all over the world. Rainbow trout are prized as freshwater game fish as well as a preferred foodfish. Much of the early aquaculture of trout was developed in order to stock and restock natural waters to enhance fish populations to support recreational trout fishing. There are records of aquaculture production of rainbow trout from 1950 (FAO 2004).

Private, commercial production of trout for foodfish markets has grown and expanded over the years. However, the rate of growth of trout production increased in the 1980s and 1990s (Fig. 6.9). In 2002, world trout production was 555,802 MT (FAO 2004).

Capture fisheries for trout exist, but in negligible quantities. For example, in 2002, trout capture fisheries for trout were only 8,784 MT compared to the 555,802 MT from aquaculture. Aquaculture production began to surpass that of capture fisheries in 1954, and by 1960, aquaculture facilities were producing four times as much as the production from capture fisheries.

The leading trout-producing countries worldwide in 2002 were Chile, Norway, France, Turkey, and Italy, with 20%, 15%, 8%, 6%, and 6% market share, respectively (Fig. 6.10). Chile produced 111,681 MT of rainbow trout in 2002. Much of the Chilean trout is sold to markets in Japan.

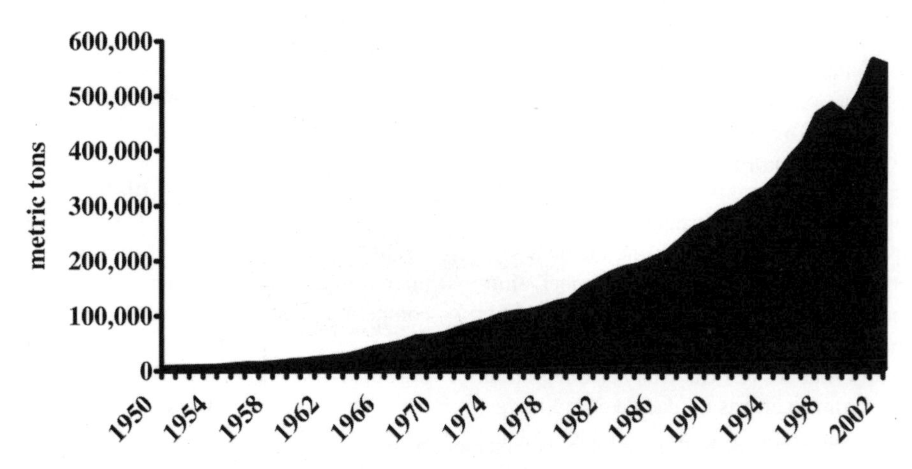

Figure 6.9. Growth of world aquaculture production of trout. (Source: FAO 2004.)

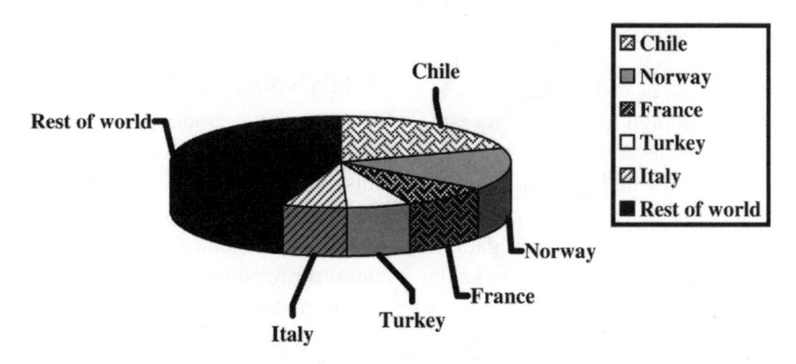

Figure 6.10. Leading trout-producing countries worldwide. (Source: FAO 2004.)

Trout are raised primarily in raceways located in areas with high volumes of high-quality surface water, typically springs. However, trout are also raised in cages in places such as Lake Titicaca in Bolivia and elsewhere. Trout are served traditionally as a whole fish with the head on. In some restaurants, butterfly fillets are served, but there is a variety of traditional preparations and forms served around the world.

The largest markets for trout are the United States, the European Union, mainly France, United Kingdom, and Italy, the Russian Federation, and Japan. There has been an increasing trend toward expansion of production in Chile for export to the major markets in the United States, European Union, and Japan.

Trout production in the United States continues to provide fish for angling as well as for the food market. Figures 6.11 and 6.12 show the distribution of foodsize (30.5 cm) trout and stocker (15- to 30-cm) trout in the United States. Although 68% of the food-sized trout produced are sold to processing plants, another 20% are sold to fee fishing businesses to provide recreational fishing opportunities to anglers. Of the stocker trout sold, 52% were sold to fee fishing businesses with another 14% being sold to the government, primarily for stocking programs in natural waters.

The U.S. foodfish trout industry has moved into value-added product development in recent years with products such as breaded, stuffed, and finger-food portions with a variety of recipes, flavors, and preparations. Many of the products are easy and quick to prepare.

The trout industry has had to cope with increasing regulations related to its discharge of waste products

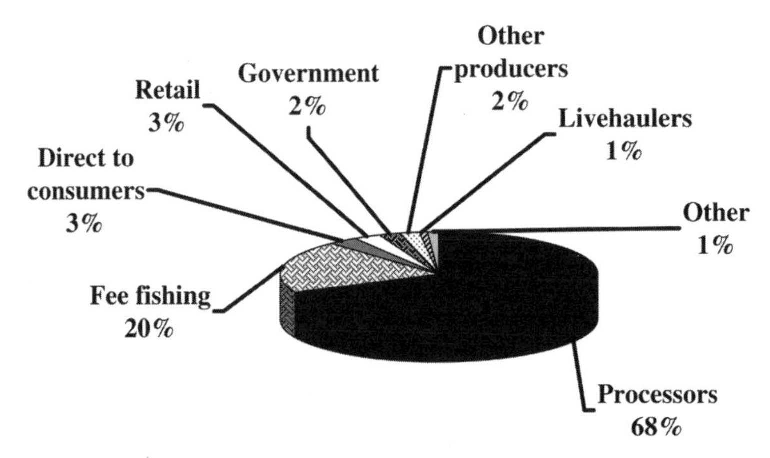

Figure 6.11. Distribution of 30.5-cm trout to various market channels, United States. (Source: USDA 2004.)

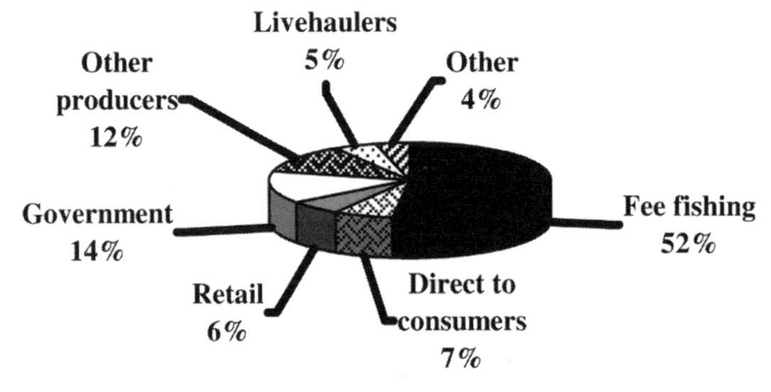

Figure 6.12. Distribution of 15- to 30-cm trout to various market channels, United States. (Source: USDA 2004.)

into the environment. New treatment technologies, new feeds, and increased monitoring have been adopted as the industry has adapted to these changing demands.

SUMMARY

A marketing channel (also called channel of distribution) is a combination of interrelated intermediaries (individuals and organizations) who direct the physical flow of products from producers to the ultimate consumers. It can be very simple and direct, as with direct sales, or can be very complex, comprising an array of brokers, sales agents, traders, distributors, wholesalers, foodservice operators, and importers. The complexity often depends on the type of seafood and whether the nation's economy is developed or is subsistence in nature. In most western economies, seafood-marketing channels consist of a wide variety and a high number of actors including importers, agents, traders, wholesalers, processors, retailers, and restaurants. In the United States, the physical flow of agricultural commodities through the marketing channel varies with commodity groups.

Evidence from the food distribution system in the United States indicates differences in relative importance of specific commodity flows, how channel agents facilitate commerce, and price-discovery mechanisms. Price discovery for different commodities involves concepts such as market structure, market behavior, market information, and price reporting, as well as futures markets and risk management alternatives. In the livestock sector, there is less dependence on open market transactions or using a spot market basis and more dependence on forward contracts, production and marketing contracts, marketing agreements, and vertical integration. Regarding the U.S. seafood market, there is a variety of pricing mechanisms including negotiation on boat-by-boat basis at the time of landing, short-term marketing agreements, and sale on consignments. For example, marketing arrangements for shrimp involves contracts with fixed price or quantities and covered periods ranging from two months to a year. There are some fish auctions in the United States in which buyers engage in competitive bidding for seafood products.

One of the fundamental issues that processors consider is the choice of an intermediary to adopt for the distribution of their products, and the issue of control of channels lies in where bargaining power lies. This is because the distribution system involves several members pursuing different objectives, therefore some mechanism is needed to coordinate the channels. Some mechanisms include short-term exchanges, promotions, long-term contracts including franchising, and vertical integration. In the food marketing system, coordination mechanisms take the form of specialized contracts between a food processor and a wholesaler or retailer involving profit sharing or quantity discounts arrangements, which allow risks and revenues to be shared by all members. Types of channel agreements include "tying agreements" and "exclusive dealings." One of the mechanisms of coordination in the channel system is the Electronic Data Interchange (EDI) that is being utilized by various members of the channel. The EDI system is used for inventory control, stock replenishment programs, warehouse management, customer management, pricing, and financial reporting.

STUDY AND DISCUSSION QUESTIONS

1. Describe the difference between commodities produced by farmers and the products demanded by consumers. Using a specific commodity, suggest how the difference is bridged by the food system.

2. What factors determine the complexity of seafood marketing channels? Illustrate with an example.

3. What are the differences between grid pricing and formula pricing? What are the advantages and disadvantages of each pricing system?

4. What are futures and options? How can a futures market be used as a price discovery mechanism?

5. What are the various criteria for classifying wholesalers? Describe the types of food wholesalers in the U.S. food system and the role each plays in the system.

6. Suppose that you are an independent fish processor and looking for a distributor for your products. What factors will you consider in your decision process?

7. Why is market power essential in the food distribution channel? What ways do businesses adopt to be able to have market power?

8. What are the advantages and disadvantages of slotting fees?

9. Describe some specialized contracts between a food processor and a wholesaler or retailer that would involve nonmarket incentives. Give examples.

10. Give two examples of channel agreements and indicate how they differ from each other.

REFERENCES

Anderson, J.L. 2003. *The international seafood trade.* Woodhead Publishing Ltd., Abington Hall, Abington Cambridge.

Asche, F., A. Fofana, J. Hartmann, S. Jaffry, and R. Menezes. 2002. Vertical relationships in the value chain: an analysis based on price information for three seafood species in Europe. SALMAR Work Package 3, European Commission DG 14.

Chauhan, D.P. 1994. Fish marketing in India. Pages 15–23 in Infofish (1994). *Fish Marketing in Asia,* Infofish, Kuala Lumpur, Malaysia.

Dimitri, C., A. Tegene, and P. R. Kaufman. 2003. U.S. fresh produce markets: marketing channels, trade practices, and retail pricing behavior. Agricultural Economic Report No. AER825. 23 pp.

FAO. 2004. Fishstat Plus. Food and Agriculture Organization of the United Nations. Accessed at http://www.fao.org.

Glaser, L.K, G.D. Thompson, and C.R. Handy. 2001. Recent changes in marketing and trade practices in the U.S. lettuce and fresh-cut vegetable industry. Economics Research Service Agricultural Information Bulletin No. 767, U.S. Department of Agriculture, Washington, D.C.

Harris, J.M., P. Kaufman, S. Martinez, and C. Price. 2002. The U.S. food marketing system, 2002—competition, coordination, and technological innovations into the 21st Century. Economics Research Service Agricultural Economic Report No. 811, U.S. Department of Agriculture, Washington, D.C.

Homziak, J. and B.C. Posadas. 1992. A preliminary survey of tilapia markets in North America. Proceedings of the Annual Gulf and Caribbean Fisheries Institute 42:83–102.

Hussain, M.M. 1994. Fish marketing in Bangladesh. Pages 24–44 in Infofish (1994). *Fish Marketing in Asia*, Infofish, Kuala Lumpur, Malaysia.

Jaffry, S., A. Fofana, and A.D. Murray. 2003. Testing for market power in the UK salmon retail sector. *Aquaculture Economics and Management* 7(5/6): 293–308.

Johnson, H.M. 2004. 2004 annual report on the United States seafood industry. H.M. Johnson & Associates. Jacksonville, Oregon.

Johnsen, O. and F Nilssen. 2001. The future for salmon in Germany. Norwegian Institute of Fisheries and Aquaculture Research Report 10/2001, Tromsø, Norway.

Karundeng, H.H. and S. Pawiro. 1994. Some aspects of fish marketing development in Indonesia. Pages 7–14 inInfofish (1994). *Fish Marketing in Asia*, Infofish, Kuala Lumpur, Malaysia.

Lahidji, R., W. Michalski, and B. Stevens. 1998. The future of food: an overview of trends and key issues. In *The future of food: Long-term prospects for the agro-food sector*. OECD, Paris.

Legaspi, A.S. 1994. The socio-economic and quality aspects of fish marketing in the Philippines. Pages 54–64 in Infofish (1994). *Fish Marketing in Asia*, Infofish, Kuala Lumpur Malaysia.

Lence, S.H. 2000. A comparative marketing analysis of major agricultural products in the United States and Argentina. Argentina Report 2, Midwest Agribusiness Trade Research and Information Center Research, Iowa State University, Paper 00-MRP 2, Ames, Iowa.

Leyva, C. 2004. Optimizing tilapia (*Oreochromis* sp.) marketing in Honduras. Master's thesis, Department of Aquaculture and Fisheries, University of Arkansas at Pine Bluff, Pine Bluff, Arkansas.

Martinez, S. 2002. Vertical coordination of marketing systems: lessons from the poultry, egg and pork industries. Economics Research Service Agricultural Economic Report No. 807. 45, U.S. Department of Agriculture, Washington, D.C.

McBride, W.D. and N. Key. 2003. Economic and structural relationships in U.S. hog production. Resource Economics Division, Economic Research Service Agricultural Economic Report No. 818, U.S. Department of Agriculture, Washington, D.C.

Molnar, J.J., Hanson, T. and Lovshin, L. 1996. Social, economic, and institutional impacts of aquaculture research on tilapia. Research and Development Series, No. 40. International Center for Aquaculture and Aquatic Environments, Alabama Agricultural Experiment Station, Auburn University, Auburn, Alabama, U.S.A.

Muhaiyadin, M. 1994. Fisheries sector profile of Malaysia. Pages 1–6 in Infofish (1994). *Fish Marketing in Asia*, Infofish, Kuala Lumpur, Malaysia.

National Marine Fisheries Service (NMFS). 2004. Fisheries of the United States, 2003. Office of Science and Technology Fisheries, Statistics Division, Silver Spring, Maryland.

Piamsombun, S. 1994. Domestic fish marketing in Thailand. p. 50–53 in Infofish (1994). *Fish Marketing in Asia*, Infofish, Kuala Lumpur, Malaysia.

Radtke, H., and S. Davis. 2000. Description of the U.S. west coast commercial fishing fleet and seafood processors. The Research Group Report prepared for

Pacific States Marine Fisheries Commission, February 2000. http://www.psmfc.org/efin/docs/fleetreport

Rogers, J. 2000. Sea-fish marketing channels in Northern Ireland. Department of Agriculture and Rural Development, Economics and Statistics Division. A Government Statistical Publication, December 2000.

United States International Trade Commission (USITC). 2004. Certain frozen or canned warmwater shrimp and prawns from Brazil, China, Ecuador, India, Thailand, and Vietnam. Investigations Nos. 731-TA-1063-1068 (Preliminary) Publication 3672, U.S. International Trade Commission, Washington D.C.

University of Alaska, Anchorage. 2001. A planning handbook. Institute of Social and Economic Research, University of Alaska, Anchorage, Alaska.

Ward, C.E and T.C. Schroeder. 2001. Understanding livestock pricing issues. Oklahoma State University, Oklahoma Cooperative Extension Fact Sheet F-551, Norman, Oklahoma. http://pearl.agcomm.okstate .edu/agecon/marketing/wf-551.html

7
Marketing by Farmer Groups

Many observers of the food and agricultural industries indicate that there has been a phenomenal change in the structure of the food and agricultural industries since the 1980s and 1990s. One of the changes that is often mentioned is the increased concentration in the food and agricultural industries. This means that there are fewer companies competing with each other. The strongest concentration trends can be observed in the livestock industries, particularly in hog and dairy production. A related concern stems from the recent trends in mergers in the food processing industries. This has resulted in sharp increases in the sizes of processing plants. This same trend in consolidation has occurred in the retailing and wholesaling sectors. As a result of the mergers and acquisitions, store sizes and company sizes have grown with the emergence of fewer but larger corporate chains.

In many cases, these consolidations into larger plants and companies are the results of new economies of scale and competition in the market place. Economists have suggested that, when market demand for a product is growing slowly, increased consolidation can lead to increased concentration. Therefore, the structural changes occurring in the food and agricultural industries could harm small-scale and medium-scale farmers. On the other hand, increased consolidation sometimes may be beneficial for consumers and the society at large. This is because the economies of scale reduce costs, which can translate into reduced prices to consumers.

These changes occurring in the food and agricultural industries can affect farmers in a variety of ways. These can include: (1) disparity of bargaining power; (2) use of production contracts; and (3) competition from imports (Torgerson 2000). Each of these are discussed in more detail in the following sections.

DISPARITY IN BARGAINING POWER

Increased mergers and concentration of the companies that purchase farm products result in farmers facing fewer, but larger, buyers of their products. This often results in an apparent disparity in market and bargaining power between farmers and buyers. The large companies often have the greater bargaining power, and farmers can lose market access as a result. Therefore, by organizing into cooperatives or farmer groups, farmers can control a larger portion of their products, have greater bargaining power than an individual farmer would have, and can approach several different potential buyers in different regions.

The greater product resources available through farmer organizations provide opportunities to negotiate and develop larger supply contracts. Such farmer cooperatives or groups are likely to be able to negotiate higher prices for their products than an individual farmer with more limited available product supplies. With a cooperative, farmers or other marketing group can collectively exercise some influence in the market place and begin to correct for market failure. The U.S. Department of Agriculture (2004) lists 12 associations that collectively bargain for processing fruit and vegetable commodities including apricot, cling peach, Bartlett pear, processing tomatoes, olives, prunes, raisins, potatoes, peas, barley, flax, processing apples, plums, red tart cherries, asparagus, feeder pigs, raw milk, hazelnuts, sugarbeets, and perennial ryegrass. The number has declined from previous years, suggesting a possible decline in bargaining activity, probably due to increased use of contract production.

USE OF PRODUCTION CONTRACTS

Large companies and chains increasingly tend to use production contracts. For example, more than 90%

of broilers and processed vegetables have been produced according to the terms of contracts with processors for several decades. The concept of production contracting is becoming popular and has been adopted for the production of hogs and cattle as well as for the production of other commodities. Contract production involves relationships and activities between an owner of the plant or animal (often the major buyer or large company) and the services of the farmer. For example, in a typical poultry contract, Tyson Foods, Inc., provides the chicks and inputs while the farmer provides the labor, management, and facilities required to raise the Tyson-owned chicks to the appropriate processing weight. The farmer receives an agreed price per bird in addition to some performance incentives. In most coastal fisheries, some fish processing companies contract with fishermen by providing them with inputs such as nets, boats, motors, and gear while fishermen supply their catch to processors at some negotiated price minus the cost of the inputs (University of Alaska 2001).

Production contracts can be beneficial but also be disastrous for the farmer. Farmers are concerned when the market is dominated by a few, large, integrated buyers. Some farmers find that contracting limits their opportunities for growth, restricts entrepreneurship, and applies pressure to keep up with technological changes. Moreover, farmers may lack the leverage to negotiate for better contract terms. Farmer groups can effectively participate in the development of agricultural contracts and provide farmers the opportunity to work with buyers and processors to eliminate unfair or unreasonable terms. The group or cooperative can negotiate terms of sale, prices, and payment arrangements to share any financial risks between the members and the buyers.

COMPETITION FROM IMPORTS

There has been growing competition in the United States between domestically produced food products and low-cost imported products. Many of these imported food products are produced under relatively few environmental guidelines, food safety, and labor controls. Production is subsidized in many other countries, and labor costs are very low. Therefore, the price of many imported food commodities is low compared to the price of domestically produced products. The domestic seafood industry in the United States especially has faced very stiff competition from low-priced imported seafood products in recent years.

Agricultural growers have found it necessary to assume greater control of their industry by working together, developing effective cooperatives, and coordinating cooperative systems for collective actions in marketing. With pooled resources, a cooperative can provide better market information and data for members to utilize in their management decisions. The cooperative can serve as a clearinghouse for trade information, promote the product in both the domestic and foreign markets, develop partnerships with other groups in foreign countries, and serve as a voice for producers. With some market power, cooperatives can influence terms of trade on the domestic or international markets, or both. Terms of trade relate to price, timing, form, and other quality or quantity specifications. Cooperatives can provide mechanisms for resolution of trade disputes and enforcement of trade regulations and standards to ensure a fair playing field in the market place.

As discussed in Chapter 5, antitrust laws exist to promote competition, but organization of farmers into groups is not against antitrust laws. In fact, the Capper-Volstead Act of 1922 (Appendix 7.A, at the end of this chapter) provides immunity to farmers who organize into groups for purposes of developing some bargaining power to better deal with other competitors, and for addressing supply chain issues from a cooperative and coordinated position of strength. The Act essentially grants farmers the legal right to pool their bargaining and marketing resources to place them on an equal footing with the large buyers of their raw agricultural products.

MARKETING COOPERATIVES

A marketing cooperative is a farmer organization with the purpose of collectively selling members' farm products. The cooperative provides farmers the opportunity to perform joint marketing responsibilities that include assembling of products, negotiating with large buyers of farm products, exercising some degree of power in the marketplace, spreading risks and costs, and in some cases processing farm commodities. Thus, a marketing cooperative may function as a contract- and price-bargaining cooperative, or it may be involved in processing or manufacturing of specific agricultural commodities. By joining the cooperative, farmers are provided a guaranteed outlet for their farm products. Another advantage to farmers in joining a cooperative is the benefit of economies of size and scale. There is also sharing of marketing risk and costs among farmer-members. This risk sharing plays an important role in develop-

Fishermen's Cooperatives in Bangladesh

There are numerous examples of fish and seafood marketing cooperatives throughout the world. In Bangladesh, for example, Hussain (1994) reported about 4,500 primary cooperative fishermen's societies. Overall membership was estimated at 537,224 in 1994. The cooperative societies were funded with government loans and credit from Japan, Denmark, and other countries.

ment of individual farm enterprises and in developing markets.

There are four classes of marketing cooperatives based on how they are organized, membership affiliation, control, and geographic area. These classes are as follows: (1) local cooperatives; (2) centralized cooperatives; (3) federated cooperatives; and (4) mixed cooperatives. Each is discussed in the following sections.

Local Cooperatives

Local cooperatives are usually farmer groups at the local or community level. They perform a limited number of marketing activities for the group such as assembling and grading farm products. Most of the cooperatives for fruits, vegetables, specialty crops, and fisheries are local in nature because of the localized nature of the production of the commodities involved. Consequently, membership is almost exclusively restricted to farmers engaged in producing the commodity.

Centralized Cooperative

Centralized cooperatives, unlike local cooperatives or associations, operate over larger geographic areas and have members in several states. In addition to assembling farm products, they often provide more vertically integrated services such as processing. This is the common form of cooperative in agriculture. In the livestock sector, for example, several small livestock producers in Nebraska, Kansas, and Oklahoma have organized into marketing associations that ship livestock to central markets. Such cooperatives have enabled small livestock producers to pool their small sale lots for more efficient shipment to terminal markets.

Another example is the Staple Cotton Cooperative Association (Staplcotn), which is America's largest and oldest cotton marketing cooperative, based in Greenwood, Mississippi. It is owned by 2,500 cotton growers in Mississippi, Arkansas, Louisiana, Tennessee, Missouri, Alabama, Florida, and Georgia and handles about 15% of the U.S. cotton crop. Cotton growers may store product with Staplcotn but not market through the cooperative, or vice versa. Under the cooperative's Mill Sales Program, members market their cotton through two basic options. Most members choose the Seasonal Option, through which the coop makes the pricing decisions. Those grower-members who want to make their own pricing decisions can choose the Call Option, through which the grower makes the futures market decisions while Staplcotn markets the basis, the other major component of cotton pricing.

Federated Cooperative

The federated cooperative consists of local associations or cooperatives. Leaders from member local cooperatives or associations elect directors and provide general operating guidelines for the federation. Federated cooperatives perform more complex and expensive marketing activities that the member local associations or cooperative cannot perform. These can include manufacturing, involvement in financial markets, and international marketing, among others. CherrCo is an example of a federated marketing cooperative with 28 member cooperatives in the U.S. and Canada. It represents 75-80% of Michigan's tart cherry production and significant portions of the production in New York, Utah, Washington, Wisconsin and Ontario, Canada. Members range in size from producing about 272,727 kg (600,000 lb) to more than 4.5 million kg (10 million lb) annually. Ocean Spray is also a federated cooperative owned by more than 800 cranberry growers and 126 grapefruit growers located throughout the United States and Canada. It is the nation's largest cranberry marketing organization and North America's leading producer of canned and bottled juices and juice drinks. It has been the best-selling brand name in the canned and bottled juice category since 1981.

Mixed Cooperative

Mixed cooperatives serve both the local cooperatives and the individual farmer members. The structure combines the features of local, centralized, and federated cooperatives as well as individual memberships. Mixed cooperatives are not common and are usually formed to fit particular industry situations. The Dairy Farmers of America is an example of this type of cooperative. The cooperative is the largest milk cooperative in the United States, representing more than 22,924 producer-members who market their milk through the cooperative. Dairy Farmers of America was formed in 1998 as a result of a merger of four leading dairy cooperatives: Associated Milk Producers, Inc., Mid-America Dairymen, Inc., Milk Marketing, Inc., and Western Dairymen Cooperative, Inc. Other cooperative organizations joined after 1998, including the Independent Cooperative Milk Producers Association, the Valley of Virginia Milk Producers Association, and the California Cooperative Creamery. The cooperative represents 13,445 dairy farms in 49 states and markets more than 11.7 billion kg (25.7 billion lb) of milk, giving it a market share of 33% of the total U.S. milk supply. The cooperative also has nine bottling joint ventures, three manufacturing joint ventures, and 25 cooperatively owned processing plants. Brands of products produced by the cooperative are Borden Cheese, Golden Cheese, Mid-America Farms, Jacobo, Enricco, CalPro, Sport Shake, and VitalCal.

MARKETING COOPERATIVES AS MARKETING AGENTS

Most marketing cooperatives operate as a marketing agent by collecting products of members for sale, grading, packaging, and performing other marketing functions. Livestock cooperatives, milk cooperatives, and grain elevator cooperatives are examples of marketing agents. For example, CHS Cooperatives, formed in 1998, was a merger between two regional cooperatives, Cenex, Inc. and Harvest States Cooperatives. CHS markets substantial amounts of member-produced grain. However, in recent years, the trend has been toward affiliation with a global grain marketing company such as Archer Daniels Midland (Dunn et al. 2002). Some milk marketing cooperatives in Wisconsin, for example, do not process or physically market their members' milk but instead only represent members in pricing or establishing other terms of trade with processors on their members' behalf. The Alaska inshore pollock bargaining association historically has utilized exclusive delivery contracts between a surimi plant and the fleet delivering to that plant (Matulich and Sever 1999).

The Role of Associations in Developing Trademarks to Differentiate Aquaculture Products

There has been an increase in the development of collective trademarks and certification programs for aquaculture products over the last decade (Girard and Mariojouls 2003). In France, for example, the Comité National de la Conchyliculture (CNC) has taken a stronger role in developing new brands for groups of shellfish growers. Although shellfish growers traditionally have been very independent and individualistic, farming associations under a brand or trademark allows them to gain competitive advantages and improve their marketing position. Supermarkets have begun to sign contracts with groups of farmers for products certified under distinct trademarks that assure compliance with particular specifications. This is a departure from traditional emphases on buying strictly on price. New producers' organizations are emerging in France that have responsibility for developing differentiated products and organizing marketing initiatives.

MARKETING COOPERATIVES AS PROCESSING GROUPS

Some marketing cooperatives are organized to perform processing functions. This typically includes packaging products of members as well as wholesaling final products. Examples of such marketing cooperatives can be found in vegetable canning, fruit packing, and cheese and butter manufacturing. These functions are part of the overall marketing activities performed by these cooperatives in an attempt to control their products as they move to the marketplace.

Delta Pride Catfish, Inc. is a catfish marketing cooperative that controls 26,000 ha (65,000 ac) of catfish ponds in Mississippi and Arkansas, representing about 50% of the acreage of catfish farms in the United States. The cooperative is the nation's largest processor of farm-raised catfish, processing about 45.5 million kg (100 million lb) of live farm-raised catfish every year. The cooperative processes catfish of its farmer members. Farmers must own stock in the company to be eligible to sell fish to the processing plant. Farmers are entitled to sell volumes of fish in proportion to their shares of stock owned. For many years, Delta Pride paid the highest price to farmers for fish delivered to the plant. However, if the plant had not generated a profit at the end of the year, farmers were billed back for the volume of fish delivered during that year. The amount of the billback is calculated to ensure that the company breaks even for the year. The cooperative produces and markets fresh and frozen catfish products as well as value-added products of breaded, marinated, and seasoned fish. The cooperative sells its catfish products to restaurant chains, food service institutions, and other discount wholesale food outlets such as Costco and Sam's Club. The major product sold by the cooperative is catfish fillets.

FARMERS' BARGAINING GROUPS

Agricultural bargaining groups are a special type of marketing cooperative. These bargaining groups do not own, process, or market the farm commodities of farmers. Instead, they negotiate with processors or buyers on behalf of the members. The cooperative negotiates for price (including premiums and discounts), quality standards, and time and method of payment. In some cases, the bargaining group coordinates the distribution of product and timing of delivery. It may also negotiate other terms of transaction that may include grading, the duration of the contract, production rights and responsibilities, and transportation. A bargaining cooperative represents the occupational interests of farmers in the policy arena and in the marketplace. The association is mainly financed through checkoff programs. A checkoff could be a flat fee per unit of sale or some specified percentage of sale value of the products sold by members that is retained by the association. Other methods of financing include service charges to processors, annual dues, and membership fees.

A bargaining cooperative usually does not physically handle the farm produce. Members sell farm products directly to processors at the price negotiated by the cooperative. With control over large volumes and supplies of farm products, bargaining associations have more market power than do individual growers and are able to negotiate price more effectively. Bargaining associations are common in processing sectors of fruit, vegetable, specialty crop, dairy, and sugar beet industries. Iskow and Sexton (1992) conducted a comprehensive survey of all active bargaining associations in markets for processing fruits and vegetables. The authors reported that bargaining associations bargain for raw product price, the terms of trade (including time and method of payment), and quality standards. Only 25% of the associations surveyed reported negotiating for the quantity of raw product to be purchased by processor/handlers. In most cases, the total volume of raw product to be purchased was predetermined prior to price negotiations.

In the fisheries sector, perhaps one of the most successful bargaining groups is the Alaska inshore pollock bargaining cooperative. The inshore pollock processing sector is highly concentrated, with catcher vessels delivering more than 80% of inshore allocation to four onshore surimi processors while the remainder is delivered to motherships. The Alaska inshore pollock bargaining cooperative negotiates a formal contract, involving price, with each of the processors prior to each season and represents a countervailing monopolistic bargaining association (Matulich and Sever 1999).

A bargaining association could be an effective bargaining agent for farmers engaged in production contracts. A cooperative bargaining association can work legislatively toward establishing institutional rules that augment the bargaining process. These rules could include provisions for good faith negotiations, dispute resolution mechanisms, and enforcement procedures. The cooperative can effectively represent farmers negotiating marketing contracts and those negotiating production contracts.

MARKETING ORDERS

A marketing order is a legal instrument authorized by the U.S. Congress through the Agricultural Marketing Agreement Act of 1937. The Act authorizes the secretary of agriculture to establish "marketing orders" for milk, fresh fruits, vegetables, tobacco, peanuts, turkey, and specialty crops (such as almonds, walnuts, and filberts). The order establishes production quotas, allocates the quotas among producers, and forbids producers to sell more than the amounts allocated. Producers who attempt to sell more than their quota are subject to fines. The primary objective of the order is to stabilize market conditions and provide benefits to producers and consumers by establishing and maintaining orderly marketing conditions.

Many states also have parallel legislation modeled after the Federal Act (Agricultural Marketing Agreement Act of 1937) to provide for state marketing orders. The jurisdiction of federal marketing orders may be limited to an industry defined within the boundaries of one state or a subregion within a state, or it may extend to a production region encompassing more than one state. With state marketing orders, the jurisdiction is limited to individual states or subregions within the states.

The legal provisions of federal marketing orders can be classified into three broad types: (1) Quality control provisions that involve specifying standardized packages or containers and establishing uniform, mandatory quality standards, such as size, color, or minimum maturity; (2) quantity control methods that include smoothing the flow of the product to market (for example, prorate or shipping holidays) and specifying producer allotments and volume management provisions, such as allowing only a certain portion of the crop to move into specified outlets (for example, reserve pools or market allocation); and (3) market-facilitating provisions that include production research, market research and development, market information, and market promotion and advertising.

The enabling legislation of the Agricultural Marketing Agreement Act allows producers to form marketing orders that comprise elements from all three of the above types of provisions. However, in practice, commodity groups generally prefer to include only some of the provisions when designing a marketing order for their product. Most commodity groups forming federal marketing orders have tended to focus on quality regulations (such as grade, size, and packing or container regulations), research and advertising, and sometimes quantity controls.

Most of the state marketing programs have been utilized for research, promotion and advertising, or both, because of the support from farmers and policy makers. State marketing orders have been used more for quality regulations than for quantity control. Quantity controls have been controversial.

Marketing orders are established for commodities by a vote of the producers in the geographic area for which the order is proposed. After the marketing order is established, committees of producers develop the details of enforcement. The details including (1) uniform quality standards; (2) produce allotments; and (3) production and marketing research. The detailed regulations are forwarded to the secretary of agriculture and, upon approval, the order is published in the Federal Register, whereupon it becomes law and legally binding on all producers and handlers. Industries may voluntarily elect to enter into these programs and choose to have Federal oversight of certain aspects of their operations.

In Texas, the Federal Marketing Order for oranges and grapefruit contained provisions that established the grades and sizes of fruit that could be shipped, container size and pack specifications, and inspections to ensure compliance with regulations and allowed for funding of market research and development, including paid advertising. Quality regulations were implemented by establishing the grades, sizes, and containers in which citrus could be shipped in regulated trade channels. These regulations could be changed from season to season or even within a given marketing season as market conditions changed. One objective of such regulation was to stimulate demand for citrus by increasing buyers' and consumers' satisfaction and confidence in the product. Federal Market Orders also regulated the importation of some commodities. The Federal Marketing Order for Texas Oranges regulated the importation of all fresh fruit into the U.S., and the Federal Marketing Order for Florida Grapefruit regulated all fresh grapefruit imports.

GENERIC ADVERTISING OF AQUACULTURE PRODUCTS

One of the major programs of coordinated cooperative action in marketing is generic advertising. A generic marketing campaign is typically conducted to benefit a generic product or grouping of similar products without identifying brand names or product origins. Generic advertising campaigns for individual commodities have often been supported and funded by producer groups, food companies, food organizations, and sometimes state governments. State governments have engaged in generic promotion programs to enhance the state's agricultural product sales. For example, generic state promotion programs include those conducted for Washington apples, Florida citrus, and Idaho potatoes. Generic marketing campaigns are run by organizations such as the Alaska Seafood Marketing Association, the Catfish Farmers of America, the Virginia Marine Products Board, and the National Fisheries Institute. Tilapia producers and suppliers engaged, in the past, in a generic campaign to promote farmed tilapia. Successful generic campaigns have been run by non-seafood producer organizations such as the American Dairy Association, American Egg Board, Beef Industry Council, California Raisin Advisory Board, International Apple Institute, National Honey Board, National Pea and Lentil Association, National Yogurt Association, Peanut Advisory Board, Popcorn Institute, the Wine Institute, and others.

These advertising programs are designed to stimulate consumer demand for the related commodity. Consequently, in 1996, the U.S. Congress mandated that all commodity promotion programs utilizing price checkoffs be evaluated at least once every five years under Section 501-(c) of the 1996 Farm Bill. Ward and Lambert (1993) found that generic advertising increased beef demand, and their results have been used to support additional funding on generic advertising. In contrast, Brester and Schroeder (1995) and Kinnucan et al. (1997) found that generic beef and pork advertising had little effect on demand. Lenz et al. (1998) reported that the effect of advertising by New York dairy farmers on fluid milk demand with an advertising elasticity of 0.06 was minimal for New York City. Chung and Kaiser (1999) confirmed this with an advertising elasticity estimate of 0.07. For catfish, Zidack et al. (1992) reported a benefit-cost ratio of about 13:1. The ratio suggests an enormous benefit, which the authors attributed largely to the inelastic supply of catfish. However, the authors reported an advertising elasticity of 0.007. Kinnucan et al. (1995) concluded that generic advertising was always beneficial to catfish farmers in that incremental benefits exceeded incremental costs if producers and feed mills share the levy equally.

Evaluation of the effectiveness of other promotion programs includes cotton (Capps et al. 1996), soybeans (Williams et al. 1998), avocados (Carman and Green 1993), eggs (Reberte et al. 1996), and apples (Richards et al. 1997).

All generic promotional campaigns promote the generic product and do not promote one brand of product over another. Examples of generic promotional campaigns by fisheries and aquaculture-related agencies are identified in the following sections.

Advertising of Farmed Fish: The National Fisheries Institute (NFI)

The National Fisheries Institute primarily promotes the interests of the seafood industry in general in the U.S. Congress and before regulatory agencies. NFI also promotes and defends the industry and its products to the media and consumers through generic advertising of fish and seafood. The NFI sponsors advertising programs that cover various species including catfish, sea bass, cod, crab, halibut, lobster, menhaden, oyster, pollock, quahog, salmon, scallops, shrimp, skate, tilapia, and tuna, among others. It frequently provides advertising materials relating to seafood recipes, seafood safety, and the health benefits of eating fish and seafood. One of the major advertising campaigns of the NFI has been the "Eat Fish and Seafood Twice a Week" campaign. The campaign focuses on the message that fish and seafood are economical, delicious, and quick and easy to prepare. Therefore, eating fish and seafood at least twice each week can go a long way toward achieving healthful dietary goals. The campaign includes promoting that fish oil also provides significant health benefits, especially in combating heart disease.

The institute is a nonprofit trade association representing more than 1,000 companies involved in all aspects of the fish and seafood industry. Membership includes U.S. firms that operate fishing vessels and aquaculture facilities; buyers and sellers, processors, packers, importers, and exporters and distributors of fish and seafood; as well as operators of retail stores and restaurants that sell fish and seafood.

Salmon Advertising: Salmon Marketing Institute (SMI)

The salmon industry recognized the potential benefits of developing a generic advertising promotion program. Initial difficulties revolved around funding for the program and whether the program should include both farmed and wild salmon. In 1997, the Salmon Marketing Institute (SMI) was funded by salmon farmers in Chile, Canada, and Norway. SMI developed radio advertisement programs that aired in some major U.S. cities at various times of the year and promoted consumption of fresh salmon. However, when the United States filed an antidumping and countervailing duty case against Chilean salmon farmers in mid-1997, the Chilean farmers terminated their funding to the institute. The reduction in funding eventually led to the collapse of SMI.

Catfish Advertising: The Catfish Marketing Institute (TCI)

The Catfish Institute (TCI) was formed to develop generic promotion and public relations programs to encourage consumption of U.S. farm-raised catfish (Fig. 7.1). TCI is a nonprofit organization established in 1986 by catfish farmers and feed manufacturers to raise consumer awareness of the positive qualities of US farm-raised catfish. It is a producer-controlled organization that is funded from catfish

Figure 7.1. The Catfish Institute logo.

feed mills located in Alabama, Arkansas, Louisiana, and Mississippi through a voluntary $5.00 checkoff per ton of catfish feed sold.

TCI's activities include advertising, public relations, and providing services to foodservice operators. The Institute runs a summer mobile marketing and sampling program that targets consumers at the retail seafood counter. The program visits grocery stores and food fairs across the nation during the summer months. The foodservice and marketing program is designed to educate chefs and food service operators about the use of U.S. farm-raised catfish. Activities include workshops at culinary schools and sponsoring booths at chefs' and caterers' conferences.

The advertising focus of the Institute has been to enhance the image of U.S. farm-raised catfish as a versatile, high-quality, convenience, and mild-flavored fish. Early advertising themes by TCI focused on the quality of U.S. farm-raised catfish, its availability, versatility, low cost, taste, and its being part of the new American cuisine. Later programs highlighted varieties of preparation methods. A "Made in America" theme emphasized the stringent food quality regulations in the United States. A "Spice it Up" campaign was developed in collaboration with a spice company. The emphasis was on demonstrating grilling recipes and promoting summer sales of catfish. TCI has emphasized primarily the print media. Some spot and network radio advertising has been done along with one national and cable television campaign.

The Catfish Farmers of America (CFA) is a national association of catfish farmers that is also engaged in some promotional activities. Much of CFA's advertising activities are offered through TCI programs. The association also provides some promotional materials on catfish at its Web site as well as in its monthly publication, "The Catfish Journal." Besides these established agencies involved in generic advertising and promotion of catfish, individual catfish processors do advertise and promote their brands and product lines of catfish.

Tilapia Advertising: The Tilapia Marketing Institute (TMI)

The Tilapia Marketing Institute (TMI) was formed to develop generic advertising programs in the United States similar to those of TCI. However, the TMI is not well established and has not become as functional as TCI. It was started in 1997 as a consortium of producers and suppliers of goods and services to the tilapia industry. The founding members provided the initial funding for advertising programs to increase U.S. consumer awareness of tilapia.

The early emphasis was on a marketing communications program. Some of the earlier activities of TMI included working with journalists and a variety of print media to create familiarity and awareness of tilapia to U.S. consumers. TMI worked actively to obtain coverage of food stories that included recipes and food reviews, business stories that covered the growth of the tilapia industry, technology stories that discussed production practices, and travel stories that enlightened consumers about the international status of tilapia. Much of TMI's generic campaign focused on tilapia's mild flavor, recipe versatility, and widespread availability. The campaign used media such as food magazines and newspapers, well-known television personalities, and respected chefs. They also sponsored events at conferences of chefs.

Initial funding for TMI activities was for two years. The lack of funding since 2000 has prevented TMI from continuing any meaningful advertising campaigns. However, individual tilapia companies have developed brand advertising programs by promoting their brands and products in the seafood marketing trade literature.

There also is the American Tilapia Association (ATA), which engages in a minimal amount of promotion of tilapia. Advertising of tilapia by ATA is mainly in the form of providing information on tilapia including production, supply, prices, trade, markets, and recipes on the Internet.

Trout Advertising: United States Trout Farmers Association (USTFA)

The USTFA is the main voice of the trout industry. The major objective of the association is to promote all aspects of the trout industry and especially to establish a high-quality image of trout products in the marketplace. Membership is offered to all individual farmers and companies engaged in or associated with the trout industry, including major suppliers of products or services. The association promotes trout in the form of providing trout information through its Web site as well as through a 40-page book of recipes. The book contains more than 80 complete recipes plus an additional 10 recipes for sauce and stuffing for trout. General information about trout, its excellent nutritional qualities, tips on handling, best basic preparation methods, and step-by-step instructions on how to bone a trout, whether cooked or uncooked, is included in the recipe book.

AQUACULTURE MARKET SYNOPSIS: SEABASS AND SEABREAM

Seabream (*Sparus aurata*) and seabass (*Dicentrarchus labrax*) culture technologies have been developed and converted into successful aquaculture industries in the Mediterranean Sea (Gasca-Leyva et al. 2003). Culture of the silver seabream (*Pagrus major*) has also emerged in Japan as a major supplier of a similar species (FAO 2004).

The world supply of seabream and seabasses from capture fisheries has increased over time (Fig. 7.2). Major increases in supply were recorded in the late 1990s, possibly from new reports from China and other countries (FAO 2004). Total capture fisheries of this group of fishes reached 670,016 MT in 2002. This list included 22 different individual species along with others that are reported as undifferentiated breams and seabasses.

The major species caught from the wild is the golden threadfin bream (*Nemipterus virgatus*) (Fig. 7.3). This species composes 46% of the total world catch of seabreams and seabasses. The next largest category in the FAO database is an

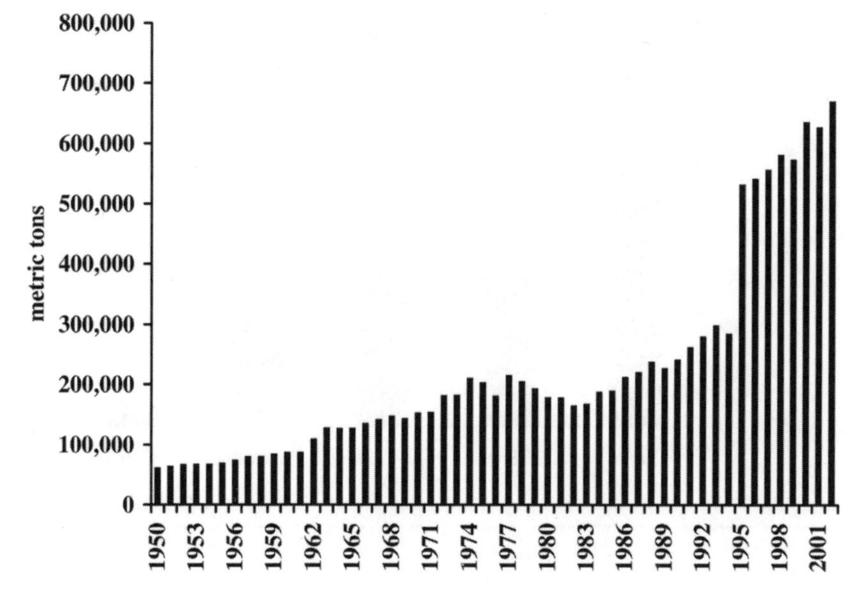

Figure 7.2. Volume of worldwide catches of seabasses and seabreams, 1950–2002. (Source: FAO 2004.)

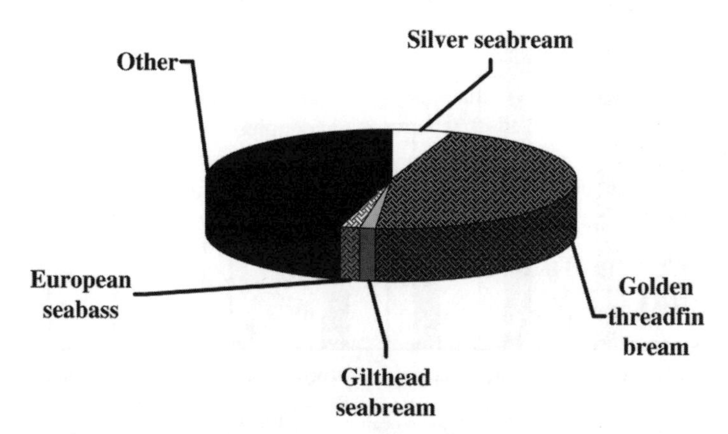

Figure 7.3. Major types of seabass and seabream supplied to world markets from capture fisheries. (Source: FAO 2004.)

undifferentiated category of threadfin breams (37%) (FAO 2004). A distant third is the silver seabream (5%), followed by the Japanese seabass (2%), the European seabass (1.5%), and the gilthead seabream (1.4%).

China is the major supplier of the golden threadfin bream, with 99% of the world catch of this species (FAO 2004). The silver seabream is a species that is also caught primarily in Asia. Japan is the leading supplier of silver seabream from the wild and is followed by Taiwan, New Zealand, Australia, and Korea (Fig. 7.4.) France and Italy are the major suppliers of wild-caught European seabass with additional supplies coming from Egypt, the United Kingdom, and Spain (Fig. 7.5). Wild-caught gilthead seabream are supplied primarily (in descending order) by Italy, Egypt, Spain, Tunisia, and Turkey.

Aquaculture of this group of fishes has concentrated primarily on the gilthead seabream, European seabass, and the silver bream, in spite of the fact that

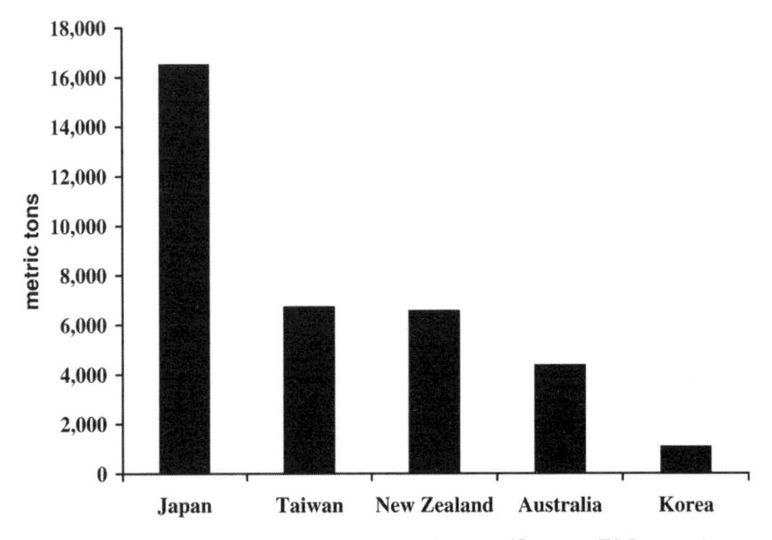

Figure 7.4. Major countries supplying wild-caught silver seabream. (Source: FAO 2004.)

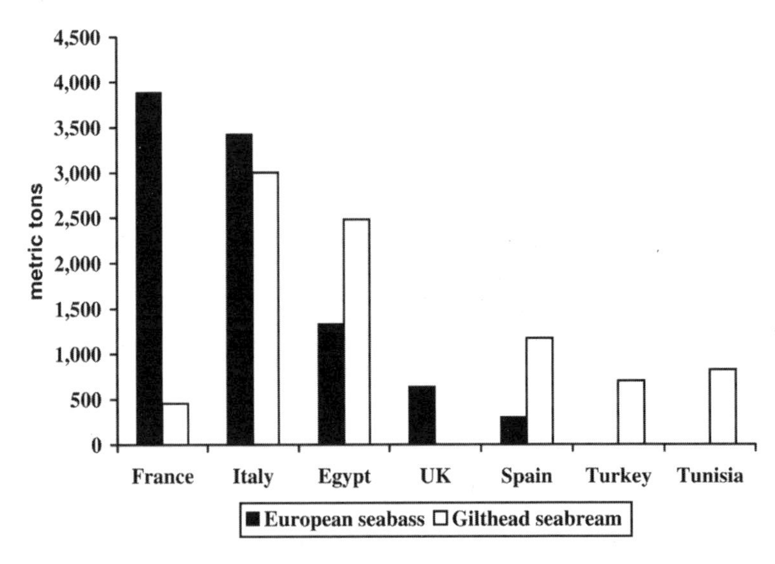

Figure 7.5. Major countries supplying wild-caught seabass and seabream to the world market. (Source: FAO 2004.)

these are not the major wild-caught species. The greatest aquaculture production is of the gilthead seabream, followed by the silver bream and the European seabass (Fig. 7.6).

The gilthead seabream and European seabass are familiar and preferred species of fish in southern Europe (Shaw and Curry 1989). Seabass are sold at sizes of up to 10–12 kg. The larger sizes are sold both as fillets and as whole fish. The gilthead seabream sold tend to be smaller, from 0.5–3.0 kg, and are more commonly sold as a whole fish. Fish of both species are commonly prepared by poaching, grilling, or baking. The meat of these fish is considered to be delicate and best purchased fresh, not processed.

The strongest demand for these fish is in Italy, particularly during holidays such as Christmas and, to a lesser extent, the Lenten period preceding Easter (Shaw and Curry 1989). Spain is the second most important market for seabass and seabream. Seabass has tended to be regarded as a higher-valued species in Europe as compared to seabream (Josupeit 1994). However, Italy has been the exception in that seabream has been sold at prices comparable to those of seabass. In recent years, the prices of seabream throughout Europe have approached those of seabass.

Greece has emerged as the dominant supplier of both gilthead seabream and European seabass to the European markets (Figs. 7.7 and 7.8). Total production from aquaculture in Greece in 2002 reached 37,944 MT of seabream and 23,860 MT of seabass. Thus, aquaculture production in Greece alone of these species is four times greater than the wild catch of seabream, and two and a half times greater than the wild catch of seabass. Much of the seabream production in Greece is sold in domestic markets, whereas nearly 80% of the seabass production is exported (Karagiannis and Katranidis 2000). This dramatic increase in the supply from aquaculture has had a negative effect on market prices that has, in turn, resulted in efforts to improve efficiencies and productivity in order to reduce costs of production (Gasca-Leyva et al. 2003; Karagiannis and Katranidis 2000). Turkey, Spain, Italy, and Israel (in descending order) are the next largest producers of seabream. For the European seabass, the next largest

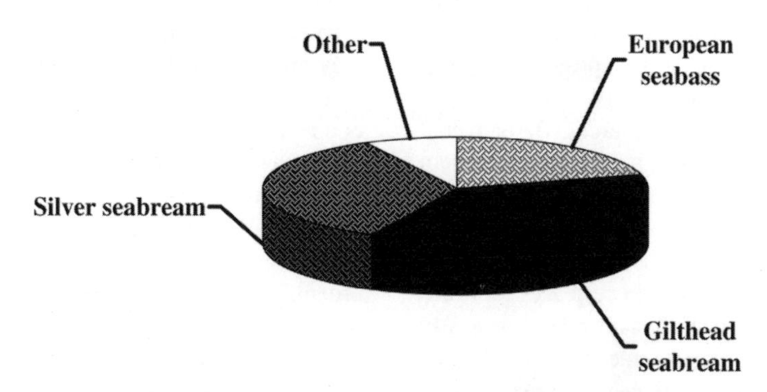

Figure 7.6. Major species of farm-raised seabass and seabream. (Source: FAO 2004.)

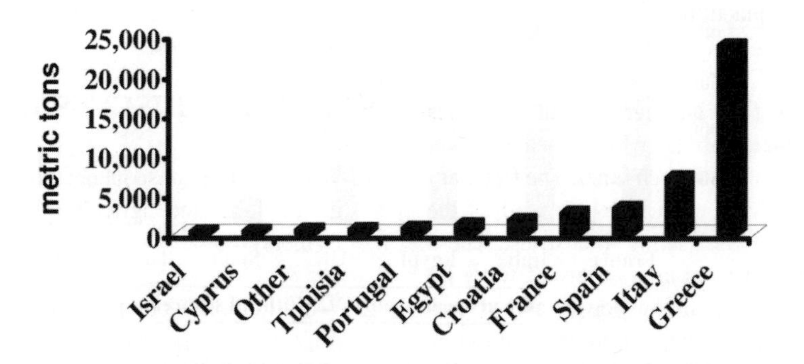

Figure 7.7. Major countries supplying farm-raised European seabass. (Source: FAO 2004.)

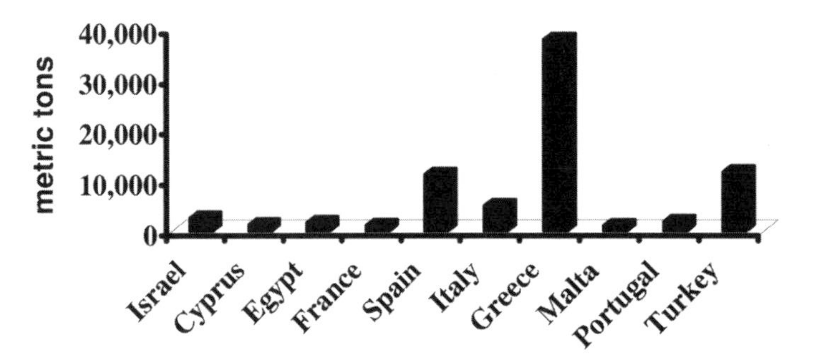

Figure 7.8. Major countries supplying farm-raised gilthead seabream. (Source: FAO 2004.)

producing countries after Greece (in descending order) are Italy, Spain, France, and Croatia. Japan dominates culture of the silver bream and produces 98% of all the silver bream cultured worldwide. There is some additional production from Korea and Taiwan.

Seabass and seabream are raised in net pens. In the Mediterranean, culture for many years was extensive. Wild-caught juveniles were confined in coastal lagoons before cages began to be used in semi-intensive production in protected areas (Arias 1976). The engineering innovations in cage and net pen production technologies have provided opportunities for intensive production in offshore areas with less competition for coastal resources (Barbato et al. 1993; Blakstad et al. 1996).

Seabass and seabream cage production faces constraints similar to those of other net pen–based industries. Regulations limit the amount of sea area that can be used by farms. Greece places a priority on water quality, particularly as it relates to nearby beaches and how the tourist trade might be affected. Also, the industry must continue to improve productivity to maintain profit levels at the lower prices that have accompanied the increase in supplies that have occurred through aquaculture.

SUMMARY

Trends in the U.S. food and agricultural industries point toward concentration, which means fewer companies competing with each other. The trend applies to fish processing plant sizes, as well as the food retailing and wholesaling industries. These changes can affect farmers in a variety of ways including disparity of bargaining power, use of production contracts, and competition from imports. Larger companies often have the greater bargaining power, and farmers can lose market access as a result. Therefore, by organizing into cooperatives or farmer groups, farmers can control a larger portion of their products, and have greater bargaining power than would an individual farmer.

The cooperative provides farmers the opportunity to perform some joint marketing responsibilities including assembling of products, negotiating with large buyers of farm products, exercising some power in the marketplace, spreading risks and costs, and in some cases processing farm commodities. There are four classes of marketing cooperatives based on how they are organized, membership affiliation, control, and geographic area: local cooperatives; centralized cooperatives; federated cooperatives; and mixed cooperatives. Most marketing cooperatives operate as a marketing agent by collecting products of members for sale, grading, packaging, and performing other marketing functions. Other cooperatives are organized to perform processing functions, or negotiate with processors or buyers on behalf of the members. One of the major programs of coordinated cooperative action in marketing is generic advertising. Marketing orders help to stabilize market conditions, and provide benefits to producers and consumers by establishing and maintaining orderly marketing conditions.

STUDY AND DISCUSSION QUESTIONS

1. What are the consequences of structural change in the food and agricultural industries on fish farmers?

2. Why should cooperatives become so much more important for the marketing of aquaculture products in the United States?

3. What are the differences between centralized cooperatives and federated cooperatives? Give examples of each type of cooperative.

4. Catfish farmers in the Mississippi delta have been trying for several years to form a bargaining association. Is this a good idea, and why?

5. Discuss the factors that will make a seafood marketing cooperative become very successful.

APPENDIX

THE CAPPER-VOLSTEAD ACT

(Public-No. 146-67th Congress)

An Act to Authorize Association of Producers of Agricultural Products

Be it enacted by the Senate and House of Representatives of the United States of America in Congress assembled, that persons engaged in the production of agricultural products as farmers, planters, ranchmen, dairymen, nut or fruit growers may act together in associations, corporate or otherwise, with or without capital stock, in collectively processing, preparing for market, handling, and marketing in interstate and foreign commerce, such products of persons so engaged. Such associations may have marketing agencies in common; and such associations and their members may make the necessary contracts and agreements to effect such purposes; Provided, however, That such associations are operated for the mutual benefit of the members thereof, as such producers, and conform to one or both of the following requirements:

First. That no member of the association is allowed more than one vote because of the amount of stock or membership capital he may own therein, or,

Second. That the association does not pay dividends on stock or membership capital in excess of 8 per centum per annum.

And in any case to the following:

Third. That the association shall not deal in the products of nonmembers to an amount greater in value than such as are handled by it for members.

Sec. 2. That if the Secretary of Agriculture shall have reason to believe that any such association monopolizes or restrains trade in interstate or foreign commerce to such an extent that the price of any agricultural product is unduly enhanced by reason thereof, he shall serve upon such association a complaint stating his charge in that respect, to which complaint shall be attached or contained therein, a notice of hearing, specifying a day and place not less than thirty days after the service thereof, requiring the association to show cause why an order should not be made directing it to cease and desist from monopolization or restraint of trade. An association so complained of may at the time and place so fixed show cause why such order should not be entered. The evidence given on such a hearing shall be taken under such rules and regulations as the Secretary of Agriculture may prescribe, reduced to writing and made a part of the record therein. If upon such hearing the Secretary of Agriculture shall be of the opinion that such association monopolizes or restrains trade in interstate or foreign commerce to such an extent that the price of any agricultural produce is unduly enhanced thereby, he shall issue and cause to be served upon the association an order reciting the facts found by him, directing such association to cease and desist from monopolization or restraint of trade. On the request of such association or if such association fails or neglects for thirty days to obey such order, the Secretary of Agriculture shall file in the district court in the judicial district in which such association has its principal place of business a certified copy of the order and of all the records in the proceeding, together with a petition asking that the order be enforced, and shall give notice to the Attorney General and to said association of such filing. Such district court shall thereupon have jurisdiction to enter a decree affirming, modifying, or setting aside said order, or enter such other decree as the court may deem equitable, and may make rules as to pleadings and proceedings to be had in considering such order. The place of trial may, for cause or by consent of parties, be changed as in other causes.

The facts found by the Secretary of Agriculture and recited or set forth in said order shall be prima facie evidence of such facts, but either party may adduce additional evidence. The Department of Justice shall have charge of the enforcement of such order. After the order is so filed in such district court and while pending for review therein the court may issue a temporary writ of injunction forbidding such association from violating such order of any part thereof. The court may, upon conclusion of its hearing, enforce its decree by a permanent injunction forbidding such association from violating such order or any part thereof. The court may, upon conclusion of its hearing, enforce its decree by a permanent injunction or other appropriate remedy. Service of such complaint and of all notices may be made upon such association by service upon any officer or agent thereof engaged in carrying on its business, or any attorney authorized to appear in such proceeding for such association, and such service shall be binding upon such association, the officers, and members, thereof.

Approved, February 18, 1922 (42 Stat. 388) 7 U.S.C.A., 291-192

REFERENCES

Arias, A. 1976. Sobre la biología de la dorada, *Sparus aurata*, L., de los esteros de la provincial de Cádiz. *Investigaciones Pesqueras* 40(1):201–22.

Barbato, F., A. Fanari, F. Meloni, and R. Savarino. 1993. Cage culture of gilthead sea bream (Sparus aurata) in two Sardinian coastal lagoons. Pages 95–104 in G. Barnabé and P. Kestemont (eds.), European Aquaculture Society, Special publication 18. Proceedings of the International Conference Bordeaux Aquaculture ‘92, Bordeaux, France.

Blakstad, F., A.F. Fagerholt, and D. Lisac. 1996. Cost of bass and bream production: comparisons between land based and cage facilities. Pages 245–248 in B. Chatain, M. Saroglia, J. Sweetnan, and P. Lavens (eds.), Proceedings of seabass and seabream culture: problems and prospects. European Aquaculture Society, Verona, Italy.

Brester, G.W., and T.C. Schroeder. 1995. The impacts of brand and generic advertising on meat demand. *American Journal of Agricultural Economics* 77: 969–79.

Capps, O., Jr., D.A. Bessler, G.C. Davis, and J.P Nichols. 1996. Economic evaluation of the cotton checkoff program. In: J.L. Ferrero and C. Clary (eds.), Economic evaluation of commodity promotion programs in the current legal and political environment, NEC-63 and NICPRE (p. 119–55). Ithaca, NY: Cornell.

Carman, H.F., and R.D. Green. 1993. Commodity supply response to a producer-financed advertising program: the California avocado industry. *Agribusiness* 9:605–621.

Chung, C and H.M. Kaiser. 1999. Measurement of advertising effectiveness using alternative measures of advertising exposure. *Agribusiness* 15(4):525–537.

Dunn, J.R., A.C. Crooks, D.A. Frederick, T.L. Kennedy, and J.J. Wadsworth. 2002. Agricultural cooperatives in the 21st Century. Rural Business-Cooperative Service, Cooperative Information Report 60, United States Department of Agriculture, Washington, D.C.

FAO. 2004. Fishstat Plus. Food and Agriculture Organization of the United Nations, Rome, Italy. Accessed at www.fao.org

Gasca-Leyva, E., C.J. León, and J.M Hernández. 2003. Management strategies for seabream *Sparus aurata* cultivation in floating cages in the Mediterranean Sea and Atlantic Ocean. *Journal of the World Aquaculture Society* 34(1):29–39.

Girard, S. and C. Mariojouls. 2003. French consumption of oysters and mussels analysed within the European market. *Aquaculture Economics and Management* 7(5/6):319–333.

Hussain, M.M. 1994. Fish marketing in Bangladesh. Pages 24–44 in Infofish (1994). *Fish Marketing in Asia,* Infofish, Kuala Lumpur, Malaysia.

Iskow, J. and R. Sexton 1992. Bargaining associations in grower-processor markets for fruits and vegetables. Research Report 104. Rural Development Agency, U.S. Department of Agriculture, Washington, D.C.

Josupeit, H. 1994. Markets in the European Union for turbot, seabream and seabass. Globefish Research Programme, Volume 31. Food and Agriculture Organization of the United Nations, Fishery Industries Division, Rome, Italy.

Karagiannis, G. and S.D. Katranidis. 2000. A production function analysis of seabass and seabream production in Greece. *Journal of the World Aquaculture Society* 31(3):297–305.

Kinnucan, H.W., H. Xiao, C.J. Hisa and J.D. Jackson. 1997. Effects of health information and generic advertising on U.S. meat demand. *American Journal of Agricultural Economics* 79:13–23.

Kinnucan, H.W., R.G. Nelson, and H. Xiao. 1995. Cooperative advertising rent dissipation. *Marine Resource Economics* 10:373–384.

Lenz, J., H.M. Kaiser, and C. Chung. 1998. Economic analysis of generic milk advertising impacts on markets in New York state. *Agribusiness* 14:73–83.

Matulich, S.C., and M. Sever. 1999. Reconsidering the initial allocation of ITQs: the search for a pareto-safe allocation between fishing and processing sectors. *Land Economics* 75(2):203–19.

Reberte, J.C., T.M. Schmit, and H.M. Kaiser. 1996. An ex-post evaluation of generic egg advertising in the U.S. In: J.L. Ferrero & C. Clary (eds.), Economic Evaluation of Commodity Promotion Programs in the Current Legal and Political Environment, NEC-63 and NICPRE (p. 83–95), Cornell University, Ithaca, New York.

Richards, T.J., P. Van Ispelen, and A. Kagan. 1997. A two-stage analysis of the effectiveness of promotion programs for U.S. apples. *American Journal of Agricultural Economics* 79:825–827.

Shaw, S.A., and A. Curry. 1989. Markets in Europe for selected aquaculture species: salmon, trout, seabream, and seabass. Globefish Research Programme, Volume 1. Food and Agriculture Organization of the

United Nations, Fishery Industries Division, Rome, Italy.

Torgerson, R. 2000. Cooperative marketing in the new millennium. Rural Cooperatives, Washington, D.C.

United States Department of Agriculture (USDA). 2004. Directory of farmer cooperatives. Service Report 22. Rural Business-Cooperative Service, U.S. Department of Agriculture, Washington, D.C.

University of Alaska, Anchorage. 2001. A planning handbook. Institute of Social and Economic Research, University of Alaska, Anchorage, Alaska.

Ward, R.W. and C. Lambert. 1993. Generic promotion of beef: Measuring the impact of the U.S. beef checkoff. *Journal of Agricultural Economics* 44(3): 456–465.

Williams, G.W., C.R. Shumway, H.A Love, and J.B. Ward. 1998. Effectiveness of the soybean checkoff program. Texas Agricultural Market Research Center Commodity Market Research Report No. CM-2-98, Texas A&M University, College Station, Texas.

Zidack, W., H. Kinnucan, and U. Hatch. 1992. Wholesale- and farm-level impacts of generic advertising: The case of catfish. *Applied Economics* 24: 959–968.

8
Wholesaler Marketing

Chapter 6 demonstrated the importance of distributing products through intermediaries or middlemen in enhancing efficiency in the distribution system. Wholesalers generally perform the functions of purchasing, transporting, assembling, storing, and distributing. Thus, wholesale marketing also improves efficiency in the distribution system and reduces cost. For example, wholesalers are able to get volume discounts that individual retail companies are not able to obtain themselves as individuals. Food retailers, foodservice establishments including hotels and restaurants, hospitals, government institutions such as schools, prisons, and other government catering operations, and other wholesalers handle a large assortment of products. A significant portion of sales at the wholesale level goes to the food retail food sector, accounting for about 40% of total wholesalers' grocery and related products sales (Fig. 8.1). Wholesalers play an important role in the timely delivery of such assorted products from many different companies and sources to these institutions and establishments. For many retailers and establishments, purchasing through wholesalers is a more convenient way of purchasing a diverse range of products. For such retailers, dealing with one major supplier rather than several supplier company accounts reduces administrative costs.

Several large retail chains perform their own wholesaling functions. Some independent retailers band together in the form of a cooperative to provide their own wholesaling, or they may contract with a wholesaler. These are termed self-distributing retailers.

STRUCTURE OF WHOLESALER MARKETS

The U.S. Census Bureau identifies the wholesale trade sector as comprising establishments engaged in wholesaling merchandise, generally without transformation, and rendering services incidental to the sale of merchandise. Based on this definition of the sector, the Census Bureau classifies wholesalers into three major segments: (1) merchant wholesalers that buy and take title to the goods they sell; (2) manufacturers' sales branches and offices that sell products manufactured domestically by their own company; and (3) agents and brokers who collect a commission or fee for arranging the sale of merchandise owned by others. In 1997, merchant wholesalers accounted for about 56% of total wholesale sales, manufacturers' sales branches and offices accounted for 25%, and agents and brokers accounted for 19% (U.S. Department of Commerce—Bureau of the Census 1997).

The food wholesaling business in the United States consists of that part of the food marketing system in which goods are assembled, stored, and transported to customers, including the grocery retail food store sector, foodservice sector, other wholesalers, government, and other types of food businesses. The food wholesaling business in 1997 was estimated to be a $589 billion industry (Harris et al. 2002).

There are a wide variety of different types and sizes of seafood wholesalers. For example, Engle (1997) showed that seafood wholesalers in Atlanta, Chicago, Los Angeles, New York, and San Francisco ranged in size from less than $20 million to more than $100 million in annual sales. Seafood wholesale companies tended to specialize in either finfish or shellfish, but were equally likely to sell fresh and frozen product. The companies that sold tilapia tended to be either those in the smallest or largest size categories. Most of the tilapia was sold to retail grocers, primarily Asian and Hispanic, or to independent restaurants. A very few large wholesale companies had very high sales of tilapia, primarily of frozen, whole tilapia. Fresh tilapia products were purchased more frequently and in lower average purchase amounts than other types of seafood.

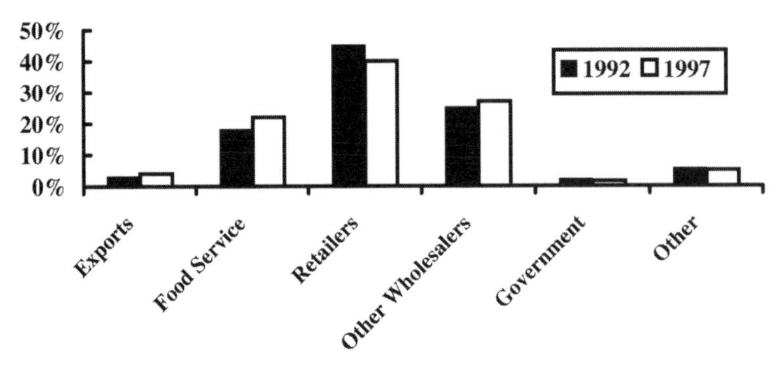

Figure 8.1. Food wholesale sales by type of customer. (Source: Harris et al. 2002.)

The U.S. Census Bureau classifies companies as merchant wholesalers if they primarily buy groceries and grocery products from processors or manufacturers, and resell to food retailers, institutions, and other businesses. Merchant wholesalers often repackage loads of larger-sized product into smaller units, or case sizes, for sale to clients. Examples of this class of wholesalers are given under other classifications that follow.

The sales branches and offices of grocery manufacturers and food processors function as wholesale divisions that market the company's products. Several seafood processing companies operate their own wholesale/distribution divisions. For example, Inland Seafood, the fourth largest seafood wholesaler/distributor in the United States in 2003, is a seafood processor that purchases seafood from fishing ports and aquaculture farms to produce fresh, frozen, smoked, and specialty seafood products including salmon, lobster, shrimp, tilapia, tuna, red snapper, catfish, rainbow trout, scallop, crab, and clams. The company produces value-added seafood products under the brand name Bimini Island Seafood®. With distribution facilities across five southern states, the company operates its own wholesaling/distribution functions. The company operates its own fleet of refrigerated trucks and utilizes air freight to deliver fresh and specialty seafood products to its restaurant, hotel, and grocery retail customers in the United States.

Large retail chains such as Kroger, Albertsons, Wal-Mart, Safeway, and Ahold also own their own distribution centers and are becoming a more significant part of the wholesaling sector. Figure 8.1 shows a decline in the percentage of total wholesale sales to the retail sector from 1992 to 1997. This suggests the increasing integration by large retailers into food wholesaling and distribution, in which they deal directly with food manufacturers.

Grocery wholesalers can also be classified by the type of merchandise they handle. Grocery wholesalers can be classified as general line, specialty, or miscellaneous. Specialty food distributors specialize in the distribution of items such as frozen foods, dairy products, poultry products, seafood, meat and meat products, and fresh fruits and vegetables. Among the three groups, specialty grocery wholesalers account for 43% of all sales, the highest percentage among these classifications (U.S. Department of Commerce-Bureau of the Census 1997).

Miscellaneous distributors are primarily engaged in the wholesale distribution of a narrow range of dry groceries such as canned foods, coffee, bread, or soft drinks, accounting for 32% of grocery wholesale sales in 1997 (Harris et al. 2002). General-line wholesalers are distributors and are sometimes referred to as broad-line or full-line distributors. Examples include SuperValu, Fleming, C & S Wholesale Grocers, and Nash Finch. Miscellaneous distributors handle a broad line of dry groceries, perishable food products, health and beauty products, and household products. General-line wholesalers accounted for about 25% of grocery wholesale sales in 1997.

Just as large retail chains such as Kroger, Albertsons, Wal-Mart, and Safeway have integrated into food wholesaling and distribution, larger general-line wholesalers such as SuperValu, Fleming, Giant Eagle, and Nash Finch have also ventured into food retailing. In 2001, sales derived from retail operations accounted for 45% of SuperValu's $20.9 billion total sales, 64% of Giant Eagle's $4.5 billion total sales, 15% of Fleming's $15.6 billion total sales, and 25% of Nash Finch's $4.11 billion total sales value (Harris et al. 2002). However, not all wholesalers that have ventured into retailing have been successful. In 2003, Fleming and its operating subsidiaries filed voluntary petitions for reorganization under Chapter 11 of the U.S. Bankruptcy Code. Consequently, most of the Fleming's retail stores were sold to competing retailers, and C & S Wholesale Grocery acquired Fleming's wholesale grocery business. Spartan Stores is the seventh largest general-line grocery wholesaler in the United States, with retail sales accounting for 40% of its $3.5 billion total sales in 2001. Spartan Stores divested a number of its retail stores in 2003 to focus on its core business of wholesaling.

RETAIL FOOD STORE WHOLESALERS

In 1997, sales of wholesale grocery to retail food operators, such as supermarkets and other grocery outlets, were estimated at about $236 billion (Harris et al. 2002). Distribution to retail food stores follows the structure outlined previously. Examples of merchant food wholesalers include Supervalu, Fleming, and Nash Finch. Merchant wholesalers have distribution centers in strategic locations from which to serve thousands of independent grocery stores as well as stores owned by the company. Merchant wholesalers account for 38% of wholesale distribution to retail food stores (Harris et al. 2002); they generate profits on the price spread and on the services provided. In 2004, Supervalu owned 24 wholesale/distribution facilities with approximately 14 million square feet of warehouse space; Fleming owned 32 wholesale/distribution centers. The buying power of these merchants allows them to obtain volume discounts and acquire leverage in food auctions by food companies.

The merchants use business-to-business (B2B) information technology gateways for ordering, invoicing, tracking, and customer service. An example of this technology is the Electronic Data Interchange (EDI) introduced in Chapter 6. The technology is Internet-based and involves computer-to-computer exchange. Merchant wholesaling requires some demand/supply planning and collaboration, distribution accounting for lead times and constraints, network optimization involving markets to serve and what products to serve, and general planning to reduce costs, management, accounting, evaluation, and reporting.

Retail food store wholesaling also involves direct-store delivery by grocery manufacturers and food processing/manufacturing companies. Typical examples are Coca Cola and Frito-Lay. Direct-store deliveries account for 28% of distribution to retail food stores (Harris et al. 2002). Typically, the vendors deliver products directly to individual retail stores and arrange products on display shelves for retailers. One of the ways grocery outlets and grocery/food manufacturers streamline the supply chain and reduce inventory is through the adoption of scan-based trading.

Self-distributing by grocery and food retailers is an increasing trend in the wholesaling industry. Food retail giants such as Kroger, Albertsons, Wal-Mart, Safeway, and Ahold buy directly from grocery and food manufacturers and producers, which then deliver the products to the wholesale/distributing centers of these retailers. Self-distributing food retailers account for about 34% of all food distribution (Kinsey 1999). In 1999, 47 out of the 50 largest food retailers in the United States were self-distributors (Harris et al. 2002). This type of wholesaling is beneficial to retailers because it reduces labor and general operating costs. The proportion of labor cost to sales at inventory for self-distributors is 0.9 percentage points lower than similar costs for merchant wholesalers, and their nonlabor costs are 1.3 percentage points lower (Kinsey 1999).

Warehouse clubs or cash-and-carry establishments such as Costco, BJ's, and Sam's Club are emerging as a significant segment of the wholesaling industry. Their activities are a blend of wholesaling and retailing of grocery food and other nonfood items. These establishments require membership for shopping at the outlets, and members include both individuals and small businesses, including businesses in the hospitality industry. Though warehouse clubs are wholesalers, prices charged are slightly above bulk wholesale.

FOODSERVICE WHOLESALERS

The food service sector has grown rapidly in recent years and has averaged 5.5% annual growth in sales. The number of food service locations has also increased over the last decade. This rapid growth in both sales and locations is in response to the growing trend of increasing away-from-home food consumption (U.S. Department of Commerce—Census Bureau 2004). Food service distributors are a form of wholesalers that carry a full range of seafood products. Food service distributors purchase seafood from processors and other wholesalers and sell primarily to restaurants. Cash-and-carry wholesalers typically supply small retail fish stores. Store owners travel to the cash-and-carry wholesalers to purchase fish, pay cash, and transport the fish back to their store.

Wholesale sales to food service institutions such as hotels, restaurants, and other hospitality and catering companies accounted for about 22% of all sales of groceries and related products by all wholesalers (Harris et al. 2002). There are important differences between wholesale distribution to the foodservice industry and other types of food retail outlets, such as supermarkets and grocery stores. Foodservice wholesalers fall under the categories of general line, specialty, or miscellaneous wholesalers.

General-line or broad-line foodservice wholesalers typically purchase a wide range of food products from manufacturers and stock them at distribution centers for distribution to their clients. The centers can carry up to 10,000 stock keeping units (SKU) and set prices competitively using economies of scale as leverage (Friddle et al. 2001). Prices may be negotiated or may be set with cost-plus pricing. The major foodservice broad-line distributors include Sysco Corporation, U.S. Foodservice, Alliant Foodservice, Performance Food Group, Gordon Food Service Incorporated, and Food Services of America. Sysco Corporation is the largest broad-line as well as seafood distributor in the United States. In 2003, Sysco Corporation, U.S. Foodservice, Performance Food Group, and Gordon Food Service, respectively, ranked first, second, third, and eighth among the top seafood wholesalers/distributors in the United States (Fig. 8.2).

Most broad-line foodservice distributors offer more than just distribution services. Many also offer value-added services tailored to the needs of their customers. Sysco Corporation and U.S. Foodservice offer a variety of services and proprietary food product lines in addition to food manufacturer brands. Sysco Corporation owns a number of brands, which include Buckhead Beef and Newport Pride (beef products) as well as Sysco Natural and FreshPoint (fresh produce).

Specialty wholesale distributors usually do not handle a wide range of products but focus on special products and markets or niche markets. For example, a specialty wholesaler may service Asian niche markets and handle Asian foods or customer segments such as convenience stores. McLane Company, for example, is one of the nation's largest wholesale food distributors to convenience stores, drug stores, quick-service restaurants, and movie theaters. Wholesale specialists may handle particular types of products such as seafood, dairy, meats, produce, ice cream, and so on. Some of the specialty distributors among the top ten seafood distributors in 2003 include Inland Seafood (fourth), East Coast Seafood (fifth), Supreme Lobster and Seafood Company (sixth), Morey's Seafood International (seventh), and Southstream Seafoods (tenth). Inland Seafood, for example, handles more than 1,000 seafood products involving species such as salmon, lobster, shrimp, tilapia, tuna, red snapper, catfish, rainbow trout, scallop, crab, and clams. It has the largest inland holding facility for lobsters in the United States and sells about 35,000 kg (77,000 lb) of salmon a week (personal communication). East Coast Seafood specializes in fresh lobster (*Homarus ameri-*

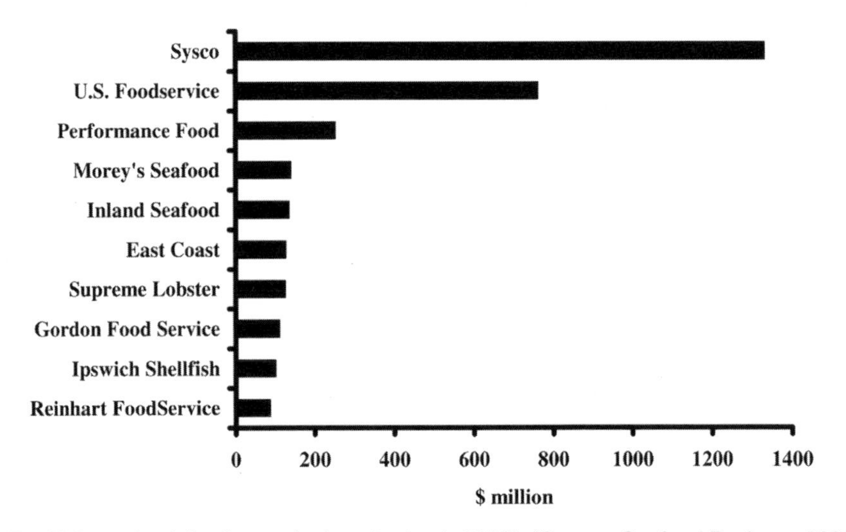

Figure 8.2. Top U.S. seafood distributors (value of sales in 2003). (Source: Seafood Business 2003.)

canus), dogfish (*Sqalus acanthias*), monkfish (*Lophius americanus*), skate (*Raja sp.*), scallops (*Placopecten megallanicus*), squid (*Loligo Pealei*), and whiting (*Merluccius Bilinearis*).

Wholesale specialists may also be warehouse club establishments such as Costco, BJ's, and Sam's Club. Membership in these clubs includes many small businesses in the foodservice industry that obtain particular supplies from the club outlets.

Miscellaneous wholesalers are also known as system distributors. This category of foodservice wholesalers serves a customer base that includes chain restaurants with centralized purchasing and menu development. The leading miscellaneous wholesalers in the food service industry are Ameriserve/McLane, MBM Corporation, Martin Brower, Golden State, and Marriott Distribution Services (Directory of Foodservice Distributors 2002).

A great deal of consolidation has occurred in the food service sector over the last decade. Sysco Corporation, the nation's largest distributor, represented more than 40% of the total food service business in 2003 (Seafood Business 2003). In 2001, the second leading seafood distributor, U.S. Foodservice, bought the third leading seafood distributor, Alliance Foodservice, further consolidating this sector of the marketing chain. By 2002, Sysco, with $1.3 billion in sales, together with U.S. Foodservice sales of $760 million, accounted for 65% of seafood distributor sales in the United States (Seafood Business 2003). The four-firm concentration ratio was 79% in 2002.

ROLE OF BROKERS IN SEAFOOD AND AQUACULTURE PRODUCTS WHOLESALING

Chapter 6 outlines the functions and compensation modes of seafood brokers. The National Oceanic and Atmospheric Administration (U.S. Department of Commerce—National Oceanic and Atmospheric Administration 2004) lists about 571 major seafood brokers in 2003. These brokers seek out information on the species, package sizes, and price from seafood suppliers and then offer these products for sale at a certain price to prospective buyers. Brokers also seek out buyers, identify specification needs of buyers, and then look for suppliers who can supply products according to those needs. Food broker companies typically operate in regional market areas instead of national markets, but the global and competitive nature of the seafood business makes it necessary for them to have a worldwide sourcing network to obtain supplies of quality products. Broker companies have a number of sales associates responsible for contacts with corporate headquarters of suppliers, warehouses for receiving samples, test kitchens, and conference areas for presentations by clients.

Food brokers can also be classified as broad-line or specialty brokers. However, the majority of food brokers fall into the category of broad-line because of the number of products that they handle. For example, Asmussen Waxler Group L.L.C. is a broad-line broker that handles a variety of products from different food manufacturers including Chicken of the Sea International (tuna products), Contessa Food Products (raw and cooked shrimp products), Country Select Catfish (farm-raised catfish products), Dean Foods/Land O'Lakes Milk (lactose-free milk products), Fishking Processors, Inc. (value-added shrimp, scallops, oysters, salmon, surimi, and lobster products), Icelandic USA, Inc. (fresh and frozen fish and seafood products), Orca Bay Foods, Inc. (salmon, swordfish, tuna, halibut, mahi and crab products), and Tyson Foods, Inc. (chicken products and branded concepts). The Asmussen Waxler Group operates in the Chicago area. Buzz Crown Enterprises, Inc. handles a similar variety of product lines as the Asmussen Waxler Group but the market area includes Washington, D.C.; Baltimore, Richmond, Roanoke, Virginia Beach, and Charleston across three states. ACH Food Service, Inc. operates in the Charlotte, Greensboro, and Raleigh areas in North Carolina as well as in Columbia in South Carolina. Food Sales West, Inc. is a major food broker that serves major cities in the west including Bakersfield, Fresno, Los Angeles, Sacramento, San Diego, and San Francisco in California, Las Vegas and Reno in Nevada, and Salt Lake City in Utah.

Transactions of brokers have traditionally taken place through phone contacts, but the Internet now plays a major role. Transactions usually do not involve contracts but rather one-time purchases on a day-to-day or week-to-week basis. It involves some specified quantity, price, and shipping arrangements. The supplier usually delivers products to the buyer. There are exceptional cases in which the broker pays some of the shipping costs if the demand is high but supply is limited.

CONSOLIDATION AND GROWTH OF WHOLESALERS/DISTRIBUTORS

The food distribution industry has experienced continued mergers and acquisitions with emphasis on the supply chain, optimizing its performance, and increasing sales while solidifying partnerships. A report produced in 2004 by the Unison Capital Group concluded that there were more than 6,000 small to medium-sized independent distributors with sales between 10 million and 100 million, and that, since 1996, companies such as Sysco Corporation, JP Foodservice, U.S. Foodservice, Nash, Performance Food, and others have acquired more than 200 food distribution companies (see Table 8.1).

The competitive nature of the wholesaling business has led to companies looking to economies of scale and economies of size through acquisitions and consolidation to improve efficiency and reduce cost of operations. In general, high sales volume and reduced costs of operations have enabled wholesalers to operate at lower gross margins. In 1997, the average gross margin for all types of grocery wholesalers was 15.3% (U.S. Department of the Treasury 1997). This was an increase from 14.3% in 1990. The gross margin for merchant wholesalers of grocery products was 17.2% in 1997 and, by 1999, it had risen to almost 18% (Harris et al. 2002).

Among retail food store wholesalers, merchants were the most concentrated. In 1997, the top four general-line merchant grocery wholesalers accounted for about 41% of sales, and the top eight accounted for 58% (U.S. Department of Commerce—Bureau of the Census 1997). The leading general-line merchant wholesalers in 1997 were Supervalu and Fleming. A major reason for this concentration was that besides serving thousands of independent

Table 8.1. Wholesale Acquisitions, 1994–2001.

Company	Acquisitions
Foodstore Wholesalers	
Nash Finch	Super Food Services, Erickson's Supermarkets, Military Distributors of Virginia, Super Food Services, T.J.Morris, United-A.G.Cooperative Inc., Hinky Dinky Supermarkets, Inc., K&N Meats (producer of beef and other meat for fine dining establishments)
Unified Western Grocers (result of merger between Certified Grocers of California and Unified Grocers)	Market Wholesale Grocery Company
Richfood Holdings	Super Rite, Farm Fresh, Inc., Shoppers Food Warehouse
Spartan Stores	Family Fare, Ashcraft's Markets, Seaway Food Town, Inc., Prevo's Family Markets, Inc. (retailer)
Fleming	Jitney-Jungle retail stores, Minter-Weisman (convenience store distributor), seven Food4Less retail stores from Whitco Foods, Inc.
Supervalu, Inc.	Richfood Holdings, Shop 'N Save, Randall Stores, Inc.
Roundy's	Mega Marts, Ultra Mart, Copps Corp.(retailer and wholesaler)
C&S	Grand Union retail stores
Associated Food Stores	Macey Inc., Lin's A.G.Foodstores
Foodservice Wholesalers	
Ahold (managed under U.S. Foodservice name)	U.S. Foodservice, GFG Foodservice, PYA/Monarch, Parkway Foodservice, Mutual Distributors, Inc. (broad-line distributor), Mutual Wholesale Co., Alliant
JP Foodservice, Inc. (managed under U.S. Foodservice name)	Mazo-Lerch, U.S. Foodservice, Squeri Food Service Inc., Arrow Paper Supply and Food Co., Valley Food Distributors, Parkway Food Service, Stock Yards Packing Co., Inc. (custom meat processor)

grocery stores, the large wholesalers were also vertically integrated, owning several chain grocery outlets. For example, in 2003, Supervalu was the nation's tenth largest supermarket retailer and owned more than 1,400 stores, including more than 800 licensed locations (Tarnowski and Heller 2004.). The company owned grocery chain stores including Bigg's, Save-A-Lot, Cub Foods, Scott's Foods, Farm Fresh, Shop 'n Save, Hornbacher's, Shoppers Food Warehouse, and Deals.

Besides the wholesaling and distribution business, Fleming Companies, Inc., had a growing presence in grocery retailing in 2001. The company served approximately 3,000 supermarkets, 6,800 convenience stores, and more than 2,000 supercenters, discount, limited assortment, drug, and specialty businesses. Fleming was the wholesale distributor of food and consumables for K-mart stores and owned grocery retail outlets, including Rainbow Foods, Food-4-Less, and Yes!Less.

Nash Finch is a food wholesale company that supplies products to independent supermarkets and military bases in approximately 30 states. The wholesale business accounts for about 75% of company sales. In addition to wholesaling, the company owns and operates approximately 85 retail supermarkets throughout the Midwest.

In the foodservice wholesale/distribution business, the top four distributors accounted for 23% of sales in 2000, compared with 14% in 1995 (Friddle

Table 8.1. Wholesale Acquisitions, 1994–2001 (*continued*).

Company	Acquisitions
U.S. Foodservice	Clark Foodservice, Goode Foodservice, Fort Myers Meat & Seafood, CP Foodservice
Rykoff-Sexton (managed under U.S. Foodservice name)	U.S. Foodservice
Clayton, Dubilier and Rice (investment firm) (managed under the Alliant Foodservice name)	Alliant (formerly Kraft Foodservices), Belca Foodservice, ACME Food Atlantic Food Services, Leone Foodservice, K-B Foods, Inc., K&N Meats (producer of meats for tablecloth dining establishments)
Performance Food Group	Fresh Express, AFI Food Service Distributors, W.J. Powell Company, McLane Foodservice (certain assets), NorthCenter Foodservice, Carroll County Foods, Inc., Dixon Tom-A-Toe Companies, Inc., State Hotel Supply Co., Nesson Meat Sales (certain assets)
McLane Co. (subsidiary of Wal Mart)	AmeriServe Food Distribution, Inc.
AmeriServe Food Distribution, Inc.	ProSource, Inc., PepsiCo Food Systems
Sysco Corporation	Strano Foodservice, Fresh Point Holdings, Watson Foodservice Inc., Malcolm Meats (specialty meat cutter), Buckhead Beef Co. (specialty meat cutter), Newport Meat Co. (Specialty meat cutter), Doughtie's Foods, Jordan's Foods, Beaver Street Fisheries, Inc., 5 specialty meat operations from Freedman Food Service, North Douglas Distributors, Ltd.(Canada), Albert M.Briggs Co.(specialty meat supplier), HRI Supply Ltd.(Canadian foodservice distributor), Fulton Provision Co.(specialty meat company)

Source: Harris et al. (2002)

et al. 2001). Acquisitions by major broad-line distributors of other broad-line and specialty distributors are partly responsible for this consolidation and the growth in the sector (Table 8.1). For example, Sysco Corporation, the nation's largest foodservice distributor, accounted for 13% of total foodservice distribution sales in 2001. Since 1994 Sysco has been active in acquiring other foodservice wholesalers and distributors, including specialty wholesalers. U.S. Foodservice, the second largest foodservice distributor, accounted for 10% of total foodservice distribution sales in 2001 and, since 1994, has also been active in mergers and acquisitions. Royal Ahold, the fourth largest grocery retailer in the world, now owns U.S. Foodservice.

AQUACULTURE MARKET SYNOPSIS: COMMON CARP

The common carp (*Cyprinus carpio*) was the first finfish species to be cultured. There are records of common carp farming that date as far back as 2,400 years ago in China and 1,900 years ago in Japan (Suzuki 1979). Common carp today is the third most important cultured finfish in the world, behind silver carp (*Hypophthalmichthys molitrix*) and grass carp (*Ctenopharyngodon idella*) (Fig. 8.3).

Statistics on aquaculture production collected by the Food and Agriculture Organization of the United Nations date back only to 1950. Figure 8.4 demonstrates the growth of farmed production of common carp from 1950 to 2002. Total production has generally increased over time across the world, but the rate of increase accelerated dramatically in the 1990s, reaching a total volume of 3.2 million MT. In contrast, the total volume of wild-caught common carp worldwide was only 66,884 MT in 2002. Thus, farmed production of common carp was nearly five times greater than the wild catch of common carp in 2002. The average increase in farmed production of common carp was 11% per year from 1992 to 2002.

China is by far the leading carp-producing nation in the world. China alone produced 2.2 million MT, or 69% of the world production of common carp in 2002. The other leading carp-producing countries are (in descending order of volume produced): India, Indonesia, Brazil, the Russian Federation, and Myanmar (Fig. 8.5; China was excluded from the graph because the scale of production in China did not allow comparisons among the other countries on the graph.). World carp production has shifted to some degree over time. In 1972, the Union of Soviet Socialist Republics led the world in common carp production and was followed, in descending order of importance, by China, Indonesia, Romania, Japan, and Hungary.

HUNGARY

Hungary was a leading producer of freshwater fish for the European carp market for decades. The history of fisheries in Hungary dates back to the sixteenth and seventeenth centuries, when fish production in all river basins and wetlands was common in the country. Much of the land in Hungary was drained for agriculture. To compensate for the decreasing capture from natural waters, an extensive fish pond construction program was started at the turn of the century with the establishment of the first fish farms in Hungary in the 1890s. As a result of a second new fish pond construction program after World War II, the total fish pond area reached 22,000 ha (55,000 ac) in Hungary by 1975.

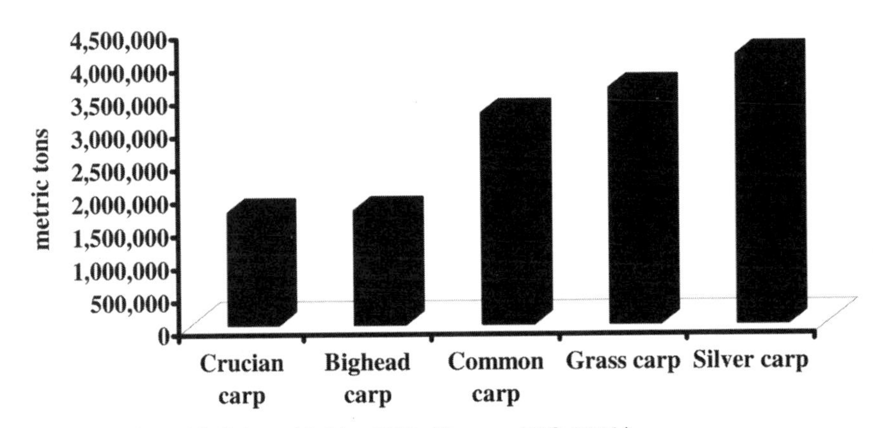

Figure 8.3. Top five cultured finfish worldwide, 2002. (Source: FAO 2004.)

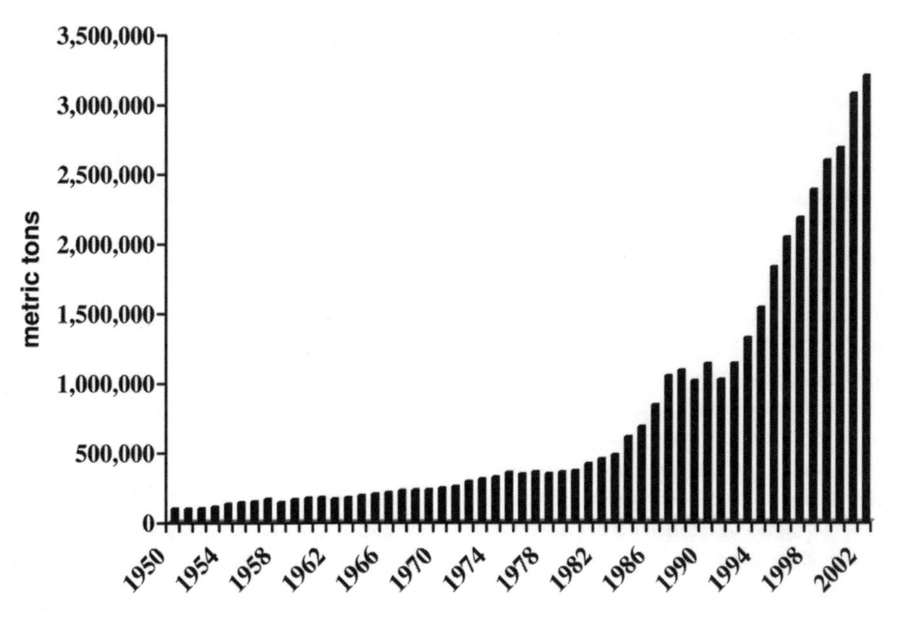

Figure 8.4. World aquaculture production of common carp, 1950–2002. (Source: FAO 2004.)

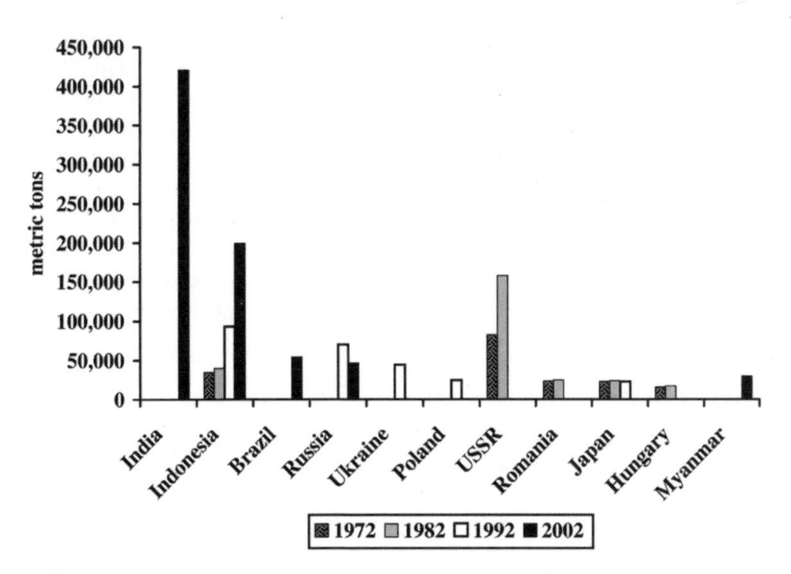

Figure 8.5. Top common carp-producing countries 1972, 1982, 1992, 2002, excluding China. (Source: FAO 2004.)

The most important cultivated species in Hungary has been the common carp, which represents 74% of total fish production. However, Hungarians traditionally eat little fish other than carp during the Christmas holidays. Much of the carp production in Hungary has been destined for export to other European countries. Hungary, along with Romania and Italy, have historically been the leading carp producers supplying farm-raised carp to European consumers. European consumers prefer carp to many other fish. The traditional Christmas Eve dish in Germany is carp, and dried or salted carp are sold as snacks in Europe. Carp is also sold as fresh or frozen whole fish, as fillets, or in cans. Carp destined for canning may be fried, smoked, or pickled prior to canning.

The economic transition from a centrally planned to a market-based economy in Hungary created a dynamic situation in the 1990s with rapidly changing conditions as market structures began to emerge. Total production of fish decreased by more than 50% from 43,000 MT in 1983 to 22,866 MT in 1995 (Varadi and Tahy 1996). The decline in farm-raised carp production in these countries produced a scarcity of farm-raised freshwater fish on the market. Aquaculture production decreased from 37,500 MT in 1989 to 28,000 MT in 1992. Nearly all (including live common carp) of the seafood currently marketed in Hungary is imported.

The market size of common carp is about 1 kg (2.2 lb). Under normal conditions, it takes about three years to reach market size. Average yield of 1.5–2 tons per ha (2.47 ac) can be reached in a well-managed fish farm in a 150-day growing season. More than 70% of table-sized fish are harvested and sold during Christmas time.

Market channels for freshwater fish in Hungary are complex (Fig. 8.6). Fish farmers sell to large distributors or to joint venture producer-owners who in turn sell to recreational fishing outlets or to small distributors to sell to grocery stores. Market channels for frozen fish and seafood are more direct, with limited numbers of firms emerging as retail wholesalers, distributors, and food service distributors. Canned fish products are primarily imported by small import companies and sold through small grocery stores.

The Hungarians have a series of carp dishes such as paprika fish soup that are considered traditional Hungarian cooking (Kouka and Engle 1997). The majority of seafood available is freshwater fish, predominantly of the carp family. Common carp fillets are priced at levels similar to those of good cuts of beef and pork. Live common carp (held in aquaria in supermarkets) are priced similarly to organ meats that take up the majority of supermarket shelf space for meats. There are numerous canned fish products available in Hungary, ranging from canned sardines and tuna to canned fish livers, caviar, krill and shrimp.

Germany

Carp is a traditionally important species in Germany facing decreased domestic supply (Lombardi 1997). Carp is traditional fare that is consumed in Germany around the Christmas and Easter holidays. It is customarily sold live (freshly killed) to customers during the holiday season. Imports of freshwater fish are on an upward trend. In 1994, approximately 74% of the total supply was imported (Infofish Trade News 1996). Germans generally buy their fresh fish at fish/seafood markets or in department store gourmet food sections. Department stores tend to offer the widest selection of live, fresh, frozen, canned, smoked, ready-to-eat, and other processed seafood products. Though fresh fish is traditionally not sold in grocery stores, it is common to have a fresh fish vendor located outside the grocery store, usually in the store parking lot, selling fresh fish and ready-to-eat seafood products from a small trailer.

Live carp is the most preferred product form in Germany (Lombardi 1997). Whole carp was the sec-

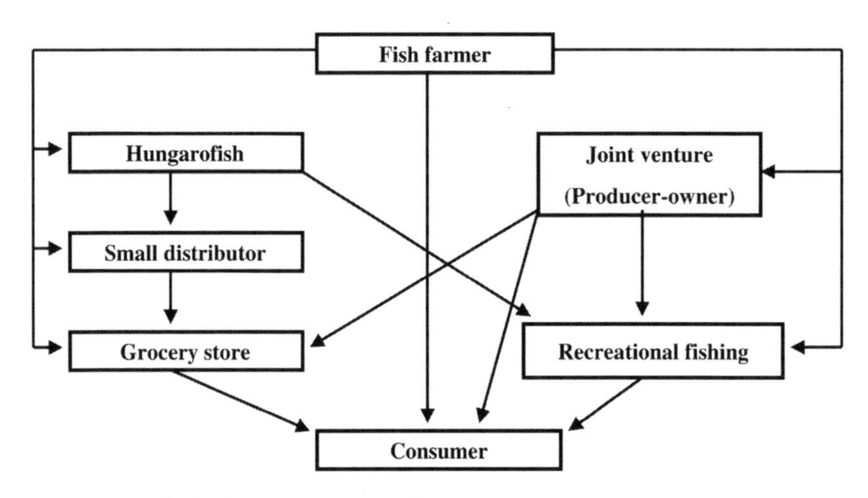

Figure 8.6. Market channels for live common carp, Hungary, 1995.

ond most popular form. German sales of carp are seasonal, with most sales occurring during the Christmas and Easter holiday season. Retailers indicate that approximately 67% of carp sales occurred between October and December.

The trend toward ready-to-eat meals is evident in European markets as in markets in the United States. The challenge for carp growers is whether to continue to supply only the traditional forms of carp for which seasonal markets exist or to try to move into the newly emerging new product markets. Lombardi (1997) provided some evidence of a degree of substitutability among freshwater fish products in Germany. The continued growth of carp production and sales may come more from developing nations in Asia than from the historically important Central European areas.

SUMMARY

Wholesalers generally perform the functions of purchasing, transporting, assembling, storing, and distributing at reduced costs because they are able to get volume discounts that individual retail companies are not able to obtain themselves as individuals. They service food retailers, foodservice establishments including hotels and restaurants, hospitals, and government institutions such as schools, prisons, and other government catering operations. The food retail food sector accounted for about 40% of total wholesalers' grocery and related product sales in 1997. Wholesalers are classified into three major segments: merchant wholesalers, manufacturers' sales branches and offices, and agents and brokers. Merchant wholesalers mainly serve the grocery retail, and foodservice sectors can be classified into general-line, specialty, or miscellaneous wholesalers. There is an increasing trend toward integrating the wholesale business into other aspects of the food marketing system. Larger restaurants and retailers are dealing directly with food manufacturers and handling their own wholesaling functions. Large wholesalers are in turn operating food retail stores and therefore handle their own wholesaling functions.

Food agents and brokers also play a major role in wholesaling and distribution. They seek out information on the species, package sizing, and price from seafood suppliers and then offer these products for sale at a certain price to prospective buyers. They also seek out buyers and their specification needs and look for suppliers who can supply products according to those needs.

STUDY AND DISCUSSION QUESTIONS

1. Identify the various classes of food wholesalers. What makes them differ from one another?

2. Describe how integration operates in the wholesaling business. What are the advantages and disadvantages of integration in the wholesale business?

3. With the competitive nature of the wholesaling business, discuss how specialty wholesalers are able to survive the competition.

4. Concentration has increased in the food wholesaling industry. In your opinion, how does that benefit the consumer?

5. Food brokers play a vital role in the food wholesaling business. Describe how they operate and make profits.

REFERENCES

Directory of Foodservice Distributors. 2002. SeaFood Sourcebook 21(7).

Engle, C.R. 1997. Marketing of the tilapias. In: Costa-Pierce, B. and Rakocy, J. (eds.), *Tilapia Aquaculture in the Americas,* Volume 1. World Aquaculture Society, Baton Rouge, Louisiana.

Food Institute Report. 2001. Supercenter and wholesale club growth continues.

Friddle, C.G., S. Mangaraj, and J.D. Kinsey. 2001. The food service industry: trends and changing structure in the new millennium. The Retail Food Industry Center Working Paper 01-02. The Retail Food Industry Center, University of Minnesota, St. Paul, Minnesota.

Harris, J.M., P. Kaufman, S. Martinez, and C. Price. 2002. The U.S. food marketing system, 2002-Competition, Coordination, and Technological Innovations Into the 21st Century. Economics Research Service Agricultural Economic Report No. 811, U.S. Department of Agriculture, Washington, D.C.

Infofish Trade News. 1989–1996. Kuala Lumpur, Malaysia.

Kaufman, P.R., C.R. Handy, E.W. McLaughlin, K. Park, and G.M. Green. 2000. Understanding the dynamics of produce markets: consumption and consolidation grow. Economics Research Service AIB-758, U.S. Department of Agriculture, Washington, D.C.

Kinsey, J.D. 1998. Concentration of ownership in food retailing: a review of the evidence about consumer impact. The Retail Food Industry Center Working

Paper 98-04. The Retail Food Industry Center, University of Minnesota, St. Paul, Minnesota.

Kinsey, J.D. 1999. The big shift from a food supply to a food demand chain. Minnesota Agricultural Economist, No. 698. University of Minnesota, St. Paul, Minnesota.

Kinsey, J. 2000. A faster, leaner, supply chain: new uses of information technology. *American Journal of Agricultural Economics* 82:1123–29.

Kouka, P.J. and C.R. Engle. 1997. The emerging seafood market in Hungary. Final report submitted to USDA, University of Arkansas at Pine Bluff, Pine Bluff, Arkansas.

Lombardi, W.M. 1997. The market for freshwater aquaculture products in Germany: a focus on carp and catfish. Major Paper, University of Rhode Island, Kingston, Rhode Island.

National Oceanic and Atmospheric Administration. 2004. U.S. seafood dealer list 2003. Southeast Fisheries Science Center. Accessed at http://www.sefsc.noaa.gov/seafooddealers.jsp.

Seafood Business. 2003. North America's top 25 seafood suppliers. *Seafood Business* 22(5):26.

Tarnowski, J. and W. Heller. 2004. The super 50. *Progressive Grocer.*

Suzuki, R. 1979. The culture of common carp in Japan. Pages 161–166 in Advances in Aquaculture, FAO Technical Conference on Aquaculture, Kyoto, Japan. 653 p.

Varadi, L. and B. Tahy. 1996. The change of the fish production sector in Hungary during the transition period into market economy. Fish Culture Research Institute, Szarvas, Hungary.

United States Department of Commerce, Bureau of the Census, Census of Wholesale Trade, various issues.

United States Department of the Treasury. 1997. Corporation Source Book, Internal Revenue Service, U.S. Department of the Treasury, Washington, D.C.

9
Market Trends

THE ROLE OF IMPORTERS IN SEAFOOD WHOLESALING

Imported seafood has become a significant part of the U.S. seafood trade (Fig. 9.1). The total value of imported seafood in 2003 was $10.9 billion compared to $10.4 billion in 2002. The total value of imported seafood has grown by about 60% over the past decade. The major species of seafood imported are crustaceans including shrimp, lobster, and crab. Shrimp accounted for about 32.3% of total imported seafood in 2003; crab, 8.9%; and lobster, 8.3%, making the total volume of imported crustaceans almost 50% of all imported seafood products in 2003 (Fig. 9.2). The major sources of seafood imports in 2003 were Canada, Thailand, China, Chile, Vietnam, India, Indonesia, Mexico, Ecuador, and Russia. In 2002, the top six seafood exporting countries to the United States, that is, Canada, China, Thailand, Chile, Ecuador, and Vietnam accounted for approximately 63% of imported seafood. In 2002, the United States imported seafood from an estimated 160 countries and 13,000 foreign processors (National Marine Fisheries Service 2004).

More than 80% of seafood consumed in the United States was imported in 2002 (National Marine Fisheries Service 2004). Figure 9.3 compares total U.S. seafood consumption and total seafood imports between 1993 and 2002. In 2002, the United States imported about 2 billion kg (4.4 billion lb), more than 80% of the seafood consumed in 2002. In addition, U.S. seafood consumption rose approximately 25% between 1980 and 2002, from 5.7 kg to 7.1 kg (12.5–15.6 lb) per person.

The increased dependence on seafood imports to meet domestic consumption needs has made the role of seafood importers important in the seafood distribution system. In 2003, NOAA listed about 1,171 major seafood importers, which includes some of the major seafood companies and distributors in the United States. These include: Morey's, Inland Seafood, East Coast Seafood, Supreme Lobster, Alaska Seafood International, Bumble Bee Seafoods, Chicken of the Sea, Slade Gorton, Contessa Food Products, Red Chamber Co., ConAgra Foods, and Trident Seafoods.

The seafood trade is competitive and, thus, seafood companies continually strive to develop a global network of sources for seafood. Most of the major seafood companies source product globally and report networks of representatives around the world for sourcing, marketing, and distribution of seafood products. For example, East Coast Seafood, Inc. is one of the largest distributors of live lobster in North America. The company boasts an integrated network of international subsidiaries that assist the company in worldwide sales, marketing, distribution, and customer service. East Coast Seafood established East Coast Europa, a seafood sales and distribution operation in Europe, with offices in Paris, Madrid, Milan, Frankfurt, Brussels, and London. Inland Seafood, the fifth largest seafood distributor in the nation, also has representatives across Europe and Asia who look for seafood products for the U.S. market. Most of the importers continue to search for new sources of seafood and appear to be the major agents who are developing the seafood market in the United States. In 1998, Seafood Connection, a seafood importer and distributor in Honolulu, Hawaii, was featured in *Pacific Business News* as the fastest growing independent seafood importer and distributor in mainland Hawaii. The company handles seafood products such as lobster, scallops, caviar, salmon, and crab and was known to import more exotic seafood products such as Russian caviar from the Caspian Sea, lobster tails from South Africa, and a variety of unique premium seafood items from Chile, Australia, and Africa for Hawaii's upscale restaurants and hotels (*Pacific Business News* 1998).

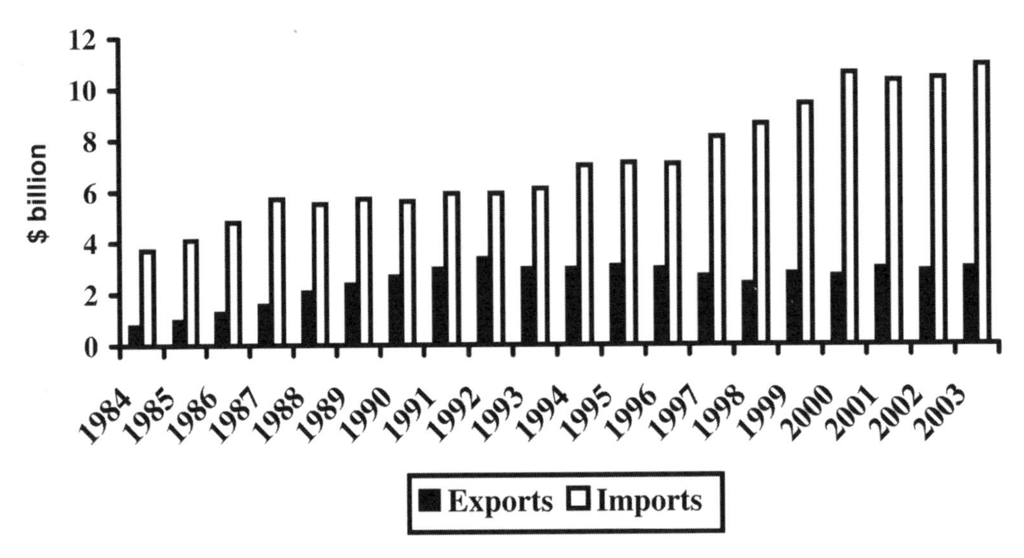

Figure 9.1. Value of U.S. seafood exports and imports, 1984–2003. (Source: US Department of Commerce.)

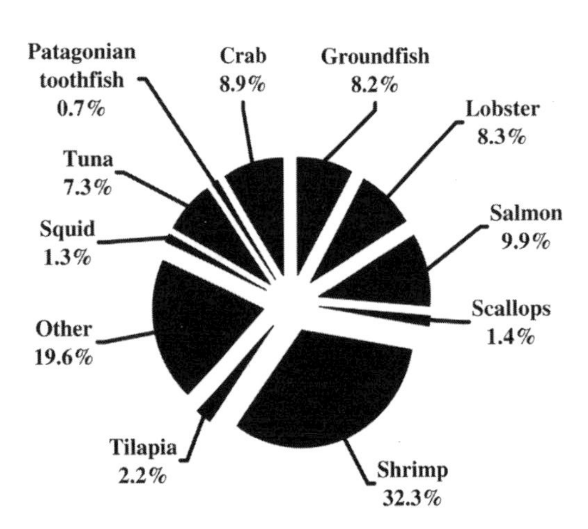

Figure 9.2. Composition of major U.S. seafood imports (total imports: $10.9 billion). (Source: US Department of Commerce 2003.)

Preferences of Asian Consumers in Retail Markets in the United States

Most Asian consumers in the United States prefer to purchase live seafood (Chan 2004). "Fresh" typically means live and swimming, and buyers prefer to select their fish and crustaceans from aquaria maintained in grocery stores. Asian consumers are accustomed to eating a wide variety of seafood products that can range from freshwater to marine fish, pelagic to groundwater fish, shellfish, shrimps, lobsters, seaweeds, sea cucumber, sea urchins, and mollusks such as octopus and squid. Asians also eat more different portions of the fish than do other consumers in the United States. Fish heads are used for soups, and roe skin, various fish organs, and cheeks are used in a variety of dishes. However, the Asian food industry is fragmented and not homogeneous. There is considerable variation in preferences among Asian consumers due to home country traditions.

WHOLESALE-RETAILER INTEGRATION

Mergers and acquisitions continue to change the structure of food wholesaling. Many food wholesalers and distributors are acquiring retail food operations. This process diminishes the share of retail food distribution accounted for by traditional third-party wholesalers. As discussed in Chapter 8, leading broad-line wholesalers such as Supervalu also rank among the top ten national leaders in grocery retailing.

Another trend that has developed in the past two decades is food retailing by nontraditional food retailers selling food and nonfood grocery products. Among these nontraditional food retailers are mass merchandisers such as Wal-Mart, Kmart, and Target, and warehouse clubs such as Costco, Sam's Club, and BJ's. Most of these are self-distributing wholesalers that have established a growing presence in

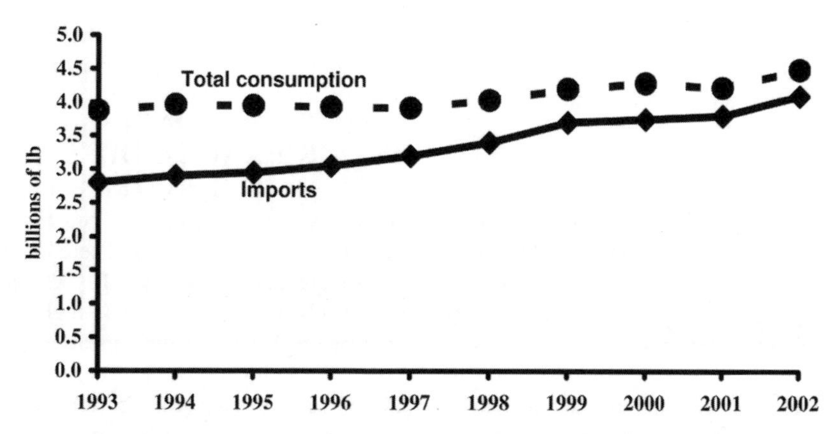

Figure 9.3. U.S. seafood consumption and imports, 1993–2002. (Source: NMFS, National Oceanic and Atmospheric Administration.)

Table 9.1. Supermarket Number and Sales, by Type.

	1980	1990	2000
Number of stores		*'000s*	
Supermarkets	26.8	24.5	24.6
Conventional	21.5	13.2	9.9
Superstore	3.2	5.8	7.9
Warehouse	1.7	3.4	2.4
Food/drug combination	0.5	1.6	3.7
Superwarehouse	–	0.3	0.5
Sales		*$ billion*	
Supermarkets	157.0	261.7	337.3
Conventional	114.7	92.3	63.4
Superstore	27.8	87.6	142.4
Warehouse	6.6	33.1	22.0
Food/drug combination	6.3	29.3	81.8
Superwarehouse	1.6	12.6	17.4

Source: Harris et al. (2002)

food retailing (Table 9.1). Wal-Mart is now the top grocer in the United States and accounted for 15.2% of all supermarket sales in 2003, up sharply from 11.7% in 2002 (Tarnowski and Heller 2004). These nontraditional food retailers accounted for 24.5% of total food sales in 2000 (U.S. Department of Agriculture—Economic Research Service 2002).

FOOD TRACEABILITY

Food wholesalers and distributors have been required to maintain full traceability of food items they handled as they made their way throughout the supply chain. This has come about largely because of concerns about food-borne illness and diseases. There are thousands of prepared, perishable, and packaged food products being offered to the consuming public; therefore, handlers of food products in the supply chain are required to have control and sanitation procedures. One of these new requirements is the establishment of written Hazard Analysis and Critical Control Point (HACCP) programs that are mandatory for all processors and handlers of meat, poultry, seafood, and fruit/vegetable juices.

There have been a number of initiatives to enhance traceability. One is the Public Health Security and Bioterrorism Preparedness and Response Act of 2002 (a law that required all food companies to develop compliance plans and register with the U.S. Food and Drug Administration); mandatory country-of-origin labeling (COOL); and the U.S. Animal Identification Plan, which would require premises-identification for livestock operations. The primary goal of this program is rapid containment of animal disease.

COUNTRY-OF-ORIGIN LABELING (COOL)

In the 2002 Farm Act, the U.S. Congress amended the Agricultural Marketing Act of 1946 and required retailers to use country-of-origin labeling (COOL) for certain covered commodities. COOL requires record-keeping and tracking systems to verify country-of-origin. However, the main purpose of COOL was to allow consumers at the retail level to make more informed decisions when making their purchases by enabling them to know the country of origin of food products. COOL was developed as a mandatory traceability system to facilitate and monitor trace-back to enhance food safety, address consumer information about food safety and quality, and protect consumers from fraud and producers from unfair competition (U.S. Department of Agriculture—Economics Research Service 2002).

The COOL rules identify two broad categories of entities that have responsibilities under COOL, suppliers and retailers. The rules identify three classes of suppliers: (1) the initiating suppliers who have the responsibility of developing the country-of-origin declaration; (2) the intermediary supplier, who is any supplier other than the initiating supplier; and (3) any person engaged in the business of supplying a covered commodity to a retailer, whether directly or indirectly. The rules require that each covered commodity offered for sale individually, in a bulk bin, carton, crate, barrel, cluster, or consumer package, must bear a legible declaration of the country of origin and, if applicable, the method of processing. The responsibility for such disclosures at the retail level falls to the retailer. The purpose of recordkeeping is to ensure that a proper audit trail exists to allow the government or other enforcement authority to track the covered commodities from origin to retailer or vice versa.

The 2004 Appropriations Act delayed the applicability of mandatory COOL to some commodities until 2006. However, under the interim final rule, fish and shellfish covered commodities were required to be labeled at the retail level to indicate country-of-origin and method of production (that is, wild or farm raised). Orca Bay Seafoods, for example, is a large seafood company based in Seattle that offers a broad assortment of domestic wild-caught and imported fish and shellfish products including crab, scallops, tuna, salmon, swordfish, and halibut. Orca Bay Seafoods sells primarily to foodservice establishments, club stores, restaurants, and grocery stores. Orca Bay has printed the words "wild" and "farm-raised" as well as a list of 10 countries and a blank for a new source on all company packages (Meat and Seafood Merchandising 2004). Employees check the appropriate method of production and country of origin on each package, thereby fulfilling the COOL requirements. New sources must be written in by hand on the blank line. Under the interim final rule, food service establishments, such as restaurants, lunchrooms, cafeterias, food stands, bars, lounges, and similar enterprises, are exempt from the mandatory COOL requirements (U.S. Department of Agriculture—Agriculture Marketing Service 2004). However, some southern states have passed laws that also require COOL labeling by food service establishments such as restaurants.

ELECTRONIC DATA INTERCHANGE (EDI)

Food wholesalers and distributors handle thousands of commodities and operate complex distribution centers and delivery fleets. Therefore their operations require an optimized and synchronized system that involves labor, inventory, warehouse space utilization, tracker and trailer utilization, and customer/vendor accounts receivable. The wholesale industry is competitive and, given current industry trends, food wholesalers and distributors must have a technology focused on the unique aspects of food distribution that reduces costs and increases profitability. The electronic data interchange system allows businesses to order merchandise, streamline delivery, and reduce overall costs. Any EDI system requires that suppliers and retailers use compatible computer systems.

THE EFFICIENT CONSUMER RESPONSE (ECR)

In 1992 the food supply industry developed the Efficient Consumer Response (ECR) system, which shares information between retailers and vendors. It allows for deliveries to be based on sales, lowering storage costs. Prior to ECR was the Quick Response (QR) system, which focused on shortening the retail order cycle. The retail order cycle is the total time elapsed from the point merchandise is recognized as needed to the time it arrives at the store. Goods that once took eight weeks or more to be ordered and received were ordered and delivered on a weekly basis, hence a "quick response." The advantage gained was that the shorter the order cycle, the lower the inventory levels required, which provided significant financial leverage for a business. Order cycles were shortened through the use of EDI and bar codes to automatically identify products.

The ECR system was built on QR techniques but addressed the order cycle as well as a wide variety of business processes involving new product introductions, item assortments, and promotions. ECR uses technology to improve every step of the cycle (or business process), which results in making every step faster and more accurate (Food Marketing Institute 2004). ECR also uses collaborative relationships in which any combination of retailer, wholesaler, broker, and manufacturer works together to seek out inefficiencies and reduce costs by looking at the net benefits for all players in the relationship.

The ultimate goal of ECR is to drive the order cycle and all the other business processes with point-of-sale data and other consumer-oriented data, giving an accurate read on consumer demand (Food Marketing Institute 2004). These data are passed by way of EDI to the manufacturer so that products can be made in quantities based on actual consumer demand, and then the data are distributed to the end consumer in the most efficient manner, hence the terminology Efficient Consumer Response. The ECR system is meant for the grocery industry to focus on the efficiency of the total grocery supply system to maximize consumer satisfaction and minimize cost (Food Marketing Institute 2004).

In 1996, Wal-Mart tested a new EDI system called Collaborative Planning, Forecasting, and Replenishment (CPFR). The system involves sharing sales forecasts of the manufacturer and that of Wal-Mart, and tailoring orders and deliveries accordingly (Kinsey 1999). A modified version of CPFR that is now commonly used in the food industry is scan-based trading (SBT). SBT is also known as Pay-on-Scan (POS).

The SBT system allows food manufacturers to bill retailers for their inventory only after the goods are scanned and sold (Kinsey 1999). Inventory is therefore on consignment basis from vendors. There could be up to 30 days lag time in billing. Some advantages of the system to the grocery retailer include savings in labor cost and improvements in cash flow because capital is not tied up in inventory. For the food manufacturer or wholesaler, the store's scanner data allows the company to monitor product movement and replenish products, thus increasing sales. The scan-based trading depends on mutual trust and accurate scanning.

Other leading food companies have proposed an Internet-based platform, called UCCNet, which operates on the World Wide Web. One element of UCCNet is CPFR, involving manufacturers and retailers separately forecasting future sales and sharing these forecasts to arrange orders and deliveries.

THE EFFICIENT FOODSERVICE RESPONSE (EFR)

A comparable system that has been initiated in the foodservice sector is the Efficient Foodservice Response (EFR). The system will help improve efficiencies in the foodservice supply chain linking manufacturing plants to distribution warehouses to operator's tables. A study conducted by Computer Sciences Corporation, Consulting and Systems Integration, and the Stanford Global Supply Chain Forum of Stanford University titled "Enabling Profitable Growth in the Food-Prepared-Away-From-Home Industries," served as the blueprint for the project. The report documents $14.3 billion in annual supply-chain savings that may be achieved across five strategies: Equitable Alliances, Supply Chain Demand Forecasting, Foodservice Category Management, Electronic Commerce, and Logistics Optimization. Savings to foodservice wholesalers, in particular, would amount to $4.7 billion (Harris et al. 2002).

A study conducted by the EFR project in 2003 suggested that despite steady progress by the foodservice industry in using bar codes on cases and inner packs, the industry required more efforts in both the use and quality of bar codes to achieve real benefits through the supply chain (Efficient Foodservice Response 2003). The study reported that case coding among foodservice manufacturers had increased since 1999. Case coding among respondents was 54% in 1999, 61% in 2000, 69% in 2001, and 77% in 2002. The EFR project has an industry-wide goal of 96% use of bar coding. Although the use of bar codes increased from previous years, the survey also revealed that the quality of bar coding efforts had slipped. The 2003 data showed that 74% of case codes were scanned accurately, compared to 82% in 2002 and 89% in 2001.

The 2003 survey also revealed significant variation in the use of case coding within different product categories. Equipment and supplies had the highest rate of case coding at 83%, followed by dry grocery at 80%, frozen and refrigerated foods at 73%, and produce at 23% compliance. The variations were consistent with 2002 data among the same categories. The survey also showed that 68% of cases were marked with bar codes on at least two sides, whereas 32% had a code printed on only one side. EFR recommends placing bar codes on two adjacent sides.

The 2003 survey recorded 29,579 cases in six different distribution facilities, including three regional broadliners in the Southwest, Southeast, and Northeast regions of the United States, two national broadliners in the West Coast and Mid-Atlantic regions of United States, and a systems distributor. The survey recorded cases from 1,719 different suppliers.

Foodservice distributors and operators have been advocating for food companies to use bar codes in order to increase supply-chain efficiencies and ensure better product traceability. The EFR project believes that as more companies use bar codes, there will be a better tracking of products from manufacturer to end user will be improved, invoice discrepancies will be reduced, communications will be more accurate, and electronic capture of company and product information will be more effective.

E-COMMERCE

Technology is having a significant impact on the way businesses operate, and the food industry is no exception. The common use of the Internet and the need for speed in obtaining information has forced the food industry to reexamine how it does business. Food companies are looking to technology to decrease costs, increase service levels, and improve the bottom line; therefore, e-commerce is becoming popular. Food companies engaged in e-commerce expect to reduce costs and improve efficiency in the supply chain by reducing fragmentation in the food supply chain.

There are various forms of e-commerce in use in the food industry. Some systems simply serve as a registry of suppliers and buyers and provide a forum for business transactions. Other systems are market based, allowing for trading. These include auctions. For example, the Uniform Code Council (UCC) operates a Web-based system called the UCCnet. It is a registry and synchronization service that helps to improve the accuracy of locating member and product information among the members' supply chain. Suppliers provide product, location, and trading partner information to the UCCnet Registry service, and the system then validates the data with demand-side partners, ensuring that all trading partners are using identical UCC standards. More than 3,000 companies have signed on to UCCnet, including several food industry companies.

Foodconnex is another example of an e-commerce platform that offers services, including a catalog database for the National Fisheries Institute members, marketing products, and customized business-to-consumer or business-to-business transactions. Clients of Foodconnex include Del Monte, Campbell's Foodservice, and the National Frozen Food Association.

In 2000, some foodservice leaders, including McDonald's, Sysco, Tyson, and Cargill, teamed up to form the electronic Foodservice (eFS) Network for their own purchases across all food categories, including seafood. The network also caters to other segments of the foodservice industry. The Internet site provides a public exchange and a private exchange for confidential customer-supplier trading. All procurements are online (Seafood Business 2000).

Another example of an online market place serving suppliers and retailers is the GlobalNet Xchange. The exchange is designed to match retailers with suppliers and to cut costs. Companies that utilize this exchange include Kroger, Sears Roebuck, Carrefour, Oracle, METRO AG, and J. Sainsbury. The Worldwide Retail Exchange is another business-to-business exchange for the retail industry and includes Albertsons, H.E. Butt, Wegman's, Kmart, and Target. Subway restaurants operate an extranet, called IPCnet, that links all its suppliers with distributors and store-level operators. The system provides for tracking, invoicing, and auditing all supply-chain activities and enables Subway operators to monitor the performance of distributors and manufacturers from different parts of the country.

In 2000, a number of seafood companies from Canada, the United States, and Iceland formed an Internet-based business called Seafood Alliance. The companies include the Pacific Seafood Group, American Seafoods, Inc., SIF Group, Pacific Trawlers/Crystal Seafoods, Inc., Fishery Products International Limited, Clearwater Fine Foods, Inc., Coldwater Seafoods, a subsidiary of Icelandic Freezing Plants Corporation, the Barry Group of Companies, and High Liner Foods, Inc. The ultimate purpose of the alliance is to find an industry-specific solution to improving the financial performance of participating companies in the seafood industry. The alliance implements an independent platform that enhances business-to-business e-commerce in the interest of all seafood industry participants (Puget Sound Business Journal 2000).

At the 2000 Boston Seafood Show, several e-commerce systems were promoted. Among them were Gofish, an online seafood exchange with 400 subscribers, Fishmonger, Globalfoodexchange, Gotradeseafood, the Gofrozen exchange with more than 900 member companies, and Worldcatch. Each Web site allows commodity buyers and sellers to exchange information and conduct product exchanges over the Internet (Seafood Business 2000). However, because many major buyers and sellers of seafood are involved in other e-commerce systems, some of these seafood companies have struggled to get buyers to sign on. In 2001, Gofish eliminated its online seafood trading, begun in 1999. Globalfoodexchange also ceased operations (Seafood Business 2001).

SUSTAINABLE SEAFOOD

Concerns over the environment have begun to have a greater influence in the marketplace. A growing number of consumers are requesting sustainable seafood products. Although the word "sustainable" has a variety of meanings and interpretations, most definitions include the concepts of harvesting seafood at rates that do not exceed the system's ability to replace the stocks harvested consistently over the long term. Several prominent retailers in the United States have initiated sustainable-seafood purchasing programs (Seafood Business 2004). Ahold USA, the fifth largest grocery chain in the United States and owned by the second-largest food retailer in the world, has been sourcing sustainable seafood for five years. The parent company of Red Lobster, Darden Restaurants is the world's largest casual dining restaurant company and has had a sustainable seafood-buying program for two years. In the distribution sector, Sysco, the largest broadline distributor in the United States, is initiating a similar effort. The effort may be driven by the need to preserve the world's supply of seafood to sustain their businesses in the future. For this reason, major seafood retailers seek to counter the increase in negative seafood stories. Seafood Business (2004) reported a 78% increase in the number of negative seafood stories between 2000 and 2003. Chapter 11 provides additional detail on industry and nongovernmental organization initiatives related to certification programs to seafood.

ORGANIC SEAFOOD

Two of the major factors influencing consumer food choices are health and food safety. Concerns about chemical residue in foods and food-borne diseases are encouraging more and more consumers to turn to organic foods. Chemical-free, organic foods are perceived by consumers to be safer and more healthful than foods produced with nonorganic materials. Organic products account for only about 1% of food sales nationwide, but sales of organic foods have quadrupled since 1990. The Organic Trade Association reported that overall sales of organic products reached $4.2 billion in 1997. However in 2002, organic food sales was estimated to be $9.7 billion (Sloan 2003). The fastest-growing categories include dairy, nondairy beverages, nutrition bars, grain snacks and candy, frozen foods, and cereals. The organic food sector has moved from a niche to mainstream industry with an average of 20–25% growth in sales over the past decade in the United States (Organic Monitor 2002).

Organic aquaculture is an emerging sector in the United States. There are growing consumer concerns about conventional production methodologies in both terrestrial and aquatic farming, perceived health benefits of food raised without the use of synthetic chemicals or drugs, and desires for humane treatment of livestock. The perceptions of seafood as a healthful food, particularly the health benefits of omega-3 fatty acids, have helped to increase sales of seafood. Unfortunately, there are no official organic certification standards for the U.S. aquaculture industry. Certified organic aquaculture products are available in Europe, Australia, New Zealand, and Asia, with some products now available on the U.S. market. However, the U.S. aquaculture industry is ready to pursue organic aquaculture after official standards are put in place. Two national task forces have been formed to attempt to spur the development of national standards for organic seafood products in the United States. Although there has been disagreement on several issues, the underlying principles related to the organic designation include breeding, feed, health care, living conditions, and recordkeeping (Institute for Social, Economic, and Ecological Sustainability—University of Minnesota 2001).

CONSUMPTION AWAY FROM HOME

The share of household food dollars going for away-from-home meals and snacks has been increasing for more than a century. An average of $1,668 was spent per person on food in 1970, out of which away-from-home meals and snacks captured 36% (U.S. Department of Agriculture—Economics Research Service 2004). By 1990, the average food spending per person increased about 20% with away-from-home meals and snacks capturing 45% of the food dollar (Fig. 9.4). In 2003, about 48% of the food dollar was spent on food away from home, indicating that Americans now spend nearly half of their food dollars on meals and snacks at foodservice facilities, such as restaurants, hotels, and schools. Total away-from-home expenditures, defined to include all food dispensed for immediate consumption outside the consumer's home, amounted to $445 billion in 2003. That is about 80% greater than annual away-from-home expenditures in 1990, which totaled $248 billion. It is anticipated that households will continue to increase their spending on foodservice meals and snacks at an annual rate of about 1.2% in real (inflation-adjusted) terms (Blisard et al. 2003).

Rising incomes, the growing incidence of nontraditional households, and other demographic developments, such as smaller household sizes, were predicted to enhance the growth in away-from-home

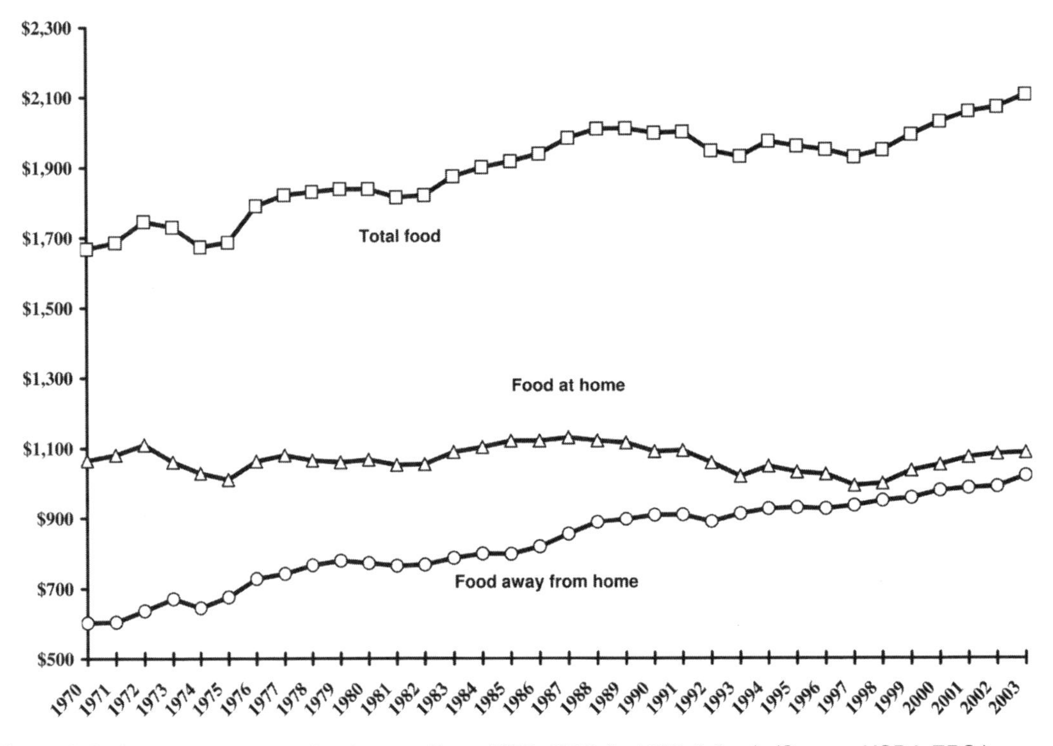

Figure 9.4. Average per-person food expenditure, 1970–2003 (in 1988 dollars). (Source: USDA-ERS.)

food expenditures in the United States (Stewart et al. 2004). Stewart et al. (2004) forecast that consumer spending at full-service and fast-food restaurants will continue to grow between 2000 and 2020. Modest growth in household income plus expected demographic developments would result in an increase in per-capita spending of 18% at full-service restaurants and by 6% for fast food between 2000 and 2020. However, the aging of the population will decrease spending on fast food by about 2% per capita.

HEALTHFUL AND WHOLESOME FOODS

The 1990s was a period of increased health awareness that resulted in consumer demand for foods and beverages that provided nourishment, health benefits, and good taste, at the right price. Consumers, in seeking to lead healthful lifestyles, have consequently recognized the appeal of fresh and particularly natural and wholesome products, with their implied benefits of safety and wellness. Increasingly, nutritionists and food manufacturers are publicizing foods as healthful and lower in saturated fats and trans fatty acids. Consequently, by January 2006, the U.S. Food and Drug Administration regulations require all food marketers to disclose the level of trans fats in their packaged products. Already, many large companies, such as Frito-Lay, Nabisco, and Tyson Foods, have eliminated or reduced the level of trans fats in their products, and many food companies will follow the guidelines.

Trends in healthful and wholesome food have focused on nutritional platforms such as fiber enriched, vitamin fortified, high protein, and omega-3. They have also focused on some general food attributes such as low fat, organic, and low carbohydrates, as well as the health benefits of food such as natural foods and functional foods. In 2002, total sales of natural products were estimated to be $36.4 billion, a 6.6% increase over sales in 2001 (Spencer and Rea 2003). About 77% of total sales of natural products was realized at the retailing and mass-market channels. Natural products retailers sold the most natural and organic foods, with 9% growth to $10.4 billion in 2002 (Spencer and Rea 2003). Food constituted 60% of total sales for natural products retailers, and 44% of the category was organic.

Consumers are searching for foods that are assured of purity and quality and that are free of chemicals. The National Marketing Institute (NMI) maintains a Health and Wellness Trends Database (HWTD) based on an annual survey of more than 2,000 U.S. consumer households. NMI reported sales of $59 billion within the consumer packaged goods health and wellness industry in 2002, representing 7.3% growth over 2001 sales (National Marketing Institute 2003). The study indicated that functional and fortified foods and beverages constituted 11% of sales, organics constituted 17%, and natural and organic personal care, 15%. Vitamins, minerals, and herbal supplements continued to thrive. About 30% of consumers indicated that they made an effort to regularly eat a meatless meal, whereas 19% considered themselves an occasional vegetarian (National Marketing Institute 2003). The study projected a 10% compound annual growth rate in consumer packaged goods in the health and wellness industry, with sales of $86 billion by 2006.

Since their discovery in the 1970s, omega-3 essential fatty acids have been the center of several studies and clinical trials. The omega-3 fatty acids have been shown to aid in the treatment of symptoms of asthma, obesity, Alzheimer's, depression, bipolar disorder, and especially the overall health of the heart and functions of the brain (Nettleton 1995). The omega-3 fatty acids benefit the hearts both of healthy people and those at high risk of or who have cardiovascular disease (Kris-Etherton et al. 2002). The American Heart Association recommends eating fish (particularly fatty fish) at least two times a week because fish is a good source of protein and does not contain the high saturated fat of fatty meat products. Fatty fish such as mackerel, lake trout, herring, sardines, albacore tuna, and salmon are high in two kinds of omega-3 fatty acids, eicosapentaenoic acid (EPA) and docosahexaenoic acid (DHA). Some wild game and grass-fed meat, and some enhanced eggs, also have high levels of EPA and DHA.

CONVENIENCE IN FOOD PREPARATION AND CONSUMPTION

The need for convenience in food preparation and consumption continues to grow among U.S. consumers as people are overwhelmed by product choices and starved for time. Convenience in home-prepared foods comes in a number of ways, including ready-to-eat, heat-and-eat, quick preparation, easy-to-cook, and packaged complete meals for on-the-go consumption. Even with the restaurant industry's off-premises market has outpaced growth in the dine-in option. The "to-go" meal sales was approximately $7.4 billion at casual dining restaurants in the year ending August 2001, which is about 12% of total dollars spent in the casual dining segment (Bailie and Bhothinard 2004).

Sales of convenient "dinner solution" meals continues to grow. It is estimated that sales of dinner solution meals grows by an eight-year compound annual growth rate of 7.5%, and breakfast solution meals by about 6.6% (Information Resources, Inc. 2002). Information Resources, Inc. (IRI), a global market research firm that has clients including Anheuser-Busch, ConAgra, Johnson & Johnson, Philip Morris, Procter & Gamble, PepsiCo, Unilever HPCE, and top retailers, reported that sales of "dinner solutions" meals added an average of more than 385 million meals sold over the previous seven years (Information Resources, Inc. 2002). In 2001, frozen entrées and meals reached retail sales of $9.3 billion, up 5.8% in supermarkets; unprepared frozen meat increased by 12.8%, ground beef by 16.8%, frozen unbreaded fish by 10.2%, frozen unbreaded shrimp by 36.9%, and other unbreaded seafood by 34.4% (Heller 2002). Other popular meals among consumers include ready-to-cook, preseasoned and prepared fresh meats, poultry, and fish/seafood, which along with precooked seafood accounted for 25% of supermarket seafood counter sales in 2001 (Bavota 2002).

It is anticipated that the success of ready-to-eat and ready-to-cook items will spawn new issues and opportunities (Sloan 2003). For example, there is the tendency to use only one appliance to prepare a meal, and it is anticipated that consumers will soon demand side dishes that can be cooked simultaneously in the microwave or oven in about the same length of time as the precooked entrée. Whether cooking is gourmet or everyday, any product that eliminates work or cleanup will likely have enormous appeal (Sloan 2003). Food products designed for easy home entertaining, such as frozen pizza, have also seen increased interest from consumers.

AQUACULTURE MARKET SYNOPSIS: GIANT CLAMS

Giant clams (*Tridacna spp.*) are listed as threatened species. Thus, trade of wild-caught giant clam products is restricted. Chapter 11 provides additional detail on restrictions under the Convention on International Trade in Endangered Species (CITES). The development of culture technologies for the giant clam provides an opportunity for the development of a new aquaculture industry that would be largely unaffected by capture fisheries that are now prohibited.

There have been several studies conducted to evaluate the market for giant clam products; these are summarized here.

Shang et al. (1991) identified markets for five types of giant clam products. These include: a food market, an aquarium market, a seedstock market, a broodstock market, and a market for giant clam shells. The greatest potential appeared to be the food market for giant clams.

Giant clams are sold as food in Okinawa, Taiwan, Australia, and the Pacific Islands (Shang et al. 1991). In Okinawa, giant clams are used in sashimi and sushi dishes. The size of the food market was estimated to be about 500 MT. However, the preferred species in Okinawa is *Tridacna crocea*, and the main species cultured are *Tridacna gigas* and *Tridacna derasa*. Preliminary taste tests indicated that the meat of *T. derasa* was too soft when compared with that of *T. crocea*. Additional research is needed to determine the degree of substitutability among species in the major food markets for giant clams. The food market in Taiwan is primarily for fresh or frozen adductor muscles of the giant clam. The Taiwanese market is in exclusive restaurants that specialize in seafood. Much of the supply is imported, given that clam fishing in Taiwanese waters is not legal. Shang et al. (1991) estimated the potential market to be about 240 MT a year. Larger muscles bring a higher price/kg. Preferences for the different clam species are related to the size of the adductor muscle produced by each species. Muscles that are whiter are also preferred and command higher prices. The giant clam market in Australia is an ethnic niche market of immigrants from the Pacific Islands and their descendents.

Shang et al. (1992) concluded that some limited market potential may exist for giant clams as aquarium specimens. This market likely would be in Japan, Australia, and the United States. Good potential was estimated for the shells of giant clams that are used in a variety of ways in Japan. They are used as serving dishes, vases in flower shops, decorator items in restaurants, and ash trays in hotels. The shells are also sold as gifts and dishes. They are sold at very high prices ($12–$188, depending on the size). If these markets were to develop, it is conceivable that markets for seedstock and broodstock would also develop.

Overall, the culture and sale of giant clam has not increased since 1989 (Fig. 9.5). Production quantities have remained fairly constant over the period of 1989–2002 in both Palau and Samoa, the two primary sources of giant clam supply.

The restrictions on trade of wild-caught giant clams pose some restrictions on the potential for developing the market. Permit requirements and certification that the products are farm raised may be required. Some type of tag will be required to distinguish farm-raised from wild-caught clams. Moreover, giant clams and products derived from giant clams are new to many markets. Taste tests and other forms of market testing will be required along with persistent marketing efforts for this market to develop.

SUMMARY

The increased dependence on seafood imports to meet domestic consumption needs in the United States has made the role of seafood importers important in the seafood distribution system. The seafood trade is competitive as seafood companies continually strive to develop global networks of sources for seafood and appear to be major agents in developing the seafood market in the United States. Many

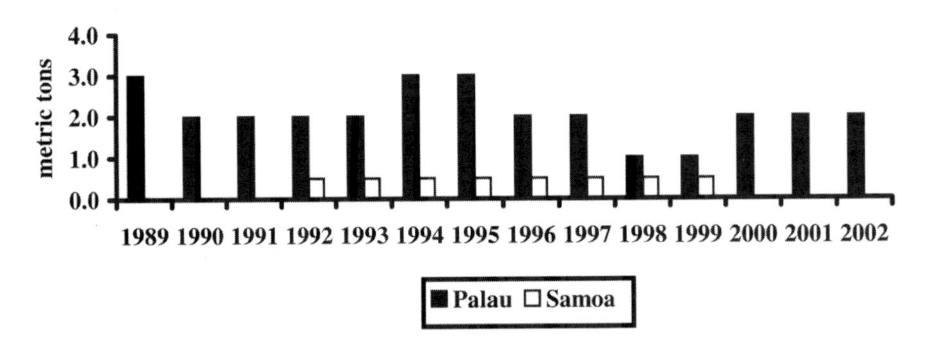

Figure 9.5. World culture of giant clams, 1989–2002. (Source: FAO 2004.)

food wholesalers/distributors and nontraditional food retailers are becoming importers of seafood and offer seafood products along with food and other nonfood grocery products to the consuming public.

Food traceability has become an important public policy issue because of concerns about food-borne illness and diseases. Handlers of food products in the supply chain are being required to have control and sanitation procedures in place such as the Hazard Analysis and Critical Control Point (HACCP) programs that are mandatory for all processors and handlers of meat, poultry, seafood, and fruit/vegetable juices. Other initiatives to enhance traceability include the Public Health Security and Bioterrorism Preparedness and Response Act of 2002 (a law that requires all food companies to develop compliance plans and register with the U.S. Food and Drug Administration); mandatory country-of-origin labeling (COOL); and the U.S. Animal Identification Plan, which would require premises-identification for livestock operations.

To optimize and synchronize the food supply chain system, technology initiatives have been adopted in the supply chain that involve labor, inventory, warehouse space utilization, tractor and trailer utilization, and customer/vendor accounts receivable. Some of these initiatives include the Electronic Data Interchange (EDI), a technology system that allows businesses to order merchandise, streamline delivery, and reduce overall costs; the Efficient Consumer Response (ECR), a collaborative relationship in which retailers, wholesalers, brokers, and manufacturers work together to seek more efficient ways of distributing manufactured food products; the Scan-Based Trading (SBT), a technological system that provides food manufacturers instant information on inventories in retailer outlets when the goods are scanned and sold; and the Efficient Foodservice Response (EFR), a technology system in the foodservice supply chain that links food manufacturers to distribution warehouses and to restaurant outlets.

The Internet is becoming a common tool for food companies in the food marketing system. Use of the Internet can decrease costs, increase service levels, and improve efficiencies in their operations. E-commerce is becoming popular.

Consumers, in seeking to lead healthful lifestyles, have consequently recognized the appeal of fresh, natural, and wholesome products, with the implied benefits of safety and wellness. It has been projected that, with a 10% compound annual growth rate in the consumer packaged goods health and wellness industry, sales will reach $86 billion by 2006. For example, the organic food sector has moved from a niche to a mainstream industry with an average of 20–25% growth in sales over the past decade in the United States. Sales of convenient, "meal solution" products continue to grow. It is estimated that sales of dinner solution meals will grow by an eight-year compound annual growth rate of 7.5%, and breakfast solutions meals by about 6.6%.

STUDY AND DISCUSSION QUESTIONS

1. Which countries were the main exporters of seafood to the United States in 2002? What was the dominant species exported to the United States?

2. The United States is a net importer of seafood. What role do importers play in the seafood distribution system?

3. Outline two initiatives that the U.S. federal government has implemented in recent years to enhance traceability in the food chain.

4. What is the major implication of COOL for consumers?

5. What is the Electronic Data Interchange (EDI)? How does this technology help reduce cost in the food supply chain?

6. What is SKU? Describe its use in the food marketing system.

7. What is Efficient Consumer Response (ECR), and how does it differ from Quick Response (QR)?

8. Scan-based trading is becoming popular with retailers and food manufacturers. What has contributed to its popularity? Compare the advantages and disadvantages of SBT.

9. What is Efficient Foodservice Response (EFR) and how does it differ from Efficient Consumer Response (ECR)?

10. What are the various forms of e-commerce in use in the food industry? Describe two of them.

11. Discuss three reasons that consumers are turning to organic seafood.

12. Discuss three factors that are predicted to enhance the growth in away-from-home food expenditures in the United States.

REFERENCES

Bailie, K. and T. Bhothinard. 2004. The 'to-go' movement in casual dining. Industry Trends, NPD Foodworld, http://www.npdfoodworld.com/foodServlet?nextpage=trend_body.html&content_id=133.

Bavota, M.F. 2002. Is ready-to-eat safe to eat? *Progressive Grocer* 81:13–58.

Blisard, N., J. Variyam, and J. Cromartie. 2003. Food expenditures by U.S. households: looking ahead to 2020, Economic Research Service AER-821, U.S. Department of Agriculture, Washington, D.C.

Chan, W. 2004. Selling seafood to Asians. *Meat and Seafood Merchandising,* Shawnee Mission, Kansas.

Efficient Foodservice Response. 2003. Standard product ID and bar coding in the foodservice supply chain. The Efficient Foodservice Response annual benchmarking survey. Survey Results, http://www.efr-central.com/barcode/surveys.html.

Food Marketing Institute. 2004. Supply Chain, ECR-Efficient Consumer Response. http://www.fmi.org/supply/ECR/, accessed 9/28/2004.

Harris, J.M., P. Kaufman, S. Martinez, and C. Price. 2002. The U.S. food marketing system, 2002—Competition, Coordination, and Technological Innovations Into the 21st Century. Economic Research Service Agricultural Economic Report No. 811, U.S. Department of Agriculture, Washington, D.C.

Heller, W. 2002. 55th annual consumer expenditure study. *Progressive Grocer* 81:13.

Information Resources Inc. (IRI). 2002. Dollar Sales: IRI's combined supermarket/drug store/mass merchandiser, including Wal-Mart reviews database.

Institute for Social, Economic, and Ecological Sustainability. 2001. National Organic Standards Board Aquatic Animal Task Force Final Recommendations, The University of Minnesota, St. Paul, Minnesota.

Kinsey, J.D. 1999. The big shift from a food supply to a food demand chain. Minnesota Agricultural Economist No. 698. University of Minnesota, St. Paul, Minnesota.

Kris-Etherton, P.M., W.S. Harris, and L.J. Appel. 2002. Fish consumption, fish oil, omega-3 fatty acids, and cardiovascular disease. Circulation: *Journal of the American Heart Association* 106:2747–2757.

Meat and Seafood Merchandising 2004. Mandatory country-of-origin labeling. *Meat and Seafood Merchandising,* Shawnee Mission, Kansas.

National Marine Fisheries Service (NMFS). 2004. U.S. seafood trade: 1984–2003. http://www.nmfs.noaa.gov/trade/2003ustrade.pdf.

National Marketing Institute (NMI). 2003. The 2003 health and wellness trends report (HWTR). Natural Marketing Institute, Harleyville, Pennsylvania.

Nettleton, J. 1995. *Omega-3 Fatty Acids and Health.* Chapman and Hall, New York.

Organic Monitor. 2002. The global market for organic food and drink. http://www.organicmonitor.com/700140.htm

Pacific Business News. 1998. Dealer goes to great depths for patrons. January 30. http://pacific.bizjournals.com/pacific/stories/1998/02/02/smallb2.html

Puget Sound Business Journal 2000. Seafood companies join in E-commerce venture, October 27.

Seafood Business. 2001–2004. Buyers navigate sustainable seafood. *Seafood Business,* Portland, Maine.

Shang, Y.C., C. Tisdell, and P.S. Leung. 1991. Report on a market survey of giant clam products in selected countries. Center for Tropical and Subtropical Aquaculture Publication #107. Department of Agricultural and Resource Economics, University of Hawaii, Honolulu, Hawaii.

Shang, Y.C., P.S. Leung, and J. Brown. 1992. Test marketing of giant clam as seafood and as aquarium specimens in selected markets. Center for Tropical and Subtropical Aquaculture, Honolulu, Hawaii.

Sloan, A.E. 2003. Top 10 trends to watch and work on: 2003. *Food Technology* 57(4):30–50.

Spencer, M.T., and P. Rea. 2003. Market overview: sales top $36 billion, The Natural Foods Merchandiser, June.

Stewart, H., N. Blisard, S. Bhuyan, and R.M. Nayga, Jr. 2004. The demand for food away from home full-service or fast food? Economic Research Service Agricultural Economic Report No. 829, U.S. Department of Agriculture, Washington, D.C.

Tarnowski, J. and W. Heller. 2004. The super 50. *Progressive Grocer*, May.

U.S. Department of Agriculture, Economic Research Service (ERS). 2004. Food CPI, Prices, and Expenditures: Expenditure Tables. ERS/USDA Briefing Room. www.ers.usda.gov/briefing/CPIFoodAndExpenditures

U.S. Department of Agriculture, Agricultural Marketing Service (AMS) 2004 AMS News Release No. 172-04, USDA Issues Regulatory Action on Mandatory Country of Origin Labeling for Fish and Shellfish, Sept. 30, 2004.

U.S. Department of Agriculture, Economic Research Service. 2002. Traceability for food marketing & food safety: what's the next step? *Agricultural Outlook*, January-February.

10
The International Market for Seafood and Aquaculture Products

THE BASIS FOR TRADE

International trade is the exchange of goods and services between two countries. Trade can occur due to: (1) differences in technology between countries; (2) differences in resource endowments; (3) differences in consumer demand; (4) existence of economies of scale in production; or (5) existence of government policies (Suranovic 1997–2004).

Trade is based on the benefits gained from specialized production and the relative advantage that the country has in the production of certain goods. For one nation to produce all the goods and services that its citizens desire would mean that some of the products are produced less efficiently than if they were produced in other nations that have a particular advantage and have specialized in their production. Moreover, most countries do not have the climate or all the resources needed to produce all the goods and services that its citizens might want.

The theory that has been developed as a basis for analyzing and understanding international trade shows that costs of production alone do not explain whether trade occurs. The lowest-cost producers may still not be competitive in international markets, and some countries may still benefit from free trade if it is less efficient than all other countries. In fact, less efficient companies can compete with foreign companies depending upon the relevant price ratios, the size of the countries involved, and the difference in domestic demand for the relevant products (Suronovic 1997–2004).

If free trade conditions exist, countries import those goods for which the domestic supply in the country is less than its demand (Anderson 2003). A country will also look to export its excess supply. If price of the product on the world market is above the equilibrium price in that particular country (price that results in all the domestic quantity supplied clearing the market because it is being purchased by domestic buyers), then this "excess" supply would be exported. The importing country would import that quantity that corresponds to price levels on the world market that are below the importing country's equilibrium price. In other words, if the world market price is higher than the domestic country's market price, it will look to import those products for which the world market price is lower than their domestic market price. Thus, the excess supply from one country interacts with the excess demand curve of another, and trade results if the price relationships are favorable. The volume of trade depends on the quantity-price relationships involved (the reader may wish to review the demand, supply, and price determination sections in Chapter 2). Changes in any of the factors that affect the excess supply of the exporting country and the excess demand of the importing country will result in changes in the price and quantity of the product traded.

Anderson (2003) demonstrated that seafood product supplies from open-access fisheries may or may not conform to traditional trade models, depending upon the shape of the fish supply curve from open-access fisheries. Seafood products supplied from aquaculture, however, will likely trade in a manner more similar to that of agricultural products due to property rights and well understood production practices. Thus, products in seafood sectors that are becoming more dominated by aquaculture products are likelier to trade in a way that conforms to traditional trade models. On the other hand, aquaculture products in markets dominated by open-access fishery products may behave differently in international trade. In these cases, careful analysis and development of appropriate trade models will be necessary to understand the impacts on trade of various policy measures.

DIMENSIONS OF THE INTERNATIONAL MARKET

International markets are more variable and difficult to predict than domestic markets. The volume of seafood that is traded internationally is very large, $60 billion in 2003 (FAO 2004). Nearly 200 countries traded more than 800 species of fish in 2003. Moreover, the volume of trade in fish and seafood products has increased, particularly over the last 10 years (Fig. 10.1; FAO 2004).

Fish products were the second most important food group traded worldwide in 2002 (Anderson 2003). Across the world, fruit, nuts, and vegetables were the only food group with a trade value greater than that of fish products. Trade in fish products is particularly important in Asia, where the all-fish products group is the largest group of food products traded. Japan is one of the largest seafood markets in the world. Its trade in food products is nearly twice as great as its trade in the all-meats food group category, the category with the next highest value.

The value of trade in fish products is also important to the United States. The value of fish products traded in 2000 was nearly as great as the leading category of fruit, nuts, and vegetables (Anderson 2003).

Traded fish and seafood has played an important role throughout history. Kurlansky (1997) argued that the search for cod resulted in international trade that played a prominent role in both the exploration and development of the New World. Moreover, access to and control of cod fisheries around the world played an important role in the economies of a number of European nations. Conflicts over cod resulted in wars and influenced political strategy in several countries for several centuries.

The seafood export trade is particularly important for developing countries, whereas the developed world is heavily dependent on imported seafood. Most seafood trade flows from lesser-developed countries to more developed countries. Most of the target markets are in the European Union, Japan, and the United States. Developing countries account for 50% of world fish exports, and revenues from

Trade, Aquaculture, and the National Debt

Chapter 10 points out that much of the aquaculture and other seafood exported is from lesser-developed countries to more developed countries. This growing volume of exports generates foreign exchange earnings for the exporting countries. Increased foreign exchange reduces the national debt and contributes positively to the national balance of trade.

However, countries with high population densities may have scarce land and water resources. These resources need to be allocated to production either for export or for domestic consumption. For example, in countries such as India and Vietnam, increasing numbers of land areas are being taken out of rice production for the domestic market and are being replaced by production of shrimp or freshwater prawns for export. In some countries, national policies promote this change and offer incentives to change to export promotion. In other cases, large corporate ventures have purchased or acquired concessions to tracts of land to produce export crops. The policy dilemma is whether the increased foreign exchange will result in sufficient economic growth and development over time to compensate for the decreased food security of becoming more dependent on imported foodstuffs.

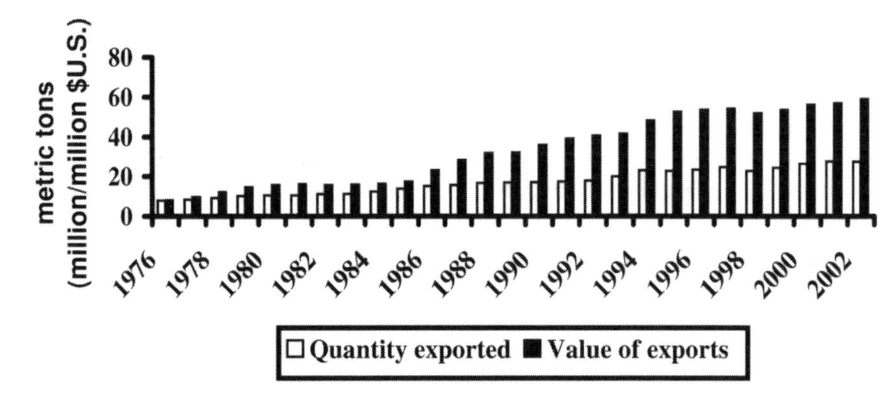

Figure 10.1. Volume of international exports of fish and seafood, 1993–2002. (Source: FAO 2004.)

fisheries are crucial for many developing countries (Lem 2004). Low-income food-deficit countries accounted for 19% of total exports (in terms of value) in 2003, and fishery products were a major source of foreign exchange for a number of countries. Net receipts of foreign exchange for fishery products by developing countries (subtracting value of exports from value of imports) more than doubled from 1981 to 2001 and exceeded that of other types of agricultural products, including rice, cocoa, tobacco, and tea. Thus, much of the developing world are net exporters of fish and seafood.

International trade in seafood was enhanced by the establishment of exclusive economic zones (EEZs). The EEZ is a 200-mile exclusive zone imposed by countries with coastal areas. Countries such as Peru, Ecuador, and Chile with important coastal fishery resources implemented EEZs in 1952 in reaction to foreign fishing vessels exploiting resources considered to belong to their nation. The United States declared its EEZ in 1976, following 37 other nations. Countries such as Spain and Japan that had relied on distant-water fishing fleets for their domestic seafood supply were cut off from rich fishing areas. Imports into these countries have increased to meet their domestic demand for seafood.

Asia as a region is a net importing region, largely due to the volume of seafood imports into Japan. When Japanese statistics are excluded, Asia is a net exporter with a large seafood trade surplus (Anderson 2003).

All countries engage in some sort of trade with other countries. Some countries, with more laissez-faire policies, engage in higher volumes of trade than do countries with more restrictive policies. However, as technologies of communications and travel have developed over time, there has been an overall increase in the volume of international trade over time (refer to Fig. 10.1). The value of seafood exports has increased faster than the volume of exports.

Approximately 38% (live weight equivalent) of world fish production was traded internationally in 2001 (Vannuccini 2003). However, the international trade in fisheries commodities increased by only 3% (from 34% to 37%) as a percent of total fishery production (both capture fisheries and aquaculture). This is due to the continued increase in total fishery production that results from the continued increase in aquaculture production.

Japan, the United States, and several European countries (Spain, France, Italy, Germany, and the United Kingdom) accounted for more than 80% of the world's imports of fishery products (Vannuccini 2003). Japan is the leading seafood-importing nation in the world, importing 26% of total world seafood production (Fig. 10.2). Japan is followed by the United States in terms of the value of seafood imported. After the United States, the following were the next most important countries: Spain, France, Italy, Germany, the United Kingdom, Hong Kong, Denmark, China, Canada, The Republic of Korea, the Netherlands, and Belgium.

The leading exporter of seafood in the world is Thailand, which exports 8% of the world's seafood (Fig. 10.3). It is followed in descending order by China, Norway, the United States, Canada, Denmark,

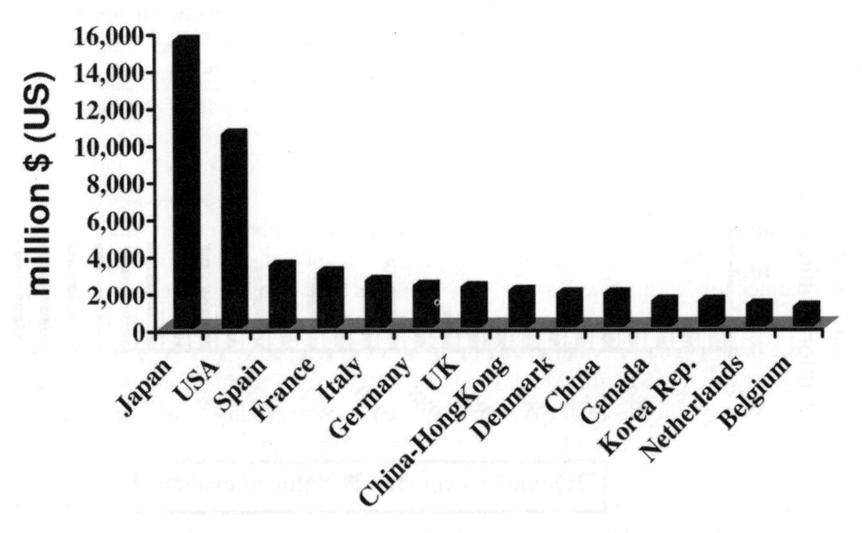

Figure 10.2. Leading importers of fish and seafood products worldwide. (Source: FAO 2004.)

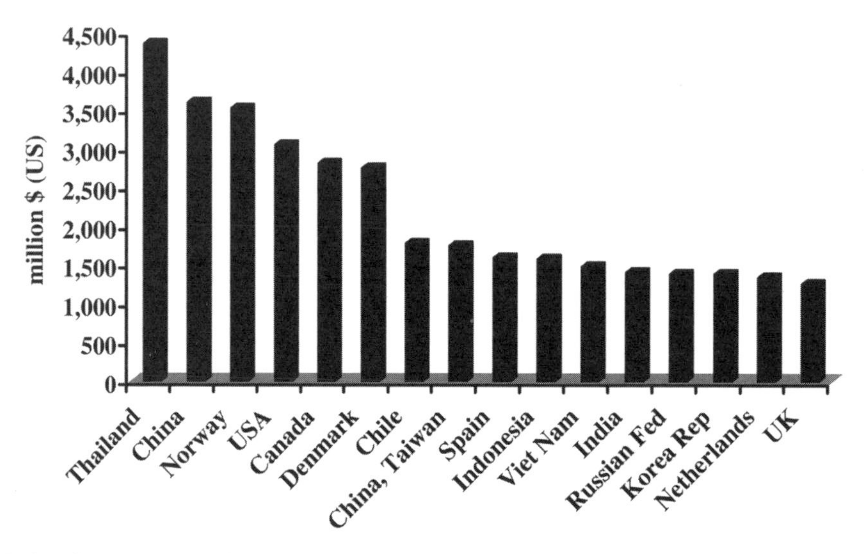

Figure 10.3. Leading exporters of fish and seafood worldwide. (Source: FAO 2004.)

Chile, Taiwan, Spain, Indonesia, Vietnam, India, the Russian Federation, the Republic of Korea, The Netherlands, and the United Kingdom. Exports from China increased by 11% in 2001 as compared to the previous year (Vannuccini 2003). Comparisons between Thailand and China show that Thai exports of fishery products were 35% of its total agricultural exports and 6% of its total merchandise trade. For China, fishery products accounted for 26% of total agricultural exports and 1.5% of its total trade in merchandise.

Anderson (2003) points out how the sources of seafood supply have shifted over the last several decades. In 1976, Japan was the leading seafood exporter (in terms of value), followed by Norway, Canada, Denmark, and Taiwan. Only Norway and Canada have continued in the top five seafood-exporting countries over time.

The most important international trading partner for North America was East and S.E. Asia ($2.8 billion), followed by North America ($2.3 billion). South America ($1.6 billion) was the third most important trading partner for North America and was followed by Central America ($1.0 billion) (FAO 2003). For the European Union, the major trading partner is the European Union ($8.0 billion) followed by Western Europe ($4.6 billion). South America ($1.7 billion) was the third most important partner and was followed by East and S.E. Asia ($1.3 billion) and North America ($0.9 billion). The major trading partners for East and S.E. Asia are: East and S.E. Asia ($0.7 billion), China ($0.4 billion), North America ($0.2 billion), and South America ($0.2 billion).

Shrimp is the major fishery commodity traded internationally and accounted for about 19% of the total value of internationally traded fishery products (Vannuccini 2003). Other main types of species exported are groundfish (cods, hakes, haddocks), tuna, and salmon. Fishmeal represented 3.6% of the value of exports; fish oil represented less than 1%. Trade from developing countries is gradually changing from exports dominated by raw products to that of value-added products.

Much of the growth in the international trade in fish and seafood has come from aquaculture, primarily shrimp, salmon, trout, tilapia, oysters, and carp (Anderson 2003). The entry of aquaculture products into the international market has provided a mechanism for countries to become major suppliers internationally. Shrimp from Thailand and salmon from Norway and Chile are good examples of countries with low levels of export until the development of significant aquaculture industries. Ecuador and Honduras have also become major traders in both shrimp and tilapia worldwide as these farm-raised industries developed.

Retrade in seafood and aquaculture products is also growing. China, for example, imports large quantities of raw product to process, repackage, and re-export. Roe herring from Alaska is imported into China, extracted, processed, and exported to Japan (Anderson 2003). China also imports frozen cod and then exports frozen fish fillets and value-added products. Moreover, re-export has been used to circumvent trade restrictions, as is discussed later in this chapter.

Trade Policy Tools

Politically, most citizens expect their government to demonstrate responsibility toward producer groups that are hurt by competition from imported products. The Trade Reform Act of 1974 (in the United States) provides for adjustment assistance to companies and workers depending on whether the circumstances meet the specific program requirements. Trade policies can be considered as either "beggar-thy-neighbor" or as a "strategic trade policy." Beggar-thy-neighbor policies benefit one country by forcing losses on its trading partners. Strategic trade polices shift profit away from international competitors or consumers.

Countries may prioritize maintaining a self-sufficient food supply to avoid dependence on another country in case of war or other threats. Some countries use trade barriers to maintain employment at home. These barriers are implemented through control of foreign exchange, lending and borrowing, licensing, or use of state trading agencies. Barriers have been used to attempt to stabilize prices by developing a more consistent supply. The U.S. import quotas on beef stimulate larger imports when the U.S. slaughter quantities are lower and reduce imports when U.S. slaughter numbers are higher. New industries tend to have higher costs initially, and protection may enable the industry to grow until it is large enough to compete effectively.

Tariffs on imported products have been used to support farm prices and incomes, whereas exporting countries may use farm support prices, export subsidies, or tariffs. Tariffs can be levied as either a specific tariff or an ad valorem tariff. Specific tariffs are based on establishing a certain cost per unit of the good imported. For example, tariffs of $0.033/kg are charged in the United States for grapefruit imported from August-September from countries without Most Favored Nation status) (International Trade Commission 2004). In the United States, the U.S. Customs Service administers and collects the tariffs.

Tariffs often hamper development of the aquaculture sector in developing countries because they protect the final consumer products industries in importing countries (Anrooy 2003). Moreover, tariffs discourage processing industries in exporting countries. Import tariffs still exist for aquatic products in many countries, including Taiwan, Vietnam, the European Union, the United States, and China.

Quotas restrict the quantity or volume that can be imported. Quotas can be either absolute or tariff rate. Limitations on the quantity imported during a specified period of time can be set to affect all imports or only those from certain countries. However, quotas can also be set to import a certain quantity at a reduced rate, or a tariff-rate quota. Quotas are more common in the international trade of agriculture products. Quotas are also generally administered by the U.S. Customs Service, with some exceptions.

Voluntary export restraints (VERs) regulate the volume of a good to be exported. VERs typically result from pressure from the importing country. With the emergence of the General Agreement on Tariffs and Trade (GATT), VERs emerged as a form of protectionism that did not violate GATT agreements. Suranovic (1997–2004) provides examples of the U.S.-Japan automobile and textile VERs with the United States and the effects of implementing these restrictions.

There are also a number of nontariff barriers that have been used to protect domestic industries: (1) government participation in trade including production subsidies; (2) customs that make it difficult to import certain products; (3) industrial, health, and safety standards that may include packaging and labeling regulations; (4) embargoes, bilateral agreements, voluntary restraints; and (5) special duties and credit restrictions (Rhodes 1993).

Export taxes and subsidies are used often to collect duties to generate revenue for governments. Malaysia, Kenya, Mali, Norway, Vietnam, and the Dominican Republic tax exports of aquaculture products (Anrooy 2003). Export taxes reduce the price received by producers as a consequence. However, if taxes are used for promotional campaigns, there may be some benefit back to the aquaculture sector. Export subsidies benefit primarily the exporting companies. If those companies were vertically integrated, primary production sectors would also benefit.

Most major fish-exporting nations have established councils to promote their export products. These councils are typically supported by public funds to enhance sales and position products in world markets. The Norwegian Seafood Export Council, Dutchfish, SalmonChile, and the Scottish Salmon Board are examples of export promotion councils for seafood.

Export Processing Zones (EPZ) have been developed in a number of countries as industrial zones to encourage foreign investment in processing (Anrooy 2003). Incentives such as free trade zones, financial services zones, free ports, duty-free imports, good infrastructure, easy market access, and others are used to attract investors to the EPZs. Companies that process aquaculture products have taken advantage of EPZs, and a number of countries have provided infrastructure to accommodate processing facilities for aquaculture products.

Exchange rate policies can affect international trade. Exchange rates determine the price received by exporters in local currency and the prices paid by importers, also in local currency. Devaluation of a country's currency results in reducing its export price and often increases overall production for export. However, devaluation raises prices of imports, which can have negative effects on sectors that are heavily dependent upon imports.

TRADE POLICY IN SEAFOOD AND AQUACULTURE

The General Agreement on Tariffs and Trade (GATT)

International trade occurred for many years within the framework of The General Agreement on Tariffs and Trade (GATT). GATT grew out of the Bretton Woods Agreement and was organized in 1948 with more than 90 countries. GATT rules defined export subsidies and included commitments from countries to reduce export subsidies with the goal of reducing or eliminating tariff and nontariff barriers to trade. An export was said to be subsidized when the export price was lower than the comparable price charged for similar products in the domestic market.

GATT was traditionally concerned with trade measures and not domestic production policies unless trade was involved. The most recent set of tariffs was implemented after the Uruguay Round of GATT negotiations. The Uruguay Round of multilateral trade negotiations was begun in 1987 (FAO 2003). It was the first round of trade negotiations in which developing countries were involved directly. The Uruguay Round Agreement on Agriculture overruled the provisions of GATT. It continued to allow export subsidies on agricultural products, but constraints were imposed. Early data on export subsidy use under the Uruguay Round Agreement on Agriculture indicated that subsidized exports for some products were small (especially as compared to a product such as wheat) and were allowed, although utilization rates for dairy products and various meats were quite high. Moreover, export subsidies on agricultural products were permitted but were subject to the potential for antidumping and countervailing duties. Overall, the Uruguay Round resulted in lower import duties for many products, including fish and fishery products.

In addition, the United States has signed free trade agreements with Canada, Mexico, and Israel. GATT also provided for the Generalized System of Preferences (GSP) for many less developed countries. The Generalized System of Preferences is a framework under which developed countries give preferential treatment to manufactured goods imported from certain developing countries. Following the Uruguay Round, the trade-weighted tariff on industrial products fell by 40% to 3.8% in 2000, and the average tariff decreased by 37% in developing countries (Ariff 2004). However, nontariff barriers began to increase over time.

World Trade Organization (WTO)

The World Trade Organization (WTO) emerged from the Uruguay Round negotiations and the Marrakech Agreement in 1995, and its membership increased to 147 in 2004. The WTO is a binding treaty that implements the GATT articles in support of free trade. However, the WTO does have greater enforcement authority than did the GATT. Russia and Vietnam were the only two major fisheries countries that did not belong to WTO in 2004, but both countries had begun negotiations to join.

Major provisions of the WTO agreement include: (1) Agreement on Application of Sanitary and Phytosanitary (SPS) Measures; (2) Agreement on Technical Barriers to Trade (TBT); (3) Agreement on Subsidies and Countervailing Measures; (4) Antidumping Agreement; (5) Agreement on Safeguards; and (6) WTO Dispute Settlement Procedures. Sanitary and Phytosanitary (SPS) measures focus on food safety. The SPS ensures that sanitary measures are not used to block imports. Import regulations that require HACCP plans are SPS applications. The United States has used this in a case against imported salmonids from Australia. Malaysian consignments of prawns and frozen seafood have been rejected by the European Commission due to bacteria counts (Ariff 2004). The European Union has set a zero bacterial count instead of the minimum acceptable level defined in international standards. Technical barriers to trade deal primarily with labeling and testing disputes. For example, in 2003, the Netherlands discovered two shipments of salmon contaminated with malachite green (www.parlamentodelmar.cl). This substance has been banned in Chile since 1997. Shipments of Chilean salmon to Japan were also detained in 2003 (Intrafish 2004). Traces of antibiotics were found in the salmon. The subsidies and countervailing measures agreement sets procedures for determining whether countries subsidize their exports. Rules for fish products typically are more stringent than those for agriculture products. The antidumping provisions have been used several times with regard to aquaculture and seafood products and have been widely publicized. Recent antidumping cases are discussed later in this chapter, and Appendix 10.A presents details of the process for antidumping lawsuits in the United States. Membership is required for dispute settlement, but most cases are settled out of court. The WTO sets defined stages with set time limits during disputes.

The most recent round of trade negotiations is referred to as the Doha Development Round, or Doha Agenda, that began in 2001. The Doha Agenda includes issues such as improved access to markets for fish and fishery products, fisheries subsidies, environmental labeling, and the relationship between WTO trade rules and environmental agreements. One Doha proposal is to eliminate all import duties on fish.

U.S. ANTIDUMPING

Countervailing and antidumping duties were authorized by the Tariff Act of 1930 in the United States. Countervailing duties are used when imports receive an unfair subsidy from the foreign government (King and Anderson 2003). Antidumping duties are used when imports are sold or likely to be sold for less than fair value. An additional measure known as safeguard remedies was authorized by the Trade Act of 1974. Safeguard remedies are used when increasing volumes of imports threaten to injure an industry in the United States.

BYRD AMENDMENT

Special measures can be taken if the products imported are being dumped. Dumping refers to selling product at prices lower than those in the home, or domestic, market. The Uruguay Round of GATT allowed countries to impose antidumping tariffs on products if it was determined that dumping had occurred.

The U.S. antidumping law (U.S. Department of Commerce 2004) is designed to provide relief to U.S. industries that are injured as a result of foreign goods being sold at unfairly low prices in the United States. Antidumping investigations are conducted in two phases, one by the United States Department of Commerce (USDOC) and the other by the International Trade Commission (ITC). The USDOC determines whether the imports in question have been sold at less than fair value in the United States. The ITC determines whether the imports in question are causing or threatening to cause material injury to a U.S. industry. An antidumping order is issued if both agencies reach positive determinations. Appendix 10.A provides additional detail on the process and procedures of antidumping lawsuits in the United States.

The Byrd amendment to U.S. antidumping law, known as the Continued Dumping and Subsidy Offset Act, was signed on Oct. 28, 2000 (Collie and Vandenbussche 2004). It provides for the distribution of revenue from antidumping tariffs imposed on foreign firms to the domestic firms that filed the dumping complaint. The justification for the Byrd amendment is that it is expected to lead to lower duties and higher welfare when compared to tax revenues if the weight on the profits of the domestic industry is sufficiently large. Also, the Byrd amendment makes it less likely that the antidumping duty will be prohibitive. However, in 2003, in response to complaints from the European Union and other countries, the WTO found that the Byrd amendment was inconsistent with the GATT antidumping agreement. When the United States did not repeal the Byrd amendment by the December 2003 deadline, the European Union applied for WTO authorization to apply sanctions in the form of higher import tariffs on products from the United States. In 2004, the European Union was approved to impose sanctions on goods from the United States in response to the Byrd amendment.

SALMON TRADE CONFLICTS

UNITED STATES AND NORWAY

Pen-raised salmon aquaculture technologies were developed on a large commercial scale primarily in Norway in the 1980s. During this same time period, the U.S. wild-caught salmon industry began to divert a larger portion of its production from canned products to the fresh/frozen salmon market in Japan (Anderson 1994). As pen-raised salmon aquaculture production grew in Norway, Norwegian exports to the United States grew rapidly.

Salmon prices declined in 1989, largely due to the increased supplies from aquaculture, and led to an antidumping petition from the Coalition for Fair Atlantic Salmon Trade (U.S.). The petition alleged that Norwegian producers had received countervailing subsidies and were dumping salmon in the United States, materially damaging the domestic industry. The U.S. International Trade Commission ruled on February 25, 1991, that the Norwegians were selling below fair market value. A countervailing duty of

2.27% and antidumping duties ranging from 15.65% to 31.81% (depending upon the company) were imposed. The magnitude of these duties caused Norway to be uncompetitive in the U.S. market. By March 1991, Norway's share of imports had sunk to less than 5% (Fig. 10.4). As Norwegian imports into the U.S. market fell in 1991, Canadian and Chilean imports began to increase. A second antidumping petition was filed subsequently in 1997 against Chilean salmon exporters by U.S. salmon producers.

United States and Chile

In 1997, the USDOC-ITC investigation found insufficient evidence of government subsidies to support imposing countervailing duties on fresh Atlantic salmon from Chile (Federal Register 1997-1998). On June 2, 1998, the USDOC-ITC investigation determined that two of Chile's largest Atlantic salmon producers had traded salmon fairly (Federal Register 1997-1998). Three other companies were found to have sold salmon at less than fair market value, and antidumping duties were set at 8.27%, 10.91%, and 2.24% with all others charged at 5.19%. The petition was later suspended by producers from the State of Maine (Sloop 2003).

European Union and Norway

In the early 1990s, Scottish and Irish farmers encouraged the European Union to impose minimum import prices for all Atlantic salmon but with a main focus on Norway (Anderson 2003). The European Union announced an antidumping and antisubsidy proceeding against imports of farmed Atlantic salmon from Norway in 1996, with antidumping and countervailing duties imposed in 1997 (Commission 2003). A subsequent decision accepted undertakings from 190 Norwegian exporters and importers, and these companies were exempted from the duties. The undertakings offered were consistent with minimum import price levels of the antidumping decision. The Norwegian government further imposed production restrictions on its farmers.

In 2004, the European Commission initiated investigation against Norway, the Faroese, and Chile. The investigations resulted from complaints by the United Kingdom and Ireland. The outcome of these investigations had yet to be determined at the time this book went to press.

BLUE CRAB CONFLICT

Imports of swimming blue crab (*Callinectes sapidus*, *Charybdis bellerii*, *Portunus pelagicus*) meat in the United States increased from 4.2 million kg (9.2 million lb) in 1994 to 12.4 million kg (27.2 million lb) in 1999 (Seafood Business 2000). The greatest concern on the part of the U.S. blue crab industry was the increase in volumes imported from Venezuela, Indonesia, Thailand, and Mexico. A number of domestic blue crab processors went out of business over this same time period. However, unlike the salmon growers, the U.S. blue crab industry could not raise the funds to file an antidumping petition. In 2000, they filed a Section 201 petition for safeguard remedies instead that required only a surge in imports thought to cause material injury and did not require proof of unfair trade practices. However, the ITC found no threat to U.S. suppliers, citing that increased imports did not result in idle plant capacity or greater underemployment.

U.S. CRAWFISH AND CHINA

For many years, the U.S. freshwater crawfish industry had experienced limited competition from imported supplies. However, in 1994, China captured 58% of the market share of crawfish tail meat in that one year. Within three years, China's market share

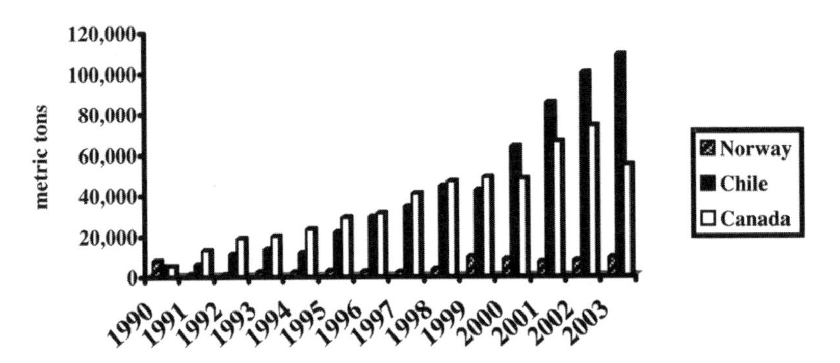

Figure 10.4. Salmon import quantities into U.S., 1990–2003. (Source: FAO 2004.)

had increased to 87% (International Trade Commission 1997). The impacts to the Louisiana industry were substantial for a number of reasons. First, the market was a small geographic market centered in southern Louisiana. Second, the Cajun French culture of southern Louisiana was linked to crawfish as a food (Roberts 2000). Thus, the U.S. crawfish market was highly localized. It was also seasonal. It operated only in the first half of the year and consisted of numerous small businesses that produced undifferentiated tail meat. Domestic supply fluctuated during this time period and was affected substantially by fluctuations in the wild catch from the Atchafalaya Basin. The combination of these factors created marketing opportunities during times of low supply, and its market characteristics made it a relatively easy target.

An antidumping petition was filed with the ITC in 1996 (Roberts 2000). The ITC found that the U.S. crawfish industry was being materially injured from crawfish tail meat imports from China being sold at less than fair value. Company-specific antidumping duties ranging from 92–123% were published in 1997. Other companies that initiated shipments following the period of investigation were assessed a tariff of 201%.

Because China was deemed to be a nonmarket economy, a normal price could not be calculated and surrogate countries were selected (see Appendix 10.A for more details on procedures of antidumping lawsuits). Spain's imported price of live crawfish from Portugal was used as the surrogate country farm price. India was selected as the processing surrogate country because it had a market economy with a large seafood processing industry that utilized hand labor. U.S. importers and representatives from China challenged this decision (unsuccessfully) because the ungraded Spanish imports bring a higher price than graded U.S. crawfish.

The ruling that resulted applied only to China. Because U.S. imports from other countries were free of the 123% import duty, China attempted to avoid the ruling by repackaging Chinese tail meat in Singapore. However, the Singapore company did not meet the substantial transformation test used by the U.S. Customs Service to determine country of origin, and duties also were levied on the shipments from Singapore. A similar increase of imports from Spain during 1999-2000 was identified.

U.S. CATFISH AND VIETNAMESE BASA

The introduction of basa and tra (*Pangasius sp.*) from Vietnam as lower-priced alternatives to U.S. farm-raised production of channel catfish (*Ictalurus punctatus*) contributed to a severe and protracted downturn in U.S. catfish prices. The quantities of imports from Vietnam increased rapidly from 2000-2001 and reached 15% of total frozen fillets in just two years (Fig. 10.5). The Vietnamese imports very

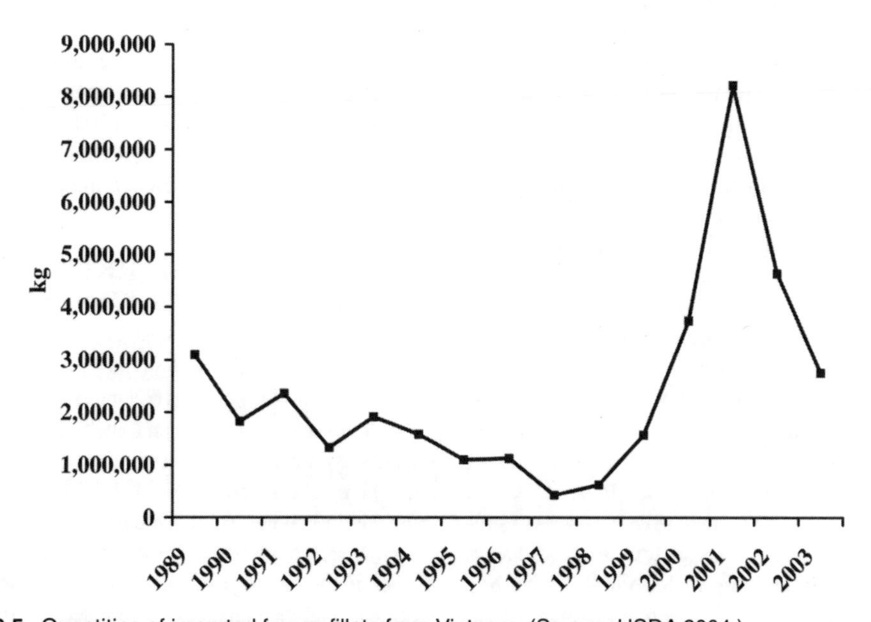

Figure 10.5. Quantities of imported frozen fillets from Vietnam. (Source: USDA 2004.)

quickly captured a noticeable portion of the most profitable and fastest growing segment (smaller frozen fillets) of the U.S. catfish market.

Within the same time period, catfish prices declined by more than 30%. Because fillets account for about 60% of the total volume of processed catfish sold, the impact of imports was considered an important factor contributing to lower price levels and price instability.

Quagrainie and Engle (2002) found that the market for domestic frozen fillets plays a significant role in the price determination of imported catfish. Thus, now that the potential for competition has been established in the U.S. farm-raised catfish market, periods of higher prices in the future may be countered by increased supplies of imported product (Quagrainie and Engle, in review). This will be particularly important during times when the U.S. dollar is strong. Ligeon et al. (1996) also concluded that the quantity of catfish imported into the U.S. will decline if the domestic price of catfish falls relative to the import price. These studies imply that, if the industry expects to see higher catfish prices, production and supply control strategies have to be pursued. Efforts to require labeling of Vietnamese fish fillets and strict inspections of imported fillets may help to reduce the quantity of imported product into the United States.

The USDOC placed an antidumping order against imports of Vietnamese frozen basa and tra fillets in August, 2004. Tariffs ranged from 36.84% to 63.88%. The average pond bank price also increased from $1.21/kg ($0.55/lb) to $1.52/kg ($0.69/lb) in July 2004. Under antidumping law, the exporters are entitled to "administrative reviews" of the tariffs that focus on the duty rate applied. Moreover, one importer has requested the USDOC to rule that live basa and tra from Vietnam processed into frozen fillets in Cambodia are not covered by the antidumping order. U.S. industry requested that this be ruled a "circumvention" of the antidumping duty order and that the Cambodian-processed fillets be covered by the antidumping order (Warren 2004).

MUSSELS CONFLICTS

Great Eastern Mussel Farms of Tenants Harbor, Maine, filed an antidumping petition in January 2001 against mussel producers in Prince Edward Island, Canada. In October 2001, the USDOC assessed preliminary tariffs on two of the four PEI producers named in the antidumping petition at 4.7% on one producer, with a 3.48% dumping margin on the other. Shortly thereafter, PEI mussel producers increased prices twice, resulting in Great Eastern Mussel Farms's withdrawing its antidumping petition against PEI mussel producers. The ITC and USDOC terminated the suit.

SHRIMP CONFLICTS

In 2003, a shrimp antidumping petition was filed in the United States from a different perspective from those filed previously in the United States. U.S. shrimp fishermen and processors specializing in wild-caught shrimp filed a dumping petition against importers who were purchasing farm-raised shrimp from China, Vietnam, Ecuador, Brazil, Thailand, and India (Seafood Business 2004). In July 2004, the USDOC imposed preliminary duties ranging from 7.67% to 112.8% on shrimp from China and Vietnam. The preliminary ruling requires importers to post cash deposits or bonds equal to the preliminary dumping margins. Other rulings will be issued against Ecuador, Brazil, Thailand, and India. In the meantime, the countries named in the suit are losing market share in the U.S. market that is being captured by Mexico, Indonesia, Venezuela, Honduras, Guyana, and Bangladesh.

THE CONVENTION ON THE INTERNATIONAL TRADE OF ENDANGERED SPECIES (CITES)

The Convention on the International Trade of Endangered Species (CITES) was adopted in 1973 with the goal of protecting species threatened by international trade. The CITES listed more than 30,000 species in 2001 and was supported by 154 member countries. CITES plays a more important role in capture fisheries than in aquaculture. However, the giant clam synopsis in Chapter 9 illustrates that development of commercial production of threatened or endangered species will require CITES permits and authorization.

AQUACULTURE MARKET SYNOPSIS: CRAWFISH

Freshwater crawfish are important segments of the aquaculture industry in the United States, Australia, Europe, and China. Crawfish have been consumed for centuries in Europe and in North America by Native Americans (Lutz et al. 2003).

There is some crawfish production in 27 states in the United States (U.S. Department of Agriculture 1998). However, 96% of the total production is located in the state of Louisiana, in the southern United States. The primary species raised in the United States is the red swamp crawfish (*Procambarus clarkia*) (Fig. 10.6). The red swamp crawfish is native to Louisiana and constitutes an important culinary tradition for the Cajun culture in that part of the United States. It is prepared in locally popular dishes such as the traditional crawfish boil and in etouffé.

Crawfish in Louisiana in the United States were sold commercially beginning in the late 1800s from wild-caught supplies (Lutz et al. 2003). Over time, the market shifted from local and household consumption to sales in urban areas of Baton Rouge and New Orleans. Growers began to reflood rice fields, woodlands, and marshland to produce crawfish in the 1950s. By the mid-1960s, a crawfish peeling industry had developed with continued increases in acreage.

Wild-caught crawfish are caught from the Atchafalaya Basin between Louisiana and Alabama. The wild catch exhibits dramatic fluctuations that are dependent on weather conditions from year to year. Wild-caught crawfish move through the same market channels as do farm-raised crawfish. Thus, price of farm-raised crawfish has been affected strongly by the fluctuations in the wild-caught supply.

Farm-raised crawfish are typically raised with some other forage crop for feeding (Fig. 10.7). The forage crop can be a flooded area with natural vegetation, a rice crop that has been harvested with the stubble remaining, or a grain crop that is planted especially to serve as forage for the crawfish crop. Crawfish emerge from burrows under the pond when the pond is flooded and begin to forage. When the pond is drained, the crawfish return to the burrows to wait for the next period of flooding.

Crawfish marketing channels in the United States include live sales, sales to processing plants, and exports of whole boiled crawfish to Scandinavian countries (Lutz et al. 2003). Crawfish are harvested by traps and specially designed boats (Fig. 10.8), graded at the pond bank, and then packed live into

Figure 10.6. Female red swamp crawfish with eggs. Photo courtesy of Dr. Greg Lutz.

Figure 10.7. Aerial view of crawfish ponds, Louisiana, United States. Photo courtesy of Dr. Greg Lutz.

onion sacks. Most crawfish farmers sell most of their product to buyers that specialize in the distribution of crawfish, although they typically will also sell a portion of their crop directly to the public, to restaurants, or to small seafood buyers.

The crawfish industry has faced several challenges in recent years. During 1999–2001, drought conditions in Louisiana resulted in yields less than half that of typical yields in previous years. Moreover, the crawfish peeling, or processing, sector has shrunk from 90–100 processors in Louisiana in 1996 to about 15 in 2003 (Lutz et al. 2003). In 2002, approximately 4.1 million kg (9.1 million lb) of frozen and peeled crawfish meat were imported into the United States from China.

The difficulties of the peeling sector are related to the lower-priced imports of crawfish from China. Crawfish in Louisiana historically were peeled and sold as fresh or frozen tail meat when crawfish harvests exceeded the demand for live, whole product. Thus, peeled crawfish tail meat served to moderate the seasonality of prices for crawfish. Peeled crawfish tail meat from China began to arrive during the period when yields had decreased and processors were having difficulty finding raw material to process. Tariffs were levied on imported Chinese tail meat in the late 1990s, and the volume imported decreased in both 2000 and 2001. The tariffs were subsequently renewed for a second five-year period. However, Chinese exporters have begun to shift to quick-frozen, whole, boiled crawfish for sale in both traditional and nontraditional markets throughout the United States.

By 2003, the crawfish industry had largely recovered from the drought years and acreage and production levels had increased substantially (Lutz and

Romaire 2003) (Fig. 10.9). However, the processing sector still had not recovered at the time this book was published. Thus, there are fewer market outlets available for U.S. growers, especially for smaller crawfish.

Although much attention is paid to the U.S. crawfish industry, crawfish of the genus *Cherax* have been cultured in Australia in a manner similar to those of the U.S. crawfish for a number of years. The Australian crawfish reach a much larger size and thus occupy somewhat different market niches. Moreover, there is a long tradition of catching and eating crawfish (*Astacus spp.*) in several European countries, including Austria and Sweden. Crawfish growers in a number of countries have targeted the European markets.

SUMMARY

This chapter presented a brief discussion of the basis for international trade and contrasted free trade and protectionist policies. The major international agreements on trade were described, including GATT, the Uruguay Round, WTO, and the DOHA Agenda. Trade policy tools of tariffs, quotas, and nontariff barriers were defined and discussed. Trade disputes related to salmon, catfish, shrimp, mussels, and crabs were described.

Figure 10.8. Harvesting crawfish, Louisiana, United States. Photo courtesy of Dr. Greg Lutz.

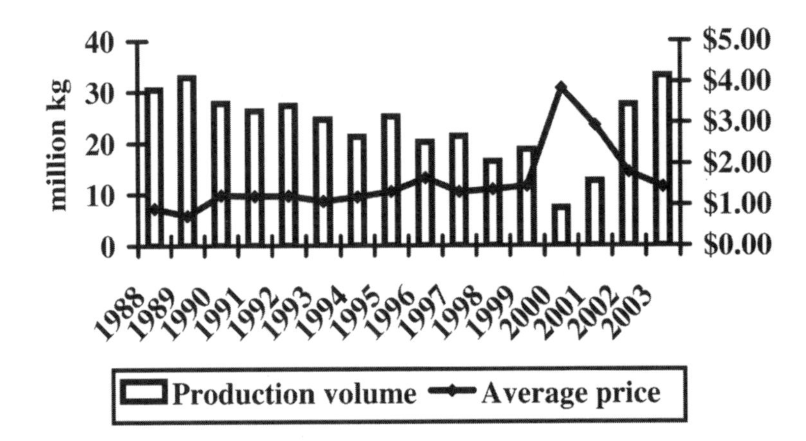

Figure 10.9. Volume produced and average price received of Louisiana farmed crawfish. (Source: Aquaculture Research Center, Louisiana State University 2004.)

STUDY AND DISCUSSION QUESTIONS

1. Explain why countries engage in trade. Is the cost of production the only factor that determines whether trade will occur?
2. How important is international trade in aquaculture products?
3. List and define the major types of trade policy tools.
4. Develop a timeline of the major international agreements on trade, with bulleted lists of the major provisions of each.
5. What is the Byrd Amendment and what does it provide for?
6. Diagram the sequence of the major trade disputes related to salmon, indicating the countries involved and the outcomes in terms of the effect on trade flows.
7. What surrogate countries were selected for the crawfish antidumping lawsuit in the United States?
8. What was different about the U.S. shrimp antidumping lawsuit as compared to the others?

APPENDIX

THE U.S. ANTIDUMPING LAW (BYRD AMENDMENT)

Both the Department of Commerce (USDOC) and the International Trade Commission (ITC) are involved in antidumping petitions in the United States. The USDOC examines whether dumping has occurred, and the ITC determines whether the U.S. industry has been harmed. For duties to be levied, both dumping and harm to the domestic industry must be proved.

The petition must include information to support the allegation of sales at less than fair value and to support the allegations of material injury, the threat of material injury, and causation. This involves obtaining information on the prices at which subject merchandise is being sold in the United States, estimated costs of production, and the estimated margin of dumping. The costs of production (or factors of production) include estimates of the labor required to produce the goods, the raw materials used for production, the energy and other utilities consumed, and capital costs. The margin is calculated by first determining the U.S. price and the normal, benchmark value.

In cases brought against nonmarket economies, a "surrogate" analysis is used for normal value. The surrogate is a market economy country that is at a level of economic development comparable to that of the nonmarket economy, and is a significant producer of the goods that are the subject of the investigation or other comparable merchandise. All factors of production, except labor, are valued using costs in the market economy country. These are taken from publicly available data.

Department of Commerce

When an antidumping petition is filed, the USDOC has 20 days to initiate an investigation. The USDOC normally bases its determination on data obtained in response to detailed questionnaires sent to foreign producers and exporters. The USDOC final affirmation is determined on day 235. The USDOC normally will compare the price at which the good is sold to the United States (U.S. price) with the price at which similar goods (the foreign like product) is sold in the foreign market (normal value). The ITC makes its final determination on day 280. On day 287, the USDOC issues its order.

An antidumping order requires an importer to post a cash deposit equal to the dumping margin. Actual duties are not paid until after there has been an administrative review. If the USDOC issues an affirmative preliminary determination, all imports that enter after that date must be accompanied by bonds or cash deposits equal to the assigned margin. The USDOC may suspend an investigation involving a nonmarket economy country if an agreement to restrict the volume of subject imports into the United States is reached. Different types of suspension agreements may be available in cases involving exports from market economy countries. A nonmarket economy country is a foreign country that does not operate on market principles of costs or pricing structures. The USDOC assumes that sales prices and costs in a nonmarket economy country cannot be used to determine the normal value of the goods.

In nonmarket cases, the USDOC will "construct" a normal value for the less than fair market value comparison. This constructed normal value is derived by: (1) identifying the cost elements involved in producing the foreign like product (factors of production); and (2) valuing those cost elements in a market economy country that is at a comparable stage of economic development as the nonmarket economy (surrogate country).

The USDOC identifies the first sale made for export to the United States to an unaffiliated purchaser in the calculation of the U.S. price. The U.S. price is compared to the normal value and the dumping margin calculated from that. The USDOC bases its determination on responses to questionnaires issued to foreign producers and exporters of subject merchandise. The questionnaires request detailed sales and cost information. Domestic producers who support the petition must represent at least 25% of total U.S. production and must account for more than 50% of the production of those domestic producers who take a position for or against the petition.

Between the preliminary and final determinations, the USDOC will conduct an on-site verification of each foreign producer. Based on the questionnaire responses, verification results, briefs filed by interested parties to the proceeding, and a formal hearing, the USDOC will make its final determination.

International Trade Commission (ITC)

The International Trade Commission (ITC) makes its preliminary determination as to whether there is a "reasonable indication" of material injury or threat of material injury to a domestic industry "by reason of" the imports in question within 45 days of filing the petition. The ITC's staff will prepare a report based on questionnaire responses received from domestic producers, foreign producers, and importers. The focus of the ITC's analysis is on the material injury or threat of material injury and the causation of injury by imports. Material injury is defined as "harm which is not inconsequential, immaterial, or unimportant." The ITC considers whether the volume or any increase in imports is significant in absolute or relative terms compared with domestic production or consumption, whether imports are underselling the domestic product, or whether they have depressed or suppressed domestic prices, and the impact of imports on the domestic industry.

The impact of imports on the domestic industry is measured by (1) the actual or potential decline in output, sales, market share, profits, productivity, return on investment, and capacity utilization; and (2) actual and potential negative effects on cash flow, inventories, employment, wages, growth, ability to raise capital, and investment. Even if the ITC determines that the domestic industry is not currently being injured by imports, it may nevertheless make an affirmative determination if it finds that the industry is being threatened with material injury.

If the ITC finds that the domestic industry is suffering from or threatened with material injury, it must then determine whether this is "by reason of the less than fair value of imports." This requires "adequate evidence to show that the harm occurred by reason of the less than fair value of imports, not by reason of a minimal or tangential contribution to material harm caused by less than fair value of goods."

The ITC's questionnaires request financial, production, shipment, and pricing information. Based on the staff report, a hearing before the staff, and briefs filed by those in favor or opposing the petition, the ITC will issue a preliminary decision. The ITC's final determination is made on day 280.

REFERENCES

Anderson, J.L. 2003. *The International Seafood Trade.* Woodhead Publishing Limited, Cambridge, England.

Anderson, J.L. 1994. The growth of salmon aquaculture and the emerging New World Order of the salmon industry. Presented at: Fisheries Management—Global Trends, University of Washington, Seattle, Washington, USA.

Anrooy, R. 2003. Policy considerations linked to better marketing and trade of aquaculture products destined for poverty alleviation and food security. Food and Agriculture Organization of the United Nations, Rome, Italy. Accessed at //www.fao.org

Ariff, M. 2004. Dismantling trade barriers. Malaysian Institute of Economic Research. http://www.mier.org.my/mierscan/archives/pdf/drariff9_8_2003.pdf

Collie, D.R. and H. Vandenbussche. 2004. Antidumping duties and the Byrd amendment. www.col lie.plus.com/byrd.pdf

Commission. 2003. Commission Decision. Official Journal of the European Union L 47/46. www.europa.eu.int/eur-lex

FAO. 2003. The Uruguay Round Agreement on Agriculture. Food and Agricultural Organization of the United Nations. www.fao.org/docrep/003/x7353e/x7353e03.htm

FAO. 2004. The Food and Agricultural Organization of the United Nations. www.fao.org

Federal Register. 1997-1998. Accessed at www.gpoaccess.gov/fr/index.html

International Trade Commission. 1997. Crawfish Tail Meat from China, Investigation no. 731-TA-752 (final), publication 3057, Washington, D.C., 1997. www.itc.gov

International Trade Commission. 2004. www.usitc.gov

Intrafish Media. 2004. Archives. Accessed at www .in trafish.com

King, J. and J. Anderson. 2003. Institutions and measures of importance to international trade in seafood. Pages 167–186 in: J. Anderson (ed.), *The International Seafood Trade.* Woodhead Publishing Limited, Cambridge, England.

Kurlansky, M. 1997. *Cod: A Biography of the Fish that Changed the World.* Walker and Co., New York, New York.

Lem, A. 2004. The WTO and the impact on aquaculture products and markets. FAO. Food and Agriculture Organization of the United Nations. www.aqua.cl/ conferencia

Ligeon, C., C.M. Jolly, and J.D. Jackson. 1996. Evaluation of the possible threat of NAFTA on U.S. catfish industry using a traditional import demand function. *Journal of Food Distribution Research* 27:33–41.

Lutz, G. and R. Romaire. 2003. Louisiana aquaculture as of 2003. Agricultural Experiment Station, Louisiana State University, Baton Rouge, Louisiana.

Lutz, C.G., P. Sambidi, and R. W. Harrison. 2003. Crawfish industry profile. Agricultural Marketing Resource Center, Iowa State University.

Quagrainie, K. and C.R. Engle. 2002. Analysis of catfish pricing and market dynamics: the role of imported catfish. *Journal of the World Aquaculture Society* 33(4):389–397.

Quagrainie and Engle. In review. Analysis of price expectations in the farm-raised catfish market. University of Arkansas at Pine Bluff.

Rhodes, V.J. 1993. The Agricultural Marketing System. 4th Edition. Gorsuch Scarisbrick, Publishers, Scottsdale, Arizona.

Roberts, K.J. 2000. Import and consumer impacts of U.S. antidumping tariffs: freshwater crawfish from China. IIFET, Corvallis, Oregon.

Seafood Business. 2000–2004. Trade news archives. www.seafoodbusiness.com/archives/00aug/ newstrade

Sloop, C.M. 2003. Chile Fishery Products Annual 2003. Foreign Agricultural Service GAIN Report (Global Agriculture Information Network, GAIN Report #C13022. //www.fas.usda.gov/gainfiles

Suranovic, S. 1997-2004. International trade theory and policy. The International Economics Study Center. http: //internationalecon.com/v1.0

United States Department of Agriculture. 1998. Census of aquaculture (1998). 1997 Census of Agriculture Volume 3, Special Studies, Part 3, United States Department of Agriculture, Washington, D.C.

United States Department of Commerce. 2004. Federal Register Notice. http://ia.ita.doc.gov

Vannuccini, S. 2003. Overview of fish production, utilization, consumption and trade. Fishery Information, Data and Statistics Unit, Food and Agriculture Organization of the United Nations. Accessed at: http://fao.org

Warren, H. 2004. Update on Vietnam antidumping case. *The Catfish Journal* 19(1):6.

11 Policies and Regulations Governing Aquaculture Marketing

Aquaculture has attracted an increasing number of regulations in recent years, particularly in the United States and the European Union. This chapter focuses on regulations that affect marketing and does not delve into regulations that affect primarily the production phase of aquaculture. Food safety concerns are summarized first, along with a description of Quality Assurance, Hazard Analysis Critical Control Points (HACCP), and the role of the U.S. Food and Drug Administration. The chapter then moves on to discuss traceability, competitive market conduct, permitting, and environmental issues.

FOOD SAFETY

Consumer concerns over food safety have increased greatly in recent decades. Recent consumer food scares in the UK alone have included salmonella, bovine spongiform encephalitis (BSE or "mad cow disease"), hormone implants, genetically modified organisms (GMOs), antibiotic residues, used cooking oil, sewage waste, polychlorinated biphenyl (PCBs), dioxin, hoof and mouth disease, chloramphenicol, nitrofurans, the mycotoxin mycophenolic acid (MPA), and nitrofen. In the United States, alar use in apples, dioxins in poultry products, BSE, *Escherichia coli* in hamburgers, and toxins in mussels have been some of the more prominent scares in recent years. Each of these resulted in dramatic decreases in sales of product that resulted in financial losses to companies producing and marketing the products affected.

Some food safety problems are caused by natural phenomena. Harmful algae blooms in natural waters can result in decreased supplies of shellfish as beds are closed and delays are incurred in reseeding the stock (Tester and Fowler 1990; Kahn and Rockel 1988; Conte 1984). Additional losses are incurred when demand for the products decreases after public announcements and public warnings appear (Brown 1969; Hamilton 1972; Sherrel et al. 1985). Public announcements that shellfish from some areas are toxic may cause consumers to fear and avoid related products (Swartz and Strand 1981). Wessells et al. (1995) distinguished between "acute" hazards that pose an immediate health hazard and those that result from a slow accumulation over a period of time. For acute hazards, it was shown that consumers based decisions on immediate, not past, news. However, in the case of a persistent accumulation of toxins, the demand impact of total cumulative information may be greater than in cases with acute effects. For example, direct losses from one farm were 8% of total average annual sales during an acute hazard event in Montreal. An additional 6.5% of total average annual sales were lost over the succeeding three months from decreased demand for the product.

Concerns over additives or residues in seafood products can prompt governments to ban their use or presence in both domestic and imported product. For example, chloramphenicol in shrimp imported into the EU from China resulted in its ban on imports from the entire country, not just from the one company where the problem was first identified. Shrimp exports from Indonesia shrank by 64%, 21% from Thailand, 39% from Malaysia, and 14% from Vietnam after imposition of the restrictions (Asia Pulse 2003). In reaction to the ban, shrimp producers in Southeast Asia and China threatened to boycott shrimp exports to the EU. They claimed that levels of chloramphenicol in meat, milk, and flour exported from the EU were of similar levels to those found in the imported shrimp. The EU removed its policy requiring shrimp from Indonesia to be free of chloramphenicol in September 2003 due to a determination that no country could comply fully with the conditions.

The EU is also expected to ban growth promoters by 2006. Supermarkets in the EU have established processes to ensure the safety of food products.

Some specific regulations have been enacted to ensure the safety of shellfish products. In the United States, for example, four states have regulations or permits for purging (depuration), transplant, and safe food handling of shellfish. In Connecticut, shellfish depuration and transplant licenses are required to operate a depuration plant and to sell processed shellfish. Transplant licenses are required to relay oysters from prohibited areas into private shellfish beds in approved areas. Florida requires a special activity license for depuration of oysters and clams in controlled purification facilities. The state of California also has shellfish safety regulations that require safe handling of shellfish; Virginia has food quality sanitation regulations that govern the inspection of food manufacturers, warehouses and retail food stores, food product sampling, and food product label review.

Industry-Initiated Programs

A number of aquaculture and food industry groups have initiated programs to ensure the quality and safety of their products. These programs are variously known as quality assurance programs or codes of practices.

Quality Assurance (QA) programs focus on medicines, additives, and safe pesticide use throughout the supply chain. Most QA programs include systems of internal and external audits that are used to inspect products to ensure safety and quality.

The Catfish Farmers of America and the U.S. Trout Farmer's Association have developed Catfish and Trout Quality Assurance programs (Brunson 1993). The Catfish Quality Assurance program was developed in 1993 as an educational program designed to maintain consumer confidence with farm-raised catfish. The program is intended for all catfish producers to ensure the safety and quality of farm-raised catfish. The Trout Quality Assurance program is organized somewhat differently and is based on the Hazard Analysis Critical Control Point (HACCP) concept. (More detail on HACCP processes is presented shortly).

The Interstate Shellfish Sanitation Conference (ISSC) is an organization of representatives from shellfish producing and consuming states, the U.S. Food and Drug Administration, the National Marine Fisheries Service, and industry representatives. It is a voluntary cooperative effort to establish uniform standards and procedures for handling shellfish. The emphasis of the ISSC is on sanitary controls on shellfish harvesting, processing, and distribution of shellfish. The states take the primary role for enforcement by monitoring waters for contamination and pollution, inspecting processing facilities, and prevention of poaching.

The U.S. Department of Commerce (USDOC) offers an optional fee-for-service Quality Assurance Inspection. USDOC inspectors will, upon request, inspect processing plants and facilities, and grade aquaculture products for quality assurance (50 CFR Part 260).

The Global Aquaculture Alliance (GAA) has established Codes of Practice for food safety with a primary emphasis on shrimp. The purpose of the Code is to maintain food safety by preventing contamination (www.gaalliance.org). Particular emphasis is placed on preventing pathogenic bacteria, chemical contaminants, and aquaculture drugs.

The Euro-Retailer Produce Working Group (EUREP) has developed a mechanism for developing production standards for commodities entering the retail trade through their outlets. The program was extended to aquaculture products in 2001 with a focus on quality, labeling, traceability, and food safety with third-party verification required. Moreover, some shellfish wholesalers have created trademarks, labels, and signs that purport to establish and certify the quality and safety of cultured products (Girard and Mariojouls 2003). France has official procedures for certifications. One such certification has been established on a local scale for mussels from the Mont St. Michel Bay region of France.

In France, the national shellfish farmers association CNC (Comité National de la Conchyliculture) established a certification list for "bouchot" mussels and has applied to the Commission Nationale des Labels et des Certifications to create a national CCP (Certification Conformité Produit) (Girard and Mariojouls 2003). Other regional professional associations are planning some type of certification applications.

One of the most recognized quality certifications programs is the Label Rouge certification program in France. The Label Rouge was created in 1965 by CERQUA (Centre de Développement des Certifications des Qualities Agricoles et Alimentaire) (Label Rouge 2004). To be approved for the French Label Rouge, the product must be demonstrated to be of superior quality as determined by appropriate taste tests.

The International Organization for Standardization (ISO) is a network of national standards institutes from 148 countries that work in partnership with international organizations, governments, industry, and business and consumer representatives (www.iso.org). The ISO 9000 series is an accreditation of the food industry in which the company sets the standards.

Regulation of Food Safety

In the United States, state health departments develop guidelines related to materials and conditions of buildings, equipment, and temperatures in processing and transportation of processed products. Local county sanitarians enforce these guidelines and have jurisdiction over sanitary conditions in processing plants. As consumer awareness and concern over food safety have grown, additional regulations by national authorities have been put in place.

The Role of the Food and Drug Administration of the United States

The U.S. Food and Drug Administration (FDA) was created from the 1906 Food and Drugs Act (www.fda.gov). It regulates the production and marketing of most food products, including fish. It is responsible for protecting the public health by assuring the safety, efficacy, and security of human and veterinary drugs, biological products, medical devices, the nations' food supply, cosmetics, and products that emit radiation. It is also responsible for advancing the public health by helping to speed innovations that make medicines and foods more effective, safe, and affordable. The FDA provides accurate, science-based information to the public as needed to issue medicines and foods to improve their health. FDA has developed regulations that deal with food production and marketing, food name and ingredients, food quality, manufacturing practices, packaging, and labeling.

Moreover, FDA specifies product labeling requirements, including the content of the product label information, the label's layout, and its size. The fundamental requirement of labeling is that the information be displayed in a prominent and visible manner.

HACCP

Hazard Analysis Critical Control Point (HACCP) programs were developed in the EU in 1996 and in the United States in 1997. In the United States, processing plants are required by the FDA to have a HACCP plan in place. The plan must identify areas with potential for product contamination or safety problems. The U.S. seafood HACCP rule covers all processors and importers, but fishing vessels, common carriers, and retailers are not required to have HACCP plans. For FDA purposes, processors are defined as seafood-related entities classified as establishments in the FDA inventory and foreign processors that export to the United States.

The HACCP rule requires every processor to conduct a hazard analysis to determine whether food safety hazards might occur. If it is deemed that no food safety hazards are likely, the processor does not need a HACCP plan, but the burden of proof is on the processor.

If the hazard analysis reveals a need, the processor must have a written HACCP plan that is specific to the plant's location and the types of products prepared. Food safety hazards that are reasonably likely to occur may include: toxins, microbes, chemicals, pesticides, drug residues, physical hazards, or decomposition. Critical control points can occur both inside and outside the processing plant and must be identified. Critical limits, or safe operating parameters, must be defined for each critical control point, monitoring procedures established, and corrective action plans developed. Verification procedures must be put in place and carried out at least annually that ensure that the HACCP plan is up-to-date and that ongoing implementation is adequate. Verification procedures may include reviewing consumer complaints, calibrating monitoring devices, and end-product testing.

A record-keeping system must be developed to document monitoring, corrective actions, and verification procedures. Records must state the name and location of the processor and the date and signature of the person making the record. Plans, HACCP records, and sanitation records must be available to FDA inspectors for review and copying. Plans and records in the possession of FDA are not available for public disclosure due to the Freedom of Information Act. Some of the HACCP functions (plan development, plan reassessment and modification, and review of HACCP records) must be performed by an individual who has been trained in HACCP through either course materials or job experience equivalents.

Importers must verify that their overseas suppliers follow HACCP rules by obtaining product from the country with which the United States has an HACCP-based agreement regarding inspection programs, developing product specifications for safety, and taking steps that might include: (1) obtaining the processor's HACCP and sanitation records; (2) third-party certification; (3) sending inspectors overseas to ensure that product meets requirements; or (4) end-product testing.

Molluscan shellfish have special requirements within the FDA HACCP rule. Shellfish must be harvested from waters approved by a "shellfish control authority." Shellfish must be purchased from harvesters in compliance with local licensing requirements, or they can be tagged.

The United Nations (UN)

The Codex Alimentarius Commission (CAC) of the U.N. Food and Agriculture Organization (FAO) and the World Health Organization (WHO) has been responsible for implementing the Joint FAO/WHO Food Standards Programme. The Codex Alimentarius is divided into two types of committees: (1) nine general-subject-matter committees that deal with general principles, hygiene, veterinary drugs, pesticides, food additives, labeling, methods of analysis, nutrition, and import/export inspection and certification systems; and (2) commodity committees that deal with a specific type of food class or group.

The World Trade Organization (WTO)

The World Trade Organization (WTO) agreement, in the Final Act of the Uruguay Round, developed an agreement on the Application of Sanitary and Phytosanitary Measures (SPS). The SPS agreement confirms the right of WTO member countries to apply measures necessary to protect human, animal, and plant life and health.

GRADES AND STANDARDS

The National Marine Fisheries Service (NMFS) administers grade and quality standards for fish. NMFS also conducts inspection and certification services. These are voluntary and are funded by fees charged to industry. For example, NMFS establishes minimum-flesh-content requirements for breaded and battered products.

TRACEABILITY: COUNTRY OF ORIGIN LABELING

Seafood products sold at retail in the United States after September 30, 2004, need to be labeled as to their origin and whether they are wild or farm raised to comply with the Country of Origin Labeling (COOL) requirements (Robinson 2004). The COOL legislation is mandated in the 2002 Farm Bill. However, beef, lamb, produce and peanuts were granted a two-year delay (until September 30, 2006) in the 2004 Appropriations Act in implementing COOL requirements. The COOL requirements are opposed by food wholesalers and retail organizations. However, COOL gives suppliers an opportunity to include product information for retailers on the backs of point-of-sale (POS) tags.

The interim final rule for USDA Country of Origin Labeling (COOL) was posted on the USDA Web site in 2004 (http://www.arms.usda.gov/COOL/ls0304ifr.pdf). Under the rule, retail products of fish and shellfish must be labeled as to their country of origin and whether they are wild or farm raised. Intermediate products (those that are used as an ingredient in other processed foods) are excluded from the mandatory country of origin labeling. Also, items that would be covered by the rule are excluded if they have undergone a change such as cooking, curing, or smoking or if they have been combined with other commodities such as with a breading or a tomato sauce. Restaurants, cafeterias, and other food service establishments are exempt from the mandatory COOL requirements. The COOL requirements outlined in the interim final rule include maintaining records in a central location for one year and store-level records as to when the product is on hand in the store. In order to have the United States country of origin label, farm-raised product must be derived exclusively from fish or shellfish hatched, raised, harvested, and processed in the United States and not have undergone a substantial transformation outside the United States.

The interim final rule takes effect six months after its publication (October, 2004), or in April of 2005. The delay was designed to allow unlabeled product to clear market channels and to give compa-

Marine Stewardship Council

The Marine Stewardship Council (MSC) is a nonprofit organization that offers a certification program that recognizes well-managed fisheries. The certification program is designed to create and promote markets for products bearing the MSC label. The MSC labeling program is based on the Code of Conduct for Responsible Fisheries of the United Nation's Food and Agriculture Organization (FAO 2004). The MSC program was developed in consultation with stakeholders. The program operates through certification first of the fishery as being well managed and sustainable, followed by issuance of a Chain of Custody Certification that guarantees conformance with standards throughout the supply chain.

nies time to make the adjustments needed to meet the requirements.

In the UK, supermarkets restored the confidence of consumers with an emphasis on traceability and assurance schemes. Beef consumption in the UK is now higher than it was in 1996, and an emphasis on traceability is credited with the increased consumption of beef.

COMPETITIVE MARKET CONDUCT (ANTITRUST LEGISLATION)

The Sherman Antitrust Act was passed in 1890 in the United States to prevent the concentration of economic power in large corporations and in combinations of business concerns. Antitrust legislation is primarily regulated by the Antitrust Division of the Department of Justice and the Federal Trade Commission. The Antitrust Act prohibits businesses that are engaged in interstate commerce from contracting, combining, or conspiring to restrain trade, or attempting to monopolize the market in a particular area of business. Violations of this Act include making contracts that unreasonably restrain trade, price fixing, group boycotts, allocating markets, and attempting to form and maintain a monopoly in an industry to injure competition.

Agriculture cooperatives are exempted under the Capper-Volstead Act of 1922. (Appendix 7.A in Chapter 7 provides additional detail on the Capper-Volstead Act). Most farmers are price takers in the economy due to the relatively large number of farmers that produce a similar product. The Capper-Volstead Act was passed to enable farmers to function cooperatively to sell their products. It was an attempt to provide farmers with additional market power in relation to the agribusinesses that frequently purchase agricultural products. Associations of farmers must be operated for the mutual benefit of their members, who must be producers of agricultural products. Cooperatives must use greater volumes of members' as opposed to nonmembers' products.

BUSINESS PERMITS, BONDING, AND REGULATIONS

Business permits, bonding, and regulations vary from country to country and within countries. Permits or licenses can be required for possession, processing, or depuration as well as other activities. Licenses may be required, fees charged, or taxes levied depending upon the specific regulation.

Marine Aquarium Council (MAC)

The Marine Aquarium Council (MAC) is a non-profit organization of the aquarium industry and includes members who maintain aquaria with pet fish as well as the collectors, wholesalers, and retailers who supply ornamental fish to this trade. Its mission is to conserve coral reefs and other marine ecosystems. MAC has developed standards, and it certifies the businesses that meet those standards. It functions by certifying independent certification companies that assess and evaluate industry participants. The key requirements for certification are the following: (1) verification that the collection area is managed as a healthy and sustainable ecosystem; (2) collection practices are consistently maintaining the health of both the ecosystem and the organisms harvested; and (3) handling practices during transport and sales outlets ensure the organisms' health and have proper documentation.

Thirty states[1] have regulations and permits involving the possession of animals for aquaculture-related activities. The definition of "possession" can vary from state to state but can include stocking, propagating, cultivating, transporting, transferring, harvesting, taking, trapping, collecting, selling, trading, storing, and purchasing. Other regulations may require additional permits or licenses to conduct aquaculture activities. Forty states and territories[2] in the United States have licensing and permitting regulations that cover all aquaculture-related activities. These include registration of aquaculture operations, education and research institutional needs, fee fishing, boat use, fish and bait dealer licenses, and marketing permits. Mississippi requires that all tilapia products offered for direct sale for human consumption have the product name specifically labeled in the manner described by the state's regulations. In Washington, regulations cover the identification requirements for products cultivated by aquatic farmers. Washington also has shellfish certification regulations, which cover shellfish sanitation practices, including a certificate of compliance, certificates of approval for shellfish growing areas, and certificates for culling, shucking, and packing facilities. In Wyoming, food safety regulations cover good manufacturing practice labeling.

Fifteen states[3] in the United States require permits for processing. These include licenses for purchasing, packing, repacking, shipping, reshipping, shucking, culling, and selling aquaculture-related products. Some states in the United States have passed regulations that require catfish processing plants to pay farmers for fish delivered to the plant within a specified period of time. Other state regulations require vendors to specify whether the product is farm raised, wild caught, or imported.

Taxes may be levied for a variety of business activities. Alabama and Arkansas in the United States both require a city privilege tax for businesses inside city limits. Some cities even have specific permits for fish markets, which would otherwise be covered by a general permit. Arkansas also requires a sales and use tax permit.

ENVIRONMENTAL ISSUES

There are several federal laws in the United States that grew out of concerns for the environment but that affect the transport and sale of live fish. These include:

- Endangered Species Act of 1973. This statute deals with any activity that might affect endangered or threatened species or their habitat.
- Lacey Act Amendments of 1981. Under this law, it is unlawful to import, export, sell, acquire, or purchase fish, wildlife, or plants taken, possessed, transported, or sold (1) in violation of U.S. or Indian law or (2) in interstate or foreign commerce involving any fish, wildlife, or plants taken, possessed, or sold in violation of state or foreign law. The Lacey Act is enforced through both civil and criminal penalties depending on the knowledge of the defendant, the type of violation, and the value of the fish involved. The Lacey Act has been invoked in situations involving shipments of fish through or into states that prohibit their entry. Although the Lacey Act was developed to protect wildlife, it is applied to farm-raised fish that are shipped across state lines.
- Migratory Bird Treaty Act. The Migratory Bird Treaty Act regulates the use of lethal control methods on migratory birds, including those that cause aquaculture crop losses. USFWS issues permits for the control of these migratory birds.

Transportation and Marketing of Nonnative Species

More than 20 states and territories[4] have regulations or permits that deal with importing or possessing nonnative species. Types of permits and regulations dealing with nonnative species include stocking licenses, general importation permits, and restrictions on possession, sale, importation, transportation, and release. Some states have special importation permits regarding specific species of aquatic animals such as grass carp (or white amur), crawfish, piranha, and rudd.

The use of bighead (*Hypophthalmichthys nobilis*), grass carp (*Ctenopharyngodon idellus*) and black carp (*Mylopharyngodon piceus*) as co-culture species in catfish ponds is regulated by some states by the respective natural resource agency under exotic species regulations. Some states prohibit exotic species whereas other states have developed "clean"

Food Standards in the European Union

www.eurunion.org/legislat/home.htm

The European Commission's Health and Consumer Protection Directorate General developed a White Paper on Food Safety in 2000 (European Union 2004). The White Paper on Food Safety contained four major initiatives: (1) creation of the European Food Safety Authority (EFSA); (2) food safety legislation; (3) a framework for monitoring the food supply chain in the European Union; and (4) food labeling rules. The European Food Safety Authority (EFSA) was established in 2002 and provides independent scientific advice on food safety. It develops and publishes opinions based on risk assessments of issues pertaining to food safety and works closely with national authorities (European Food Safety Authority 2004). The risk assessments are prepared by scientific panels convened in the following areas: food additives, substances used in animal feeds, plant health and protection, genetically modified organisms (GMOs), dietetic products, biological hazards, contaminants in the food chain, and animal welfare. Food safety legislation in the EU addresses animal feeds, animal welfare, contaminants and residues, food additives, food supplements, organic products, and packaging. The EU's Food and Veterinary Office in Dublin is charged with overseeing and monitoring food safety throughout the supply chain. Food labeling laws in the EU require the following to be included on labels: (1) name; (2) list of ingredients; (3) quantity or categories of ingredients as percentage; (4) the net quantity; and (5) date of minimum durability. There are additional labeling requirements for organic products and genetically modified organisms (GMOs).

lists of specific species that are allowed with and without permits. Other states will allow sterile triploid individuals of these species but prohibit the use of fertile diploid individuals.

Organic Standards

Organic markets for all types of products are growing rapidly in the United States and in Europe. Nearly $23 billion of organic products were sold in 2002 (Organic Monitor 2002). In the United States, organic food sales were $8 billion in 2000 (U.S. Department of Agriculture Economic Research Service 2002).

International standards for organic products have been proposed by the International Federation of Organic Agriculture Movement. However, these had not been adopted at the time this book was printed. France has adopted organic aquaculture standards that have been applied to rainbow trout for domestic sales and export. The Soil Association Certification, Ltd. and Naturland are actively certifying fish and shellfish farms.

In the United States, a National Organic Aquaculture Work Group has been formed to work toward developing national standards (Brister 2004a; 2004b). The National Organic Standards Board (NOSB) formed an aquaculture advisory group in 2000. In addition, the Independent Organic Inspectors Association has invited speakers on aquaculture to its 2002 and 2003 training courses. The growing interest in and active organization of interested individuals are expected to lead to the development of organic standards for aquaculture products in the United States.

Organic farming has developed into one of the fastest-growing segments of agriculture in the European Union (European Union 2004; European Food Safety Authority 2004). The European Commission introduced an organic logo in 2000. Producers use the logo voluntarily but must pass inspections to ensure that: (1) at least 95% of the product's ingredients were produced organically; (2) product complies with official inspections; (3) product is delivered from the producer in a sealed package; and (4) product bears the producer's name and inspection code.

Green Labeling and Standards

Two different groups have issued guides as to the "sustainability" of various types of seafood. The Monterey Bay Aquarium, as part of its Seafood Watch program, features 60 of the most popular seafood species in the United States on its Web site (www.montereybayaquarium.org). The site includes reports on each species and a pocket guide as to sustainability of each species. Sustainability is evaluated based on any bycatch, habitat damage, overfishing, pollution, or other factors. U.S. farm-raised catfish is recommended due to the diet fed and the farmer's control over water quality in the ponds where raised. Species included on the list of "best" choices are: U.S. farm-raised catfish, farmed caviar, stone crab, wild-caught Alaska salmon, tilapia and Pacific halibut. Species classified as the "worst" choices, based on the aquarium's definition of sustainability, are Atlantic and Icelandic cod, Chilean sea bass, orange roughy, swordfish, imported shrimp, red snapper, and shark. However, although wild-caught salmon is recommended, farm-raised salmon is not. The main objections to farm-raised salmon are criticisms concerning the use of fishmeal in the diet and the use of Atlantic salmon in net pens on the West coast. The Blue Ocean Institute also publishes a "Guide to Ocean Friendly Seafood" (Blue Ocean Institute 2004). Mussels, clams, oysters, Alaskan salmon, bay scallops, striped bass, mahi-mahi, and tuna (yellowfin, bigeye, and albacore) are listed in the top category. American lobster, black sea bass, stone crabs, Pacific sole, tilapia, summer flounder, farmed catfish, king crab, Pacific cod, Pacific halibut, and shortfin squid are listed in the next highest category.

The Marine Aquarium Council (MAC) has developed a certification program for ornamental fish. The MAC Certification includes certification of industry operators throughout the supply chain, including collectors, exporters, and importers (Marine Aquarium Council 2004). MAC product certification requires that marine ornamentals be harvested from a certified collection area and sold to MAC-certified buyers at the next level of the marketing chain. Key emphases are on ensuring health of the ecosystem in the collection area, and that handling procedures ensure the health of the fish being sold. In response to growing concerns over the capture fishery for marine ornamental fish, standards were developed for certification for supplying marine ornamental fish for the aquarium trade.

AQUACULTURE MARKET SYNOPSIS: MUSSELS

Mussels have been captured, eaten, and sold from the wild for centuries. The most commonly captured mussel has been the blue mussel (*Mytilus edulus*) (Fig. 11.1) with 59% of world production in 2002.

The Mediterranean mussel (*Mytilus galloprovincialis*) had the next highest wild catch, 20%, and was followed by that of the green mussel (*Perna viridis*), with 9%.

Mussels have been raised for many years in several different parts of the world (Avault 1996). In Europe, France, Spain, the Netherlands, and Sweden have long histories of mussel culture, as do the Philippines and Thailand in Asia. In fact, shellfish have been farmed for a longer time in France and in the European Union countries than any other type of aquaculture (Girard and Mariojouls 2003). Records of mussel production date back to 1235 in France (Bardach et al. 1972).

Mussel production worldwide has demonstrated a generally increasing trend over the last 10 years (Fig. 11.2). Total mussel production reached a peak of 1.3 million MT in 1999. Although total production has not reached that level since 1999, it has been at similar levels over the succeeding years. However, there are divergent trends in production over time by individual species of mussels or in groups of mussels produced (Fig. 11.3). Production of unspecified species of sea mussels, although exhibiting a high degree of fluctuation, has generally been increasing at higher rates than production of the other, more traditional species culture. In fact, the data from the FAO database show that, beginning in 1998, production of the category of unspecified species of sea mussels has exceeded that of blue mussels in every succeeding year. This may be due more to the decline in worldwide production of blue mussels. Production of green mussels and the Mediterranean mussel have generally increased over the last decade, but have declined slightly from 2001-2002.

The blue mussel is the most commonly cultured mussel in Europe. Spain is the leading producer,

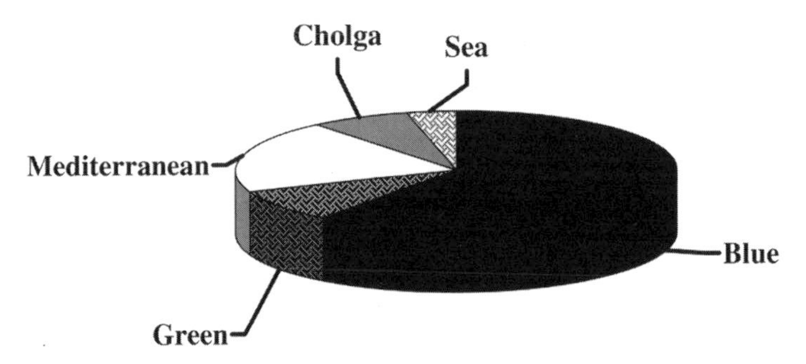

Figure 11.1. Major species of wild-caught mussels, 2002. (Source: FAO 2004.)

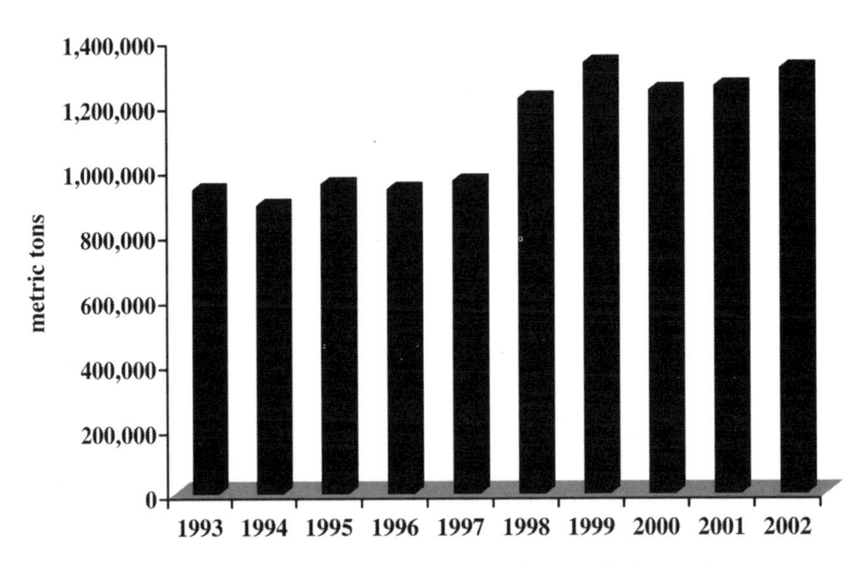

Figure 11.2. Volume of mussels cultured worldwide over time. (Source: FAO 2004.)

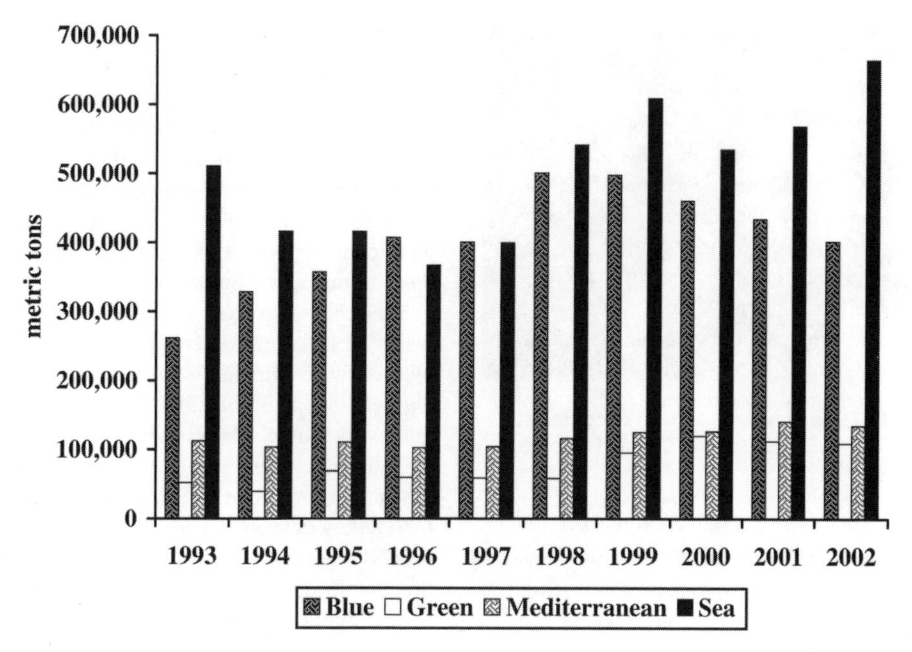

Figure 11.3. Quantity of mussels cultured over time. (Source: FAO 2004.)

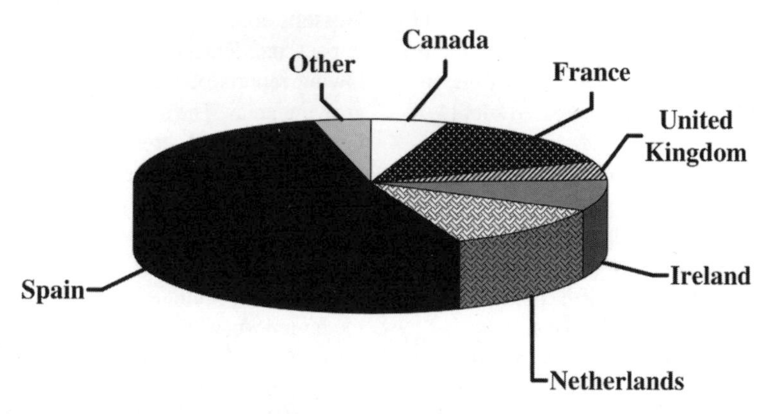

Figure 11.4. Major countries supplying cultured blue mussels. (Source: FAO 2004.)

producing 50% of the blue mussels cultured in 2002 (Fig. 11.4). France is the next largest blue mussel producer, with 14% of total production, and is followed by the Netherlands (11%), Ireland (8%), and Canada (5%). Marketable size of mussels is about 8 cm (3 inches).

Trade in mussels has increased within the European market over the last decade (Girard and Mariojouls 2003). The leading importing nations in Europe are France, Belgium, and Italy, with the Netherlands, Denmark, and Spain being the main exporters. Mussels are traded primarily as a fresh product, 80% of the total volume traded.

The majority of consumption of shellfish in Europe is in France, Italy, and Spain, countries that are also major suppliers (Girard and Mariojouls 2003). Because mussels are primarily consumed as a fresh, whole product, the proximity of the main production areas to the major markets has greatly facilitated this exchange (Fig. 11.5).

In France, 60% of the supply of mussels is from domestic production (Girard and Mariojouls 2003). The market is segmented based on the culture methods (rope-cultured mussels, "bouchot" mussels that are cultured on fixed, wooden poles, and wild mussels) and by species (between the blue and

Figure 11.5. Green mussels on sale in open-air market in Thailand. Photo by Dr. Carole R. Engle.

Mediterranean mussels). French "bouchot" mussels are considered a premium product with the highest market price. Mediterranean mussels grown in France and imported from Spain are intermediate-priced products. Mussels imported from the Netherlands ("Dutch" mussels) are the lowest-priced mussel product in France. Dutch mussels are sold primarily in supermarkets for lower prices (Paquotte 1998). However, the price of Dutch mussels increased in 2001, likely due to an increased supply of washed, debyssed, and ready-to-cook mussel products. Generally, wild-caught mussels are the lowest-priced products, with the exception of those harvested from the Basin of Marennes-Oléron (Girard and Mariojouls 2003). This region of France created a regional trademark in 1974 that has successfully resulted in higher prices for its products.

Imports of mussels into France occur mainly during the period of February to April (Paquotte 1996). This is the season of the year when French production is low, and the supply shifts to imports from the United Kingdom, Ireland, and the Netherlands.

In France, mussels are distributed primarily by large retailers (Girard and Mariojouls 2003). Most mollusks, including mussels, are consumed at home in France. Mussels are prepared as a cooked appetizer or as a main dish. However, away-from-home sales are growing. "Mussels and chip" dishes are gaining popularity in many restaurants.

The first companies to market value-added, convenience packs of mussels were Dutch companies (Girard and Mariojouls 2003). These companies developed ready-to-cook family packs of washed mussels in package sizes of 1–2 kg each. Dutch companies have continued to develop new products, and other companies, notably in Ireland and France, have followed suit. Fresh cooked dishes, precooked, vacuum-packed mussels, and intermediate products have been developed in recent years. Modified atmospheric and vacuum-packaging technologies provide opportunities for adding further value to mussel products to preserve freshness, safety, and quality of products. The new packaging technologies provide additional opportunities to add consumer convenience to mussel products.

Mussels do not close their shells when out of the water as do some other types of shellfish. Thus, long-distance shipping is more difficult with whole mussels than it is for other shellfish. Nevertheless, interest in mussels has grown in the restaurant trade as away-from-home sales of mussels has grown.

Another challenge for the mussel industry is the food safety concern related to what may have been filtered from the water by shellfish such as mussels. Shellfish beds may be closed due to contamination of the public waters where the beds are located. Contamination may result from a variety of sources that might include pathogens, harmful compounds released into the waters, or harmful algal blooms. Wessells et al. (1995) chronicles a case study of the impact on demand for mussels in Montreal following reports of harmful algal blooms in mussel-growing areas. The study documented the economic losses during and after domoic acid contamination of Prince Edward Island mussels. The effect of decreased demand on sales of mussels was calculated. In this case, losses consisted of the direct losses during a four-week ban on all mussel sales. However, loss of sales continued after the ban was lifted as media reports of the contamination event continued in the press. Those farms located outside the contamination area that had clear labels of product origin and location of the farms experienced fewer losses than farms with unlabeled product.

SUMMARY

As aquaculture industries have grown and developed, the number and type of regulations that affect the marketing of aquaculture products has grown over time. Those related to food safety have been the most comprehensive, but issues of traceability and the transport and sale of nonnative species have attracted increased regulatory attention. Several industry segments have developed industry-enforced quality assurance programs and codes of practice. National and local regulatory agencies have created a variety of permitting, licensing, and bonding requirements for all phases of the aquaculture marketing chain.

STUDY AND DISCUSSION QUESTIONS

1. What are the major areas of aquaculture marketing that are regulated?
2. What are Quality Assurance Programs? Who initiates them and what is their purpose?
3. What is the major regulatory agency in the U.S. for food and public health concerns?
4. What does HACCP stand for and what are the major components of an HACCP plan?
5. What international agencies are involved with aquaculture marketing standards or regulations?

NOTES

1. Alaska, Alabama, Arizona, California, Connecticut, Delaware, Florida, Georgia, Iowa, Idaho, Louisiana, Massachusetts, Michigan, Minnesota, Mississippi, Montana, Nebraska, New Hampshire, New Jersey, Nevada, New York, Ohio, Rhode Island, South Carolina, South Dakota, Tennessee, Texas, Virginia, Vermont, and Wisconsin.
2. Alaska, Alabama, Arkansas, Arizona, California, Colorado, Connecticut, Florida, Georgia, Guam, Iowa, Idaho, Illinois, Indiana, Louisiana, Massachusetts, Maryland, Michigan, Minnesota, Mississippi, North Carolina, North Dakota, Nebraska, New Hampshire, Nevada, New York, Ohio, Oklahoma, Pennsylvania, Rhode Island, South Carolina, South Dakota, Tennessee, Texas, Virginia, Vermont, Washington, Wisconsin, West Virginia, and Wyoming.
3. Arkansas, Arizona, California, Connecticut, Florida, Georgia, Michigan, Minnesota, New Jersey, New York, Oklahoma, Pennsylvania, South Carolina, Texas, and West Virginia.
4. Alabama, Arizona, California, Colorado, Connecticut, Florida, Guam, Iowa, Illinois, Indiana, Louisiana, Michigan, Minnesota, Mississippi, Nebraska, New Hampshire, Ohio, South Carolina, Tennessee, Texas, Virginia, and Wisconsin.

REFERENCES

Asia Pulse Pte. Limited. 2003. European Union relaxes stance on Indonesian shrimp. Asia Pulse Pte. Limited.

Avault, J. 1996. *Fundamentals of Aquaculture: A Step-by-Step Guide to Commercial Aquaculture*. AVA Publishing Company Inc., Baton Rouge, Louisiana.

Bardach, J.E., J. H. Ryther, and W.O. McLarney. 1972. *Aquaculture*. Wiley-Interscience, a Division of John Wiley & Sons, Inc. New York, New York.

Blue Ocean Institute. 2004. Guide to ocean friendly seafood. The Blue Ocean Institute. Accessed at: http://www.blueoceaninstitute.org.

Brister, D.J. 2004a. The importance of organic aquaculture. *Ecology and Farming* 35:16–18.

Brister, D.J. 2004b. Two organic aquaculture groups convene meetings. *The Organic Standard* 36:7–9.

Brown, J.D. 1969. Effect of a health hazard scare on consumer demand. *American Journal of Agricultural Economics* 51:676–678.

Brunson, M. 1993. Catfish quality assurance. Catfish Farmers of America and National Fisheries Institute, Publication 1873, Mississippi State University, Mississippi, U.S.

Conte, F.S. 1984. Economic impact of paralytic shellfish poison on the oyster industry in the Pacific United States. *Aquaculture* 39:331–343.

European Food Safety Authority. 2004. Accessed at www.efsa.eu.int

European Union. 2004. White paper on food safety. Accessed at www.eurunion.org

Girard, S. and C. Mariojouls. 2003. French consumption of oysters and mussels analyzed within the European market. *Aquaculture Economics and Management* 7(5/6):319–333.

Hamilton, J.L. 1972. The demand for cigarettes: advertising, the health scare, and the cigarette advertising ban. *Review of Economics and Statistics* 54:401–411.

Kahn, J.R. and M. Rockel. 1988. Measuring the economic effects of brown tides. *Journal of Shellfish Research* 7(4):677–682.

Label Rouge. 2004. Qui somme-nous. Label Rouge. Accessed at: //www.label-rouge.org

Marine Aquarium Council. 2004. MAC certification. Marine Aquarium Council. Accessed at: http://www.aquariumcouncil.org

Martin, R.E. and G.J. Flick. 1990. *The Seafood Industry*. Van Nostrand Reinhold, New York.

Organic Monitor. 2002. The global market for organic food and drink. Accessed at http://www.organicmonitor.com/700140.htm)

Paquotte, P. 1996. The French mussel market: evolution of supply and demand. Actes des VIIIémes Rencontres interrégionales de l'AGLIA, Talmont St. Hilaire (Vendeé), septembre 1995, Publication ADA 49:43-50.

Paquotte, P. 1998. Le marché de la moule en France: évolution de l'offre et de la demande. VIIIémes Rencontres interrégionales de l'AGLIA, 14-15 septembre 1995. *Globefish* 55:65–80.

Robinson, F. 2004. The COOL conundrum. *Seafood Business* 23(8):22–24.

Sherrel, D., R.E. Reidenbach, E. Moore, J. Wagle, and T. Spratlin. 1985. Exploring consumer response to negative publicity. *Public Relations Review* 11:13–28.

Swartz, D.G. and I.E. Strand. 1981. Avoidance costs associated with imperfect information: the case of Kepone. *Land Economics* 57(2):139–150.

Tester, P. and P.K. Fowler. 1990. Effects of the toxic dinoflagellate *Ptychodiscus brevis* on the contamination, toxicity and depuration of *Crassostrea virginica* and *Mercenaria mercenaria*. In E. Graneli, D.M. Anderson, L. Edler, and B.G. Sundstrom (eds.), *Toxic Marine Phytoplankton*. Elsevier: New York, USA.

USDA-ERS. 2002. Recent growth patterns in the U.S. organic foods market. Agriculture Information Bulletin Number 777. Economic Research Service, United States Department of Agriculture, Washington, D.C.

Wessells, C.R., C. J. Miller, and P. M. Brooks. 1995. Toxic algae contamination and demand for shellfish: a case study of demand for mussels in Montreal. *Marine Resource Economics* 10:143–159.

12
Planning Marketing Strategies (Identifying Target Markets)

"Marketing is about putting distinctive capabilities into an acceptable form and presenting them to selected market segments" (Palfreman 1999). "What is really difficult about marketing in the fish industry is actually understanding what the market requires. The characteristics of fish products can be very subtle and it requires real knowledge and experience to be aware of these."

The most successful aquaculture businesses are those that are market oriented, have diverse markets, and are committed to their customers. Many farmers want to get into aquaculture production but have little interest in spending time on a market analysis. Those who are successful in this business are those who have spent time talking to potential customers before beginning to design their production operation.

This process holds true even for growers whose primary market is a processing plant. If the plant cannot move additional farm produce, the farm will need to identify additional or alternative market outlets, such as live sales to pay lakes, different processing plants with different markets, or perhaps diversifying into production of different species of fish. Decisions on species, harvest size, and volume should be based on the market analysis, plan, and strategy.

This chapter presents background information for each component necessary in the development of a marketing strategy and plan. A sample market plan is presented at the end of the chapter.

CURRENT MARKET SITUATION ANALYSIS

Market Research

The risk associated with any business decision can be reduced by obtaining comprehensive information on the primary factors involved. However, research can be complex and expensive, and should not be done if the cost of the study exceeds the value expected from any resulting business action. For example, a small catfish farm that would generate an annual net profit of $50,000 should not accept a consultant's proposal for a $250,000 study to research the size and structure of the catfish market. This chapter includes only a short summary of the role of market research in planning and implementing market strategies. Chapter 13 provides a more detailed description of marketing research methodologies.

Research will provide the most useful information when objectives are defined clearly. Questions for research can be developed more specifically when the company is well into the planning process and has compiled detailed information on overall market conditions and trends.

Secondary information (already published) is much cheaper than generating new information. Much can be gleaned from the Internet, government reports, the U.S. Extension Service, and university resources. Thorough compilation of secondary data is an essential first step for any size of company. Although it takes time to pull the information together, the overall cost is not high. For example, information on total supply of aquaculture products worldwide is available from the Food and Agricultural Organization (FAO) of the United Nations on its Web site (FAO 2004). Total quantities produced, and their value, can be obtained for individual species or by groups of species and by country, region, and ecosystem by year to determine the overall size of the market globally. These data can also be used to identify long-term trends in supplies of species or even countries that might be competitors. Information on trade in seafood species and products can also be obtained from FAO to identify trends for future competition for specific types of export markets or to identify potential sources of competition from increased imports. Similar information can be found within individual countries. In the United States, for example, the National Agricultural Statistics Service (NASS) of the United States Department of Agriculture (USDA) publishes statistics on acreage, number of farms, quantities produced, price paid to producers, and value of the major aquaculture species produced in the United States by species and by state (NASS 2004). Some limited amount of information on imports and exports of aquaculture products is also included. Information on the overall seafood market in the United States is available from the National Marine Fisheries Service (NMFS 2004) and through its hard-copy publications. The Annotated Bibliography and Webliography in the appendices of this book include a variety of sources of this type of information.

There are a number of useful sources of information on specific fish markets that may shed light on potential competitors and their marketing strategies. Some buyers and sellers post their requirements, offers, and advertisements on Web sites. The advertisements shed light on how competitors are positioning their products, what markets they are targeting, and what their overall marketing strategy might be. Trade magazines such as *Seafood Business* and *Seafood International* provide similar information through the paid advertisements by competing businesses. Seafood shows provide an excellent opportunity to see the array of products, product forms, pricing, and marketing strategies of competitors in both the overall seafood market as well as within specific species or product type categories. In the United States, the Boston Seafood Show is still the largest, oldest, and best attended, but the newer show in San Francisco provides more of a west coast flavor in terms of both the exhibit booths and the participants. More specialized shows, such as the Fancy Foods Shows that are held several times a year in various cities in the United States, provide insight into the higher-priced, value-added, gourmet-food category. In Europe, the Anuga (Cologne, Germany, in the fall) show, the Bremen Seafood Show (Bremen, Germany, in the spring), and the European Seafood Show (Brussels, Belgium, in the spring) are the major seafood shows targeting seafood markets in Europe. Shows that target the major Asian seafood markets include the Japan International Seafood Show, China Fisheries and Seafood Expo, Singapore Seafood Exposition, and Seafood Asia (Hong Kong), among others.

Secondary information sources should be thoroughly mined before one proceeds to expend funds on direct research. However, secondary data and information should be scrutinized carefully to avoid biases. Much information on the Internet is not peer reviewed nor does it undergo any type of quality control. Individual companies promote their specific products, and trade associations represent the interests of their membership on the Web and are not obligated to provide a balanced view. Adequate efforts need to be made to ensure that information obtained represents an accurate total view of the market and its trends.

After a company has investigated secondary sources thoroughly, a decision may be made to initiate formal market research. Research can be done on a variety of levels. The first level is through direct observations; this is a necessary step in developing any research plan. Direct observations will provide many potential insights into market opportunities and can be used to develop hypotheses for subsequent, formal testing. Successful marketing requires contact with people. The yellow pages of telephone books can be used to identify contacts in any proposed, specific market. Retail markets and sellers are excellent sources for current information on their specific sales. Conversations with these individuals can provide an overall view of pricing structures, competing products, and a sense of what is most important in that market.

Direct observations provide clues as to market conditions, but their usefulness is limited to that specific situation. Identification of relationships, trends, and quantification of relationships requires more formal scientific testing and research that becomes more expensive.

Focus groups can be a cost effective means of identifying product concepts, unmet customer needs, and market opportunities. However, focus groups should be conducted by an experienced facilitator, and participants should be selected to represent the target groups.

After decisions have been made on larger questions related to products and target markets, more formal research may be required. Market experiments and surveys may be useful after very specific research questions have been developed for which secondary data are not available. Chapter 13 provides more detail on methodologies related to formal market research.

Developing a retail outlet for fish requires much advanced planning. It is important to have reliable information about the number of people passing the shop or restaurant each day, as well as the proportion of people passing by who might want to buy fish. The amount of money that each potential buyer is likely to spend on fish or fish products must be estimated. Gross margins should be estimated from these projections. External factors such as the proximity of supermarkets, the availability of fish suppliers, and relationships with wholesalers must be evaluated. Prospective development of the area, such as road widening plans, freeway construction, and other possible changes in the locality should be investigated. The business plan should include an estimate of the value of the shop in the event that the business fails.

A successful retail business will pay attention to and follow some common sense guidelines. Employees must be courteous because no one wants to return to a store or restaurant where they have been treated rudely. Prompt service provided to customers is critical to ensuring repeat business. The more convenient and easy it is for a customer to purchase from a business, the more sales will be generated. To provide service and convenience, it is essential for a business to be flexible. Each individual is different, with different tastes and preferences; with a flexible system, it will be easier to meet the needs of every customer. Prices charged must be competitive with other businesses.

Market surveys that target supermarkets can provide guidance on trends and preferences to guide fish farmers and processing plants as to which types of products will have the greatest chance of success in different types of supermarkets. For example, Olowolayemo et al. (1992) found that stores that were members of a chain, had a specialized fish market section, and had sales over $100,000 were those that had a higher likelihood of selling catfish. The study indicated that substantial potential existed for catfish market expansion if obstacles such as a negative consumer image, supply problems, freshness, off-flavor, and competition from other seafood products could be overcome. Hanson et al. (1996) found that stores that had floor space greater than 40,000 square feet, had a high-income customer base, and were part of regional chains were likelier to have seafood counters. Stores with weekly sales of $40,000 to $99,000 were more likely to have a seafood counter than were grocery stores with sales of $39,000 or less.

Competition

Open-market economies prevail throughout the world. The main defining characteristic of open markets is that there is competition among companies and products that results in the availability of choices for consumers. Successful products are those most often selected by consumers, and successful companies are those that do the best job of satisfying the needs and wants of consumers by producing products with the most desired characteristics at prices that consumers are willing and able to pay. Thus, understanding the competition is a critical first step in developing an analysis of the current market situation. It is not enough to have identified market opportunities; these opportunities must be assessed in terms of the strength of the competition (Shaw 1986). The fundamental question that business owners or managers must answer is what their business can provide to customers that is better than anything currently offered by their competitors.

The analysis of the competitive situation should include definition of the size, goals, market share, product quality, and marketing strategies of potentially competing products and companies. The company must identify those areas in which it has a particular strength and can compete successfully within the current competitive situation.

Consumer Attitudes and Preferences

It is essential to understand the attitudes and preferences of consumers in designing market strategies. Development of new markets for existing products or finding a market for a new product often follows a pattern of: (1) developing awareness by consumers; (2) increasing availability of a new product; (3) changing attitudes toward the product; (4) changing preferences for products; and (5) developing new consumption patterns. Thorough study of market characteristics and trends during the planning process should reveal to what extent the product is known, how available it and similar or competing products are, and what the prevailing attitudes, preferences, and purchasing patterns are within the market segments under consideration.

A number of studies have been conducted in the United States to identify consumer attitudes and preferences related to different types of aquaculture products. All have been based on surveys conducted at different times and in different parts of the country, and they focus on attitudes toward different cultured species.

Engle et al. (1990) conducted national surveys at restaurant, grocery store, and consumer levels to identify consumer attitudes toward farm-raised catfish. Results of this survey showed that catfish was the most preferred type of finfish in the East South

Central, West South Central, and West North Central regions (Engle et al. 1990). Important attitudes toward catfish included: no fishy odor, few bones, and low cost (Engle 1998). Catfish availability, quality, taste, price, and preparation were key attributes to stress in promotion campaigns. This survey showed that consumer attitudes toward catfish appeared to have changed in nontraditional catfish consumption areas. Consumers perceived catfish as a nutritious, high-quality product that is easy to prepare. Gempesaw et al. (1995) found that attitudes toward taste and variety in the diet had the greatest influence on decisions to purchase fresh seafood products for consumption.

Using the SRAC survey data, Kinnucan et al. (1993) showed that preferences for fish were influenced to a large degree by source availability. Fish quality and flavor perceptions were important, whereas nutrition, ease of preparation, cost, and health had little impact on preferences. Ease of preparation was an important factor affecting preference for lobster, whereas nutrition and health considerations played a role in consumer preferences for shrimp and cod. Preferences in general were found invariant to income with two exceptions: (1) high-income consumers tended to prefer lobster, and (2) low-income consumers tended to prefer catfish. Professional, clerical, and blue-collar workers tended to favor shrimp and lobster. More educated consumers favored lobster, cod, and salmon, whereas the less educated preferred catfish.

Wessells et al. (1994), in a mail survey in the Northeast and Mid-Atlantic regions, found that taste, desire to add variety to their diets, and ease of preparation were the characteristics that had the greatest influence on consumer preferences. Consumers in the New England states preferred salmon, clams, and mussels, and coastal residents tended to prefer seafood more than those residing more than 100 miles from the coast.

The Wessells et al. (1994) study demonstrated the importance that consumers are placing on the safety of seafood. Although 97% of the respondents to the survey conducted in the northeastern and mid-Atlantic region did consume seafood, only 5% were completely confident that it contained nothing harmful to their health, while 45% were somewhat confident. Respondents considered pollution as more of a problem for wild-harvested than farm-raised fish. It was also important to know the date the fish were harvested. As many as 60% of the respondents were unfamiliar with aquacultured products, but approximately one-half of those surveyed considered farm-raised products to be of higher quality, harvested in clearer water, and handled better than wild-caught seafood products (Wessells and Anderson 1992). Consumers perceived shellfish to be more of a health risk than finfish.

Engle and Kouka (1995) evaluated potential consumer acceptance of canned bighead carp in Arkansas by measuring preferences toward four attributes: taste, texture, appearance, and aroma. Probabilities estimated with logit analysis showed that canned bighead competed more favorably with canned tuna than with canned salmon. Income, region, and gender significantly affected perceptions, and taste variables affected consumers' willingness to pay as much for canned bighead as for canned tuna. The socio-demographic variables affected willingness to pay indirectly through effects of taste on preference.

Preference for nutritional value caused consumers to purchase trout, whereas perceptions related to food safety concerns, odor, and appearance of the whole fish were negative influences for nonbuyers (Foltz et al. 1999). Half of the respondents preferred farm-raised to wild trout for food safety reasons. In general, higher-income consumers liked to purchase trout. Consumers having a higher likelihood of purchasing whole-dressed trout tended to be Asian or Hispanic, had had childhood experiences eating freshwater fish, and were from smaller communities. Individuals coming from an urban background, desiring nutritious foods that are easy to prepare, and possibly having a sensitivity to fish odor and appearance preferred fillets to whole-dressed trout.

The Florida Department of Agriculture and Consumer Services commissioned a study in 2001 to understand seafood preferences of the Hispanic population in the United States (Florida Department of Agriculture and Consumer Services 2001). Respondents indicated in general that they believed that seafood tasted good, was healthful and easy to prepare, but was expensive. Shrimp, lobster, and stone crab were highly preferred, but snapper, mahi-mahi, grouper, and tilapia were also well liked. Contrary to other population segments (Mintel International Group 2002), Hispanics preferred seafood plain and not precooked or premarinated.

A 2003 study conducted in Nicaragua, Central America, showed that consumer preferences for a particular form of preparation, in this case, ceviche, can be used to create a market opportunity for an aquaculture product (Neira et al. 2003). Older restaurants, whose clientele prefer variety and ceviche appetizers, were found to be more likely to begin to include tilapia on their menu.

Overall, the seafood marketing literature clearly shows that the three most important product characteristics are typically: taste, quality, and price. Fish has been promoted in recent years for its healthful characteristics, and the emphasis on good nutrition is increasing. Nevertheless, research continues to show that the overriding factor in consumer purchase decisions is the taste of the product. Quality is a complex characteristic that includes freshness of the product, cleanliness, brand identification, brand familiarity, and brand loyalty, as well as other characteristics. If quality standards can be maintained consistently, customers will purchase repeatedly, learn to recognize the brand (brand identification), become familiar with the brand (brand familiarity), and begin to insist on buying only that brand (brand loyalty).

Brand identification has not developed widely in seafood markets. However, as aquaculture companies and industries continue to grow, increasing supplies, brand development would be expected to begin to offer some market advantages through differentiating products and developing brand loyalty.

ANALYSIS OF BUSINESS STRENGTHS AND WEAKNESSES

Careful analysis of the relative strengths and weaknesses of the business should be an integral part of the marketing plan and strategy. These strengths and weaknesses derive from both external and internal factors that can constitute either opportunities or threats to the business.

External threats to seafood businesses can come from a number of sources. Economic downturns often constitute an external threat to seafood businesses because demand for seafood decreases during economic downturns, and decreasing demand causes prices to decline. Unforeseen external shocks to the economy can cause prices to decline. For example, the September 11, 2001, bombing of the World Trade Center in New York had dramatic effects on seafood sales because restaurant and live fish sales in New York City are dependent on tourism. When tourism falls, demand for aquaculture products sold in these markets also falls. Fluctuating exchange rates of currency are another example of an external threat to businesses. A strong currency will attract imports, whereas a weak currency will create profitable export opportunities.

Demand for fish is often related strongly to income (Palfreman 1999). Consumers with rising incomes often seek to buy more fish and seafood products. The price of substitute products like other, similar, types of fish species will also affect demand. Consumers will purchase more of a cheaper type of fish if it is viewed as a good substitute. Interest rate fluctuations can affect demand for fish and seafood products because of the effect of interest rate levels on decisions to invest in aquaculture businesses and infrastructure. Expectations of higher inflation rates may provide incentives to invest in physical assets such as land, rather than cash-related assets, that may affect the availability of capital for aquaculture investment. Technological changes (computerization, and control and monitoring), political and legal changes (proposals for additional regulations), social and cultural changes (awareness of low-fat characteristics), changes in food consumption habits (fewer set family meals, and more "grazing") are important social changes that are external to the business itself but will affect the demand and, hence, market price of the product.

Other external factors in the marketplace include the following: health concerns, convenience, and product handling by buyers. For example, after fish fillets are delivered to a supermarket, there is no control over how the supermarket treats those fillets. If the fillets are stacked up high, under a light bulb, the temperature in the middle of the stack may not be adequate to preserve fillet quality. In spite of the fact that high-quality fillets may have been delivered to the supermarket, poor handling by the buyer will result in a poor-quality product.

Another type of external opportunity or threat is that pertaining to competitors, customers, distribution channels, and suppliers (Palfreman 1999). For example, competitors may have secured cheaper supplies, or customers may want a different size of box, or there may be difficulties in obtaining the required supplies in the future; these are the types of issues that can represent either a marketing opportunity or a threat to the company.

Businesses should also evaluate critically their internal strengths and weaknesses. A small company with a higher cost of production will be better poised to develop higher-valued niche markets. A business with expertise to produce certain types of fish that are difficult to spawn may develop a market as a hatchery supplier, while another business with access to large amounts of land may concentrate on growout. Internal weaknesses may include assets that are out of date, such as ponds that are old, have not been renovated, and may have become shallow. Aging staff not able to provide the physical labor required, or too many people to operate profitably, are issues that can represent a weakness of the business. A business's strength may lie in detailed knowledge of markets, excellent engineering and maintenance skills, or skill in financial analysis.

The internal analysis must include careful consideration of the financial resources available for market research and any new investment or operating capital requirements. New directions may require reallocation of company resources, and the company must have a thorough understanding of what the implications will be.

DEVELOPING THE MARKETING STRATEGY

Strategy can be thought of as the game plan to achieve the marketing and financial objectives of the business (Palfreman 1999). A strategy may be a low-cost, low-investment strategy designed to get the most out of previous investments without incurring additional capital outlays before beginning to diversify. Alternatively, if the business sees an opportunity for efficient companies to prosper, it may choose to upgrade. A processor's strategy may be to seek to be the lowest-cost producer of a commodity such as fish fingers, or be flexible with short production runs of more differentiated products that attract higher prices.

Each business should have a marketing plan of action. The market strategy should address the four P's of the marketing mix: product, price, promotion, and place. These are factors that are controlled by the business. For example, the farmer decides what species of fish and what size to raise as the product to sell. Place refers to the geographic market, or where the farmer will sell the fish. The "place" decision involves deciding whether to sell on the farm, haul to a processing plant, or sell to other farms. The farmer decides what types of advertising to use to promote the product. The farmer also decides what price at which to sell his or her product. Whether the product is then sold depends on market conditions.

When to sell the product can also be important. A baitfish farmer who has loaned money from the bank to produce a crop, grows the crop, but now looks to sell the crop in the fall because a bank payment is due will be in serious financial difficulty. The main crop of baitfish is sold in the spring, not the fall.

Before the strategy can be developed, specific marketing and financial objectives must be clearly spelled out. Goals and objectives should be specified for the short, medium, and long term. Examples of market objectives might be to increase the minimum size of fish purchased by a processing plant to reduce processing costs or to enable it to compete in a different market segment. A business may set an objective of increasing market share or to penetrate a new market segment. Specific, measurable targets could then be specified, such as: (1) reduce the percent of fish less than 0.57 kg (1.25 lb) from 25% to 10% over the next two years; (2) increase market share from 20% to 30% over the next two years; or (3) generate sales in the new market area equal to 5% of total sales within the next two years.

Financial objectives must also be defined clearly. Examples of general financial objectives may be: (1) survive and avoid bankruptcy; (2) maximize return on investment; (3) increase cash flow; or (4) reduce the debt burden. These may be refined into the following, more specific, targets: (1) within 12 months, reduce overhead expenditures by 20%; (2) undertake capital investment only if it is capable of achieving a rate of return of 15% or above; (3) increase net cash flow from $100,000 per year to $120,000/yr by the end of three years; or (4) reduce the debt/equity ratio from 50 to 30% over the next five years.

After specific marketing and financial objectives have been specified, the strategy or game plan to achieve these objectives can be developed. The following sections discuss several important considerations and decisions to be made in developing the marketing strategy.

MARKET SEGMENTATION

Markets can be segmented along many lines. Geographic regions, occupations, incomes, ages, gender, and family size are all used to varying degrees by various businesses to segment markets. The basic concept of market segmentation is to identify different segments of a market and target somewhat differentiated products to different segments at different prices. Market segments can be based on customer location, age, gender, job, special interests, lifestyles, or certain events. A key criterion for segmenting the market is that the company must be able to measure the characteristics of specific buyers. An aquaculture marketing example of market segmentation might be that of a hybrid striped bass grower targeting supplying a live fish to ethnic grocery stores in a major urban area at one price, but targeting sales of whole fish on ice to upscale restaurants in the same city at a different price. Market segmentation is common in the U.S. marketplace.

However, with a few exceptions, the seafood industry still tends to rely upon a "mass" or "undifferentiated" marketing approach. Product differentiation will generate increased sales but may also increase production, inventory, and promotion costs. Production costs can increase because production of two or more products often requires new equipment, separate processing lines, and perhaps separate packaging lines. A factory producing one item will be more cost effective than a factory manufacturing a number of different items. Inventory costs might increase because different products may require different types of storage facilities that can maintain new products at different temperatures. Different products may require different distribution systems. Moreover, the higher the number of items marketed, the greater the investment required in safety stocks that are carried by companies to guarantee adequate supplies to customers. Differentiated marketing can involve a range of marketing programs to support the various products sold. New marketing programs will increase promotion costs. Segmented markets will require different promotional programs and messages that appeal to the different types of consumers in each segment. Each advertising program will have a separate cost that may result in overall higher costs.

Given the potential for increased costs as a company diversifies production, careful analysis is required to identify the most profitable market segments for the company and to target expenditures on new product production, inventory, and promotion toward those segments with the greatest overall potential for achieving the company's objectives. A segment must be of sufficient size, have potential for further growth, not be over-occupied by competition, and have an identified need, which the company can satisfy uniquely. If the targeted market segment does not possess these characteristics, then it may be best for the business to stay with an undifferentiated product. It may choose to concentrate sales of an undifferentiated product in a particular geographic region or to a particular market segment for which it has a specific strength. Or it may choose to differentiate its products to capture sales in more than one market segment.

Products and Product Lines

The identification and selection of products and product lines for the business is an essential component of a successful business and market strategy. Product lines are a series of closely related but somewhat differentiated products. For example, several catfish processing companies have a marinated fillet product line that may include lemon pepper, Cajun, or preparations with other seasonings and flavors. The marinated fillet product line is distinct from the nugget, steak, and whole-dressed product lines. Companies with single product lines may have lower costs of production due to production efficiencies but may also have higher market risk. Differentiated product lines and multiple product lines allow a company to spread the risk associated with changing market and economic conditions.

Shrimp, for example, can be processed into many, basic product forms, such as: (1) whole, shell-on, raw, frozen; (2) whole, shell-on, cooked, not frozen; (3) whole, shell-on, cooked, frozen; (4) headless, shell-on, raw, frozen; (5) headless, cooked, peeled, frozen; (6) headless, peeled, undeveined, raw, frozen; (7) headless, peeled, deveined, raw, frozen; or (8) headless, cooked, peeled, canned. Primary markets for these different product forms vary considerably, and the choice of product forms must be made after careful analysis. A company must establish a unique identity for its product using characteristics or attributes such as price, texture, name, availability, and quality.

The selection of products and product lines must be developed concurrently with the selection of target markets in the company's market plan. A product with a high cost of production will need to be of sufficient quality to charge a price sufficiently high to be profitable. Clearly, the target market for such a product would be one in which consumers not only value the particular attributes of that product but also have high enough income levels to be able to pay the price level required. There also need to be enough consumers in that segment to have the volume of sales required to provide an adequate return on any investment in product development that was incurred.

Product Life Cycle

The decision to develop a new product needs to be planned carefully. Products go through a product life cycle (Fig. 12.1). The strategy for a new product (or an existing product being introduced into a new geographic or demographic market) is to first penetrate the target market. This is known as the product introduction phase. The product introduction stage is characterized by low sales but high marketing expenditures. The company's objective during this stage

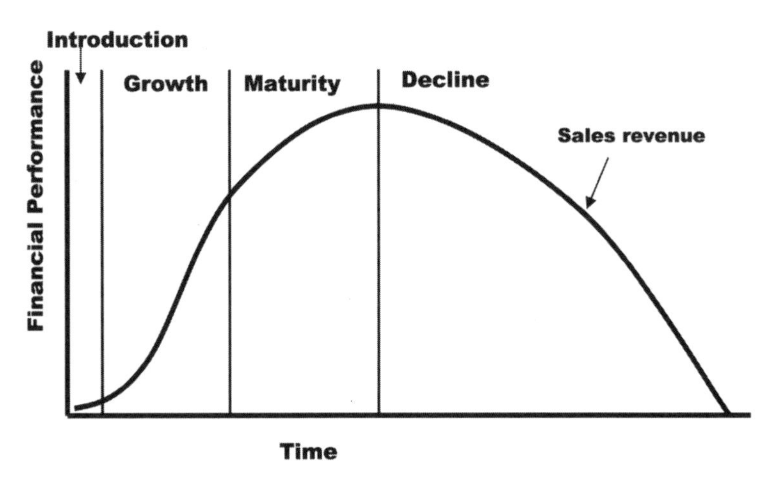

Figure 12.1. A theoretical diagram of a product life cycle indicating its various stages.

will be primarily to generate awareness of the product. Taste tests and sampling opportunities may be important strategies associated with this stage. The product may not generate profits during the introductory stage. The company should seek to generate awareness quickly and move the product into more profitable stages.

The successful product will move into the second, or growth, stage as sales increase. The growth stage is characterized by rapidly increasing sales. The company should begin to generate profit during this stage because although marketing expenditures are still high, sales are growing faster than increases in marketing expenditures. Key issues during the growth stage involve coordination of the supply chain to ensure timely deliveries and adequate quality control to guarantee quality throughout this expansion period. One key business objective during the growth stage is to saturate the market by increasing sales.

As the market approaches saturation, the product enters into the third, or maturity, stage. Sales continue to increase, but at a slower rate. The maturity stage often is characterized by increasing competition from other companies that introduce similar, competing products. At maturity, sales expansion can come only from competition, and resulting market expansion reduces profits (Chaston 1983). When a given market segment becomes saturated with that particular product, the business strategy often switches to identification of new markets with good potential for sales of that product. Thus, production costs change little, but additional promotion and distribution costs become necessary.

Sales begin to decrease in the final stage, referred to as the decline stage. It is critical to monitor and manage the stage of decline carefully. When all available markets are saturated for a particular product, then new products need to be developed. Periodic performance review will provide a basis for deciding when to eliminate a product line. Hidden costs associated with declining products must be considered. Products in the decline phase may take up too much management time, result in short production runs that increase setup time, have unpredictable sales volumes, and result in less effective advertising expenditures because fewer sales are generated for the same amount of advertising as before.

Product Positioning and Price-Quality Considerations

Businesses must make critical decisions related to positioning their product(s) in the market place. Consumers' willingness to purchase a product is related to how closely its price matches their perception of its quality. Consumers will pay very high prices for seafood that they view as being of the highest quality. This clearly holds true only for markets that include consumers with income levels that allow them to pay these prices. Conversely, they will refuse to pay high prices for a product they view as being of low quality. Moreover, price and quality need to be related for segmentation to be possible.

The aquaculture business will need to select a matching price-quality scenario for its product to be successful. To be feasible, the price will have to exceed production costs. The error committed by many aquaculture businesses has been to set prices based strictly on production costs. Businesses that do not consider the perceptions of its prospective customers related to the quality of the product and the relationship to its price are doomed from the beginning. Consumers will not pay a high price for a product perceived as being of low quality. A low price for a product promoted as high quality will cause consumers to be suspicious.

However, positioning a product as the highest quality may not always be a successful strategy. The quantity demanded for the highest level of quality might not be sufficiently high for the company to meet its revenue requirements. High-quality products frequently require additional costs related to providing and guaranteeing that level of quality that are higher than what consumers are willing to pay for that particular product, even with high-quality standards. If the financial analysis completed in the marketing plan shows that the costs of guaranteeing the highest quality exceed what consumers are willing to pay, an alternative strategy might be to target a perhaps higher-volume but lower-priced market, for which quality standards are not quite as rigid. The analyst must be careful to ensure that the expected market price will exceed costs of production. Although this seems an obvious statement, the lack of adequate effort in market analysis has caused many aquaculture businesses to fail.

Techniques that are useful to evaluate alternative product positioning strategies include: (1) a price-quality matrix and (2) a product space map. These can be developed to consider the position of the company's product or proposed product in relation to other similar or competing products. Pricing strategies should be adopted that match the price-quality positioning of the product. Table 12.1 illustrates a potential price-quality matrix for tilapia in Honduras. Production of a large 650 g (1.4 lb) tilapia for a fresh fillet product processed in an HACCP-approved plant would constitute a high-quality product. A market penetration pricing strategy would be to charge a medium price. If the targeted consumers are known to be value conscious, a low price might be required. However, in luxury markets, a premium price strategy might be pursued for lower-volume sales. An average quality product in Honduras would be a 350 g (0.77 lb) whole-dressed tilapia on ice. Charging price at the upper end of the price range for this type of product would be a market-skimming approach, with lower sales volumes. Charging a price at the lower end of the range would be an economy strategy. For low-quality, 250 g (0.55 lb) whole-dressed tilapia that is occasionally held on

Table 12.1. Price-Quality Matrix, Tilapia, Honduras.

	Price		
Product Quality	High	Average	Low
High			
650 g fresh tilapia fillet	$8.80/kg	$8.80/kg	$5.28/kg
Processed in HACCP-approved plant	Premium price strategy	Market penetration strategy	Value for money strategy
Average			
350 g whole-dressed tilapia, constantly on ice	$2.64/kg	$2.05/kg	$1.46/kg
Processed in HACCP-approved plant	Market skimming strategy	Average market position strategy	Economy strategy
Low			
250 g whole-dressed tilapia, occasionally on some ice	$1.91/kg	$1.50/kg	$0.73/kg
Several days old	Single sale strategy	Inferior goods strategy	Cheap goods strategy

Note: Price data were adapted from Green and Engle (2000); Fúnez et al. (2003a, b); and Monestime et al. (2003).

ice, selling at the upper end of the price range would likely result in only a single sale and no repeat sales.

Dover sole has been consistently viewed as a high-quality fish in the northeastern United States. Its growing scarcity has further driven its price upward. Thus, it is considered a high-quality, high-priced species (Fig. 12.2). In contrast, buffalofish is considered a low-quality, low-priced product in seafood markets in the southern United States.

Different types of products may be positioned differently even if they are of the same species. For example, small, whole, wild-caught tilapia in Nicaragua is considered a poor-quality, low-priced product. However, fresh and frozen tilapia fillets exported to the United States are positioned as being of medium-high quality and price.

After the company has analyzed carefully the current market situation, and understands consumer attitudes and preferences and the current stage of the product life cycle of its current products, including where these are positioned on the price-quality matrix, broader decisions can be made as to the numbers of product lines and the size of each product line. The size refers to the number of different products within each product line. These decisions must be based on the supply capacity of the company and the costs associated both with adding new products to existing product lines and to adding entirely new product lines. Larger companies that control greater volumes of supply and have larger processing capacity are in a better position to offer a greater degree of product differentiation than smaller companies.

Fish Species with Existing Demand

Different species of fish are frequently considered to be different products. Asche (2001) indicated that it is easier to market an aquaculture product for species that have been sold in the area. However, the business should not assume that this is always the case. If a market exists for a particular species, consumers have already developed attitudes related to its quality and the price that they are willing to pay for that quality of product. Because many of the aquaculture species in Europe are high-valued products such as salmon, trout, and sole, the existing market price for wild-caught species may be high enough to provide for profitable sales of aquaculture products. However, there are also cases in which the wild-caught species is offered in a low-quality form (small, whole tilapia with little ice) at a low price (Neira et al. 2003). In these cases, it can be difficult to create a market for a higher-quality, higher-priced aquacultured product. In the case of Nicaragua, it would not be profitable for tilapia farmers to sell farmed product at the price of wild-caught tilapia. Thus, tilapia farmers will have to overcome the im-

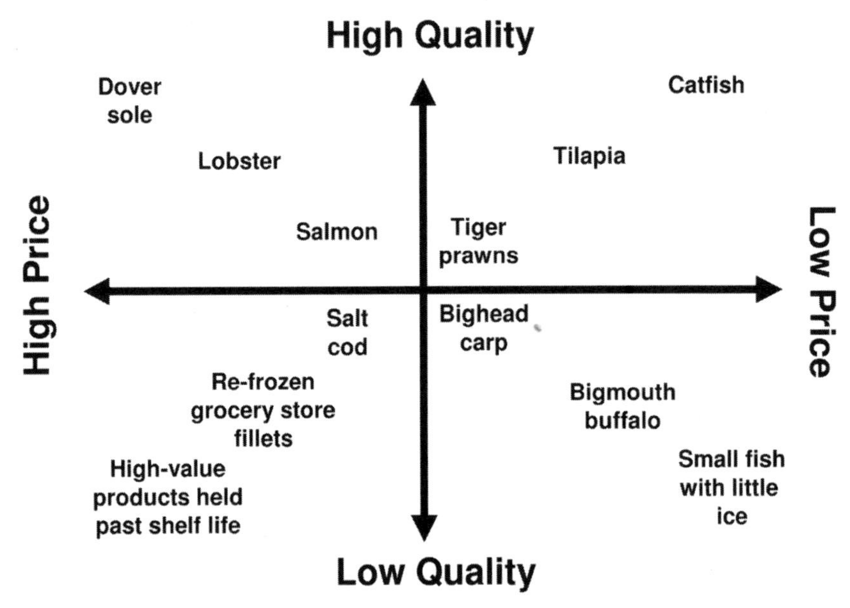

Figure 12.2. Generalized example of a product-space map with various types of seafood species. The exact position of a product will reflect not only the species but also product form, size, and handling.

age of the species and convince consumers that farm-raised tilapia is a different product from wild-caught tilapia. Marketing strategies to overcome these hurdles may include purging fish before selling, improving consistency of supply, additional education, and information for consumers to build a strong customer base for the product.

The catfish industry faced this same type of problem when it began to develop markets outside its traditional market area. Consumers along the Mississippi River have long enjoyed and consumed catfish. Although viewed as a lower-cost fish, it was consumed frequently as a major protein source by many in the areas surrounding the river and throughout the southeast. However, consumers outside this area considered catfish to be an undesirable, bottom-feeding scavenger that was not fit for human consumption. Years of generic advertising by The Catfish Institute successfully changed these perceptions in regions such as the mid-Atlantic region, and sales of catfish have increased in those areas.

New Species

Farmers who raise species for a market in which buyers have no previous experience with it will have to create and develop the market. Although this can be a long and sometimes expensive process, it is easier to develop a market for an unknown species than to overcome negative perceptions associated with a species. The companies that export tilapia fillets to the United States have successfully introduced an entirely new species into the U.S. seafood market. New products offer opportunities for price skimming and price penetration (Table 12.1).

The process of developing a market for a new species is the same as that for developing a market for a new product. The failure rate for new products is extremely high. It is critical that businesses have effective processes in place to screen new ideas to reduce the risk of failure.

Prior to investing in any new product or species, careful research is necessary. However, the total cost of the research should not exceed the potential sales value. Surveys can be conducted, but the size and scope of the survey should match the size and scope of the proposed introduction. (See Chapter 13 for details on conducting surveys). A sales curve should be forecast keeping the product life cycle in mind. The survey data should include some information on consumer attitudes and preferences from which the company can judge the possible price-quality positioning alternatives and can select promotion strategies.

Market testing is a critical step. Key parameters that should be measured in market tests include: (1) actual product trial rate; (2) level and frequency of repeat purchase; (3) relative effectiveness of various marketing plans; (4) consumer acceptance of product benefit claims; (5) reaction of the trade to the new product; and (6) potential distribution problems. The best outcome for the market test is for both trial and repeat sales to be high. This indicates that little effort (and hence, cost) will be incurred during the product introduction phase and that long-term sales potential is good. If trial sales are low but repeat sales are high, the company will need to invest more during product introduction to make consumers aware of the product, or to consider alternative product benefit claims and promotion strategies. High repeat sales still indicate favorable longer-term sales. However, high sales during the trial combined with low repeat sales would show that the promotion campaign effectively meets consumer desires but that the product is not meeting customer expectations. Careful analysis would be required to determine the specific product attributes that would need to be changed and whether it is feasible to change them. Low trial and low repeat sales indicate problems with both the image promoted of the product and with product characteristics.

Commodity Markets

Chaston (1983) defined commodity markets as "industrial markets" in which products are purchased as an ingredient or element to be used in another product that results in economic return for the buyer. A commodity is a homogeneous product produced by an industry as compared to a series of heterogeneous products with distinctive, smaller niche markets. Many commodities are sold in industrial markets as an input into a supply chain that transforms it one or more times before it reaches the end consumer. Some segments of aquaculture have grown and developed to the point where they can be considered commodities. Shrimp futures, for example, were traded for a time along with other commodities on the Minneapolis Grain Exchange. An example of a seafood commodity that is sold in an industrial market is the Peruvian anchoveta that is sold to fishmeal processors. Another example is shrimp that are sold to a manufacturer for use in a seafood entree.

Niche Markets

Some marketing experts maintain that all markets are niche markets (Palfreman 1999). Nevertheless, niche markets are commonly viewed as low-volume, high-priced specialty markets. Mass marketing is used to create products that appeal to a broad spectrum of consumers, frequently through development of a brand identity recognized across all consumer segments. Niche markets typically consist of a small segment of a large market. Sales frequently are lower in niche markets, but the strategy is to sell fewer products at a higher price. Smaller companies that successfully identify niche-marketing opportunities may have less competition from larger firms. Typically, a niche market is developed through a specific contact, and the grower uniquely supplies a custom product to that one particular market.

Small-scale aquaculture growers are often advised to seek out niche markets. Yet, there are few specific guidelines for doing so. The key component is the creativity and vision to identify a market opportunity in which a consumer need is not currently being met. Approaching an intermediary in that market line with a new concept is the first step. However, because the product is likely to be new, it is critical that the grower view this as a process of developing a relationship or partnership to develop the market. The grower will need to provide full support in terms of providing material for taste tests and sampling, point-of-sale materials, as well as to guarantee consistent product quality.

Niche markets in aquaculture typically have consisted of direct sales from the grower to the end consumer. Thus, the fish farmer performs wholesaling, distribution, and retail functions of the supply chain. In return, the grower captures the profit margins of each of these phases. However, each of these functions also entails costs, sometimes in the form of the time of the grower, in addition to costs related to holding or processing facilities, utilities, labor, advertising, transportation, and packaging (Morris 1994).

Niche marketing can be done in a cost-effective manner if basic principles are followed (Gordon 2002). The goal is to meet a unique need of the customer by tailoring the product to meet the customer's needs. It is important to understand and use the jargon of the targeted customer. What is important to a grocery store chain will be different from that of an upscale restaurant. Someone fluent in Spanish would be better positioned to approach Hispanic grocers than individuals who do not speak Spanish would be. Direct competitors must be evaluated carefully to identify how to position your product against them. Examining the advertisements, Web sites, logos, and brand names in addition to prices and delivery patterns may provide clues to needs that can be exploited. Do the customers want higher quality, lower price, more convenience, and better tasting or safer seafood products? It is important to talk to individual potential customers and look to identify how to meet a currently unmet need for that customer. Test marketing is essential to evaluate how receptive prospective buyers will be to the product. Moving cautiously minimizes risk exposure.

Growers often find it difficult to change their emphasis from production to marketing, but successful niche marketing requires a grower to spend at least 50% of their time on marketing (Kent 2002). For niche marketing to be successful, value must be added to the product either in terms of convenience, taste, or some other attribute, and it takes time and sometimes additional cost to do that. It may be difficult for growers who have made a substantial investment in a particular type of production system to switch to production of something that would move well in a particular niche market, or change to meet changing demands of that particular market.

Value-Added Products

The marketing channel comprises a value-added chain in which some type of value is added each time the product changes hands. Sometimes this value consists of the convenience offered by a large food service distributor that can supply all the food items that a restaurant needs with one telephone call or one visit to a Web site.

However, the expression "value added" more commonly refers to transformation of the product itself. In many ways, the concept of value added has been discussed under the topics of product differentiation and product lines. For example, a fresh catfish fillet product line may add value to the product and differentiate it from other fresh fillets by adding a Cajun or lemon-pepper marinade to it.

Consumers are demanding ever greater convenience, nutritional value, and variety while still purchasing based on taste. These consumer trends are creating new opportunities to add value and to differentiate products to capture these emerging market opportunities.

However, developing value-added products alone will rarely solve a particular company's economic problems. A well-developed market plan based on sound objectives and carefully analyzed strategy is the answer for struggling companies. For some companies, the move to more extensive and varied product lines may fit the company's business plan, whereas such an investment in sales force and processing and packaging infrastructure would not be feasible for others.

More than 20,000 new products are introduced in to U.S. grocery stores each year. More than 90% do not last more than three years. Thus, careful market analysis and testing are required to successfully introduce new products. The reader is referred to the sections on products and product lines, and the product life cycle as background material, for assessing the feasibility of developing a new value-added product for his or her company.

Business Organization and Contracting

Part of a marketing strategy may involve the organizational structure of the business. Many fish farmers are sole proprietorships or partnerships, but others are vertically integrated companies. Decisions related to changes in the structure of the business and their impact on the strengths and weaknesses of the business should be analyzed carefully in the marketing and business plan. Vertical integration refers to a single company that has control over several stages of the market channel or supply chain. For example, a shrimp company that owns its own farm, hatchery, and packing plant is vertically integrated. It controls its own supply chain and, thus, is in a position to be more flexible in terms of meeting customer demand throughout the supply chain. U.S. catfish farmers may own shares in processing plants and feed mills, but such a business is not truly integrated unless it is a single company involved in production through processing and final sales.

Although there are a number of examples of vertical integration in aquaculture, contract growing is not as common in aquaculture as it is in some other industries. For example, contract farming is common in the poultry industry. Growers are contracted by processors to supply a certain quantity to the plant over a given time period. Contracting companies tend to be market-oriented agri-businesses. Poultry growers bear the yield and financial risk of the growout phase.

THE MARKETING PLAN

The marketing plan should be part of an overall business plan. There are numerous books and resources available on developing business plans, and this book does not go into the broader content of the business plan. The marketing plan should focus on answering the question of why the customer should buy their fish. Characteristics such as reputation, appearance, delivery times, waiting times, and quality can be important.

Factors Essential to Effective Sales to Be Considered in Developing the Marketing Plan

All aquaculture businesses must sell products to generate revenue. Although many farmers believe that sales are marketing, this book has demonstrated that sales are only one component of marketing. Selling involves a variety of tasks that can include: (1) taking orders; (2) arranging delivery schedules; (3) delivering the product; (4) building relationships, trust, and goodwill to sustain the relationship; and (5) persuading customers to buy (Shaw 1986). Selling involves communicating the most important information to prospective customers as to what the product will do for them. The individual handling the sales must be very knowledgeable about the business and able to explain in detail the feeds fed, the quality of the water, and the post-harvest handling methods used to communicate the quality of the product. Understanding the relative production costs will also provide the seller with some flexibility in terms of negotiating changes in deliveries, packaging, and volumes and whether these changes may adversely impact costs. The seller must learn to listen well and understand the particular needs of the buyer and be prepared to meet those needs.

Table 12.2 illustrates an outline for a typical marketing plan and Appendix 12.A includes a sample, hypothetical marketing plan for an aquaculture business located in the United States. The plan typically begins with an executive summary. The first major segment is the situation analysis that includes a summary of the current market. Important subsections of the current market summary include a description of the market demographics. These include geographic information, including the number of people in targeted cities or regions. Information typically is divided into potential numbers of customers by outlet types (supermarkets, restaurants), demographic information related to age, gender, education levels, household income, lifestyle segments, and so on. Consumer needs, likes and dislikes, and buying trends by geographical area are important information. Buyers' needs must be addressed. Fish and shellfish markets are dynamic. Each market segment has its own buying patterns based on quantities purchased, product forms, price, and delivery needs.

It is important to talk to as many different buyers as possible to determine their needs. It is also useful to talk to aquaculturists and buyers in regions where the product is being sold. Substitute products sold locally should be identified and market inquiries made.

The product situation in the markets of interest would be described in terms of the recent history of sales and revenue for current products. The competition would be described by presenting the size, goals, market share, product quality, and marketing strategies of competing firms already in the markets of interest, as well as those that might be contemplating entry into those particular markets. Retail markets in the target market area should be visited to determine what the competition is. Important competitive attributes may include: price, product form, product quality, species availability, sources of competing supply, and buyer preferences. The existing distribution situation, in terms of sales through brokers, wholesalers, and retailers, should be described in detail. Finally, the macroeconomic environment of population, economic climate, and technological, legal, and social issues should be addressed.

If the target market is a processing plant, it is still important to visit the plant and identify delivery requirements. Some important types of information to obtain from a processor include: historical prices paid; dockage rates and policies; transportation charges, if any; frequency of payment to growers; seasonality trends as these affect fish deliveries at the plant; delivery volume requirements; fish size requirements; quality standards and quality control procedures; delivery quotas and scheduling; contracts; and bonding requirements.

The following section of the analysis would outline the needs in the various market segments that are obtained from information related to unmet customer wants and needs. Market trends in terms of supply and demand characteristics, market size, and past growth by geographic area and demographic segment with a particular focus on changes would then be presented and discussed in detail. Trend information is accompanied by estimates of market growth.

The second subsection within the situation analysis is the analysis of the strengths and weaknesses of the business in relation to threats facing the company from both external and internal factors and conditions. Opportunities are also identified. The competition is described along with a comparison of product offerings, pricing, volume, and distribution and customer service comparisons. From this, keys to success for the company and critical issues are identified.

Table 12.2. Outline for a Marketing Plan.

I. Executive Summary
II. Overall Market Situation Analysis
  A. Market summary
    1. Consumer Demographics
      a. Geographic areas
      b. Age groups
      c. Family structure
      d. Gender
      e. Income
      f. Education
      g. Lifestyle factors
      h. Spending habits
    2. Supermarket Demographics
      a. Geographic areas
      b. Age groups
      c. Family structure
      d. Gender
      e. Income
      f. Education
      g. Lifestyle factors
      h. Spending habits of customers
    3. Restaurant Demographics
      a. Geographic areas
      b. Age groups
      c. Family structure
      d. Gender
      e. Income
      f. Education
      g. Lifestyle factors
      h. Spending habits of customers
    4. Market Needs
      a. Product(s)
      b. Convenience/service
      c. Pricing
    5. Market Trends
      a. Supply
      b. Packaging
      c. Health consciousness
    6. Market Growth
  B. Analysis of Strengths and Weaknesses of Business
    1. Strengths
    2. Weaknesses
    3. Opportunities
    4. Threats
  C. Competition
  D. Product Offering
  E. Keys to Success
  F. Critical Issues

Table 12.2. Outline for a Marketing Plan. (*continued*)

III. Marketing Strategy
  A. Mission
  B. Marketing Objectives
  C. Financial Objectives
  D. Target Markets
  E. Distribution Channels
  F. Positioning
  G. Strategies
  H. Marketing Mix
  I. Marketing Research
IV. Financial Analysis
  A. Planned Expenses
    Sales Force Requirements
    Advertising Expenditures
  B. Sales Forecast
  C. Break-Even Analysis
V. Controls
  A. Implementation
  B. Marketing Organization
  C. Contingency Planning

Analysis of internal strengths and weaknesses should also include: (1) relationships (with buyers, suppliers, people who work in the business, and with other businesses); (2) reputation; (3) innovation; and (4) strategic assets. Relationships are key to the success of any business. Establishing and maintaining good relationships with buyers will give a business an advantage over the competition (Palfreman 1999). Special relationships with suppliers and repeated transactions may enable a business to benefit from improved services, short-term credit, improved quality, or even better prices. Within the business, a higher degree of commitment or team spirit may result in greater productivity or efficiency. Good relationships with other businesses may offer opportunities to share information and contracts or to purchase supplies at bulk prices. The reputation of the business may provide a competitive advantage or disadvantage. Companies with excellent reputations will attract more business and have greater ease of attracting sources of supply. Innovation is required to improve productivity and profits. Although innovation can be copied, it cannot be avoided if the business is to be successful. Entrepreneurs need to look deep within their own business and ask what their special abilities are and whether these can provide them with a competitive advantage in the marketplace (Palfreman 1999).

The second major segment of the marketing plan is the description of the marketing strategy itself. The strategy first needs to be articulated succinctly in a paragraph or two. Sometimes, a mission statement is included and then specific marketing and financial objectives are listed. Target markets are listed and described in detail. This is done by first determining the market area. The serviceable geographic area is defined, taking into consideration the travel distance and time. Then, the market segments are identified. Because there are different types of customers within a market area, it must be determined whether there are enough potential buyers of the product to support the proposed business. Principal customer types (market segments) for aquaculture products include: processor, wholesalers, distributors, restaurants, seafood stores, supermarkets, or consumers buying directly. The types of promotion needed are specified. The desired marketing mix is described and broken down into key categories. Any market research required is described and justified.

The third major section of the marketing plan is the financial analysis. A break-even analysis of the marketing strategy described is developed. The market potential is estimated through sales forecasts, typically on a monthly basis, by type of market outlet and target market. Annual sales goals are set. Consumer census data and business or economic development data can then be used to estimate the number of potential buyers in the targeted market area. With all the information gathered, business targets for market sales should be established by year. An example of a business target might be achieving a return on the investment of 30%/yr. Alternatively, a company could set a business target of increasing sales by 40% over the previous year. The business plan will also specify the size, type, and quality of the sales force. The level and quality of customer service should be described. The amount of advertising and sales promotion will be specified along with the amount, types, timing, and projected success of research and development needed.

The plan must also include a detailed methodology for monitoring and evaluating the company's performance in following the marketing plan. Individual contacts are invaluable in developing markets through implementing the market plan. Typically, revenue, expenses, repeat business, and customer satisfaction are categories that would be monitored to gauge performance. Contingency planning if performance does not meet expectations is a critical component of the plan.

AQUACULTURE MARKET SYNOPSIS: HYBRID STRIPED BASS

Striped bass (*Morone saxatilis*) have been a valuable species in fisheries in the northeastern United States for many years. Capture data in the National Marine Fisheries Service database date back to 1950 (National Marine Fisheries Service 2004). From 1950 until about 1981, the striped bass catch varied between about 1,900 MT to more than 6,000 MT a year. However, this catch fell to 559 MT in 1985 and remained below 1,000 MT until 1995. This decline in the striped bass fishery, primarily in the Chesapeake Bay region in the mid-1980s, created a market opportunity for a farm-raised substitute.

At the time of the collapse of the capture fisheries for striped bass, hatchery practices for production of hybrid striped bass (*Morone saxatilis x Morone chrysops*) were well developed (Fig. 12.3). Stocking programs to enhance recreational fishing had grown with the development of the hatchery techniques to produce reliable quantities of hybrid striped bass fry and fingerlings. When a moratorium was declared on the striped bass catch in the Chesapeake Bay, reported live prices of striped bass climbed to $15.40 to $19.8/kg ($7 to $9/lb). The dramatic increase in price of striped bass stimulated interest in hybrid striped bass farming in the United States as a foodfish product.

The traditional market area for wild-caught striped bass was in New York City and other urban centers in the northeastern United States. The market was primarily for a menu item in white-tablecloth restaurants that consisted of presenting the whole fish on the plate.

Measurable aquaculture production began in 1986 and grew rapidly in the early 1990s. The early production of hybrid striped bass was primarily in tanks, but tank production leveled off by the mid-1990s (Carlberg and Van Olst 2004). The subsequent increases in production have resulted primarily from increased acreage in pond production, mostly in the southern states of Texas, Mississippi, and North Carolina.

Total production declined somewhat in 2001 and 2002, primarily due to the economic downturn and the September 2001 attacks in the United States (Fig. 12.4). However, total production had rebounded by 2003 and reached an all-time high of 5, 203 MT (11.4 million lb) in 2004, a 93% increase from 1993 production levels.

Figure 12.3. Hybrid striped bass.

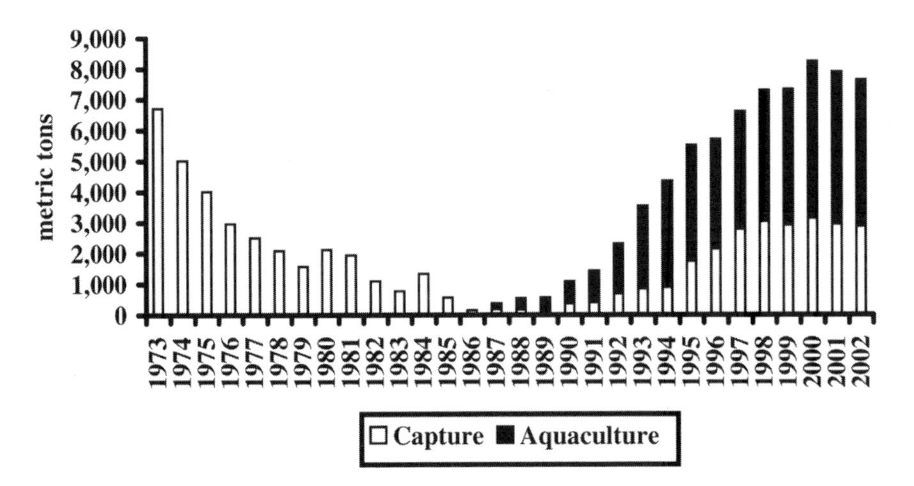

Figure 12.4. Hybrid striped bass, capture fisheries vs. aquaculture, 1973–2002. (Source: NMFS 2003.)

Hybrid striped bass are produced throughout the United States, but the largest production occurs in the western portion of the country (Fig. 12.5. The south central states and southeastern states are the other major producing regions, with some production in the upper Midwest and northeast regions.

Hybrid striped bass are marketed primarily as either a live product or as a fresh whole fish on ice. Live fish (approximately 17% of domestic production) are sold primarily to Asian distributors for retail grocery stores and for banquets in Chinese and Vietnamese restaurants. (Carlberg and Van Olst 2004). Toronto, New York, Philadelphia, Los Angeles, and San Francisco are the major market cities for hybrid striped bass. Live sales have increased at about 10% per year. The market for fresh fish on ice has grown most rapidly and reached 83% of all sales in 2003. Most fresh fish is sold as a whole, head-on, scales-on, and guts-in product to wholesale distributors that service the Asian retail market. Much smaller quantities are sold to upscale restaurants and for sushi and sashimi in Japanese and Korean restaurants. Sales of fresh fish have grown at an average rate of approximately 2% per year.

One of the challenges is to continue to develop new markets. The wholesale price for live fish has decreased over the last several years. Dressout yield for hybrid striped bass is low and will make entry into the fillet market more difficult. The meat cost will be higher with lower dressout yields for the same farm-gate price. Moreover, proposed regulations related to restricting the use of black carp (*Mylopharyngodon piceus*) in the United States may result in restricted supplies of fingerlings and reduced marketability of hybrid striped bass. Black carp are stocked in hybrid striped bass ponds to control snail populations that serve as intermediate hosts for the yellow grub. Yellow grub infestations result in mortality of hybrid striped bass fingerlings and reduce marketability of foodfish hybrid striped bass.

SUMMARY

This chapter presents specific details on the process and components of market plans and the associated marketing strategies. Techniques and information sources for developing an analysis of the current market situation provide a means to identify some potential market opportunities. Understanding the competition and consumer attitudes and preferences are keys to uncovering unmet consumer needs and wants. Analysis of the strengths and weaknesses of the business, both those that are external and those that are internal, should result in the identification of competitive advantages for the business. The marketing strategy is developed from careful analysis of the market opportunities and the competitive advantage of the business. The market plan should provide an answer to the question, What unmet consumer need can this business fulfill better than any other business? The strategy is then developed to specify the sales goals of the products and markets identified.

STUDY AND DISCUSSION QUESTIONS

1. How does one determine what scale of market research should be undertaken and whether the emphasis should be on collecting primary or secondary data?

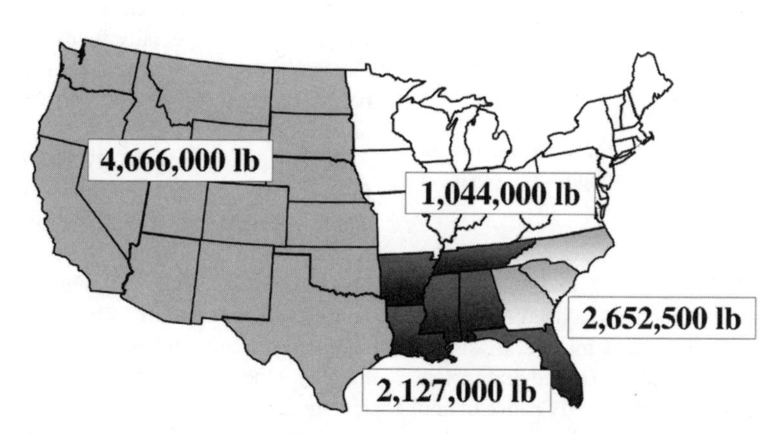

Figure 12.5. Hybrid striped bass in the United States. (Source: Kent Seatech and SBGA 2004.)

2. What is the difference between an industrial market and a consumer market? Give an aquaculture example of an industrial market.

3. Explain the costs associated with product diversification.

4. What is market segmentation? Give an aquaculture example.

5. Explain the product life cycle.

6. Explain how to successfully develop new markets.

7. Explain and give aquaculture examples of a product-space map and a price-quality matrix.

8. Think of examples of business strengths and weaknesses and how these can be used to develop a marketing strategy.

9. What are the four P's of the marketing mix? Explain and describe aquaculture examples of each.

10. List the major components of a marketing plan.

11. What are some differences in developing market strategies for species with existing demand as compared to new species?

12. What are some advantages and disadvantages of developing value-added products?

APPENDIX

Sample Market Plan (Hypothetical)

An enterprising family would like to start an aquaculture business. They live in a small city in the southern part of the United States. They need to develop a market plan to start the business off on a well-organized and well-thought-out business direction.

Executive Summary

This family-owned and operated aquaculture business will meet an unmet demand for live, fresh catfish in a small city in the southern United States. The advantage of this business is the family's love of and enthusiasm for quality fish. Our marketing challenge will be to tap into word-of-mouth advertising to be the supplier of choice to market segments that prefer very fresh fish at a reasonable price.

Vision

This family-owned farm business is based on the assumption that people will prefer to purchase live catfish due to its guaranteed freshness. The farm business will serve its clients by providing consistently on-flavor fish delivered as and when ordered to provide for their fish supply needs.

Overall Market Situation Analysis

Market Summary. Small City U.S.A. is located in the southern part of the United States. People in the area are accustomed to eating freshwater fish such as catfish, buffalofish, largemouth bass, and crappie that they have caught while fishing in the rivers and ponds in the region. Small City U.S.A. has tightly knit family groups and is a conservative town that revolves around church and family. Incomes are not high. Many rural poor looking for a better life cycle have migrated from depressed farming communities to Small City U.S.A. Educational levels generally are lower than the national average. There is a higher-income segment in the city, particularly in the areas surrounding the hospital and federal facilities located in Small City U.S.A. However, many of the higher-income residents often travel to larger cities within a few hours drive for entertainment and recreation. The city has a population of approximately 50,000 people and is roughly half white and half African-American. However, the Hispanic population is growing rapidly and there are a few Asian families in the area.

Supermarkets cater to the southern lifestyle and feature the main ingredients of southern cooking. Supermarkets located close to the hospital carry a wider variety of specialty foods and spices, but the majority of supermarkets are discount types of supermarkets that compete primarily by offering lower priced foods.

Restaurants in Small City U.S.A. include many fast-food chains and a number of Mexican and Chinese establishments. There are several catfish restaurants in the city that are popular, particularly on weekends and in the evenings, and a few barbeque houses. Other restaurants advertise plate lunches and southern cooking, and there are a few steak houses in the city. One or two offer some Cajun dishes and a few more innovative dishes, but these are few.

Given the lower-than-average income levels of residents of Small City U.S.A., pricing of products is extremely important. Restaurants that offer specialized cuisine at menu prices in excess of $10 generally do not fare well in the city. Adherence to southern lifestyles also is important, and a large percentage of the population does not have a strong sense of adventure with foods.

There has been a marked increase in foods catering to the growing Hispanic population in the city. There has also been an increase in the percentage of African-Americans in the city.

Prices of catfish in local supermarkets are priced at levels that restrict purchases. Sales of fresh fillets on ice also lack the freshness of live products. Live catfish guarantee freshness, and if they can be sold at lower per-unit prices, they may have potential in this market. Accessibility, customer service, and competitive pricing will be important.

Analysis of Strengths and Weaknesses of Business

This family has experience in raising freshwater fish, has a strong work ethic, and owns some land within 10 miles of Small City U.S.A. The family owns 25 acres of land with plentiful groundwater supplies, and a well that pumps 350 gal/min. is already in place.

The family is not from the local area and does not have strong personal ties through the family-church network of relationships. The family is also white. Because Small City U.S.A. has an increasing percentage of African-American and Hispanic residents, it may be difficult for whites to develop strong market relationships with individuals of other races, given continued racial divisions in the community.

Catfish is a well-known and desired product in the community. Prices of catfish offered in the supermarkets and restaurants are medium high as compared to chicken, beef, and other protein sources. The growing African American and Hispanic populations also offer market segments that, per capita, tend to eat more fish than do other population segments. However, prices must be reasonable and present greater value than fish products sold in supermarkets and in restaurants.

The primary competition will come from catfish restaurants, supermarkets that carry catfish (a major discount chain has begun to sell catfish in its superstore), and fish markets that carry wild-caught catfish, buffalofish, and other freshwater species. Other competition might come from larger catfish farms that might choose to sell directly to the public.

The product offered will be live catfish. The emphasis will be on the freshness and quality of the product and of exceptional service in delivering product to customers.

The keys to success will be to satisfy customers who will be carrying live fish home from the farm or from a truck parked at strategic locations. Critical issues may include the willingness of customers to drive to the farm to purchase fish or to identify locations in the city where a truck could sell fish successfully. Establishment of effective delivery routes to maximize convenience may be important to the success of the business. The willingness of individuals to clean the fish purchased may be a constraint.

Marketing Strategy

The mission of the business is to be the most preferred source of quality live fish for Small City U.S.A. The marketing objectives are as follows:

1. Sell 54,000 lb of live catfish a year.
2. Develop effective word-of-mouth advertising.
3. Become the preferred supplier of catfish for church and family reunion fish fry's.

Financial objectives are to:

1. Develop sufficient cash flow for the business to survive in year 1.
2. Beginning in year 2, reduce debt-asset ratio by 10% a year.
3. Begin to show normal profit in year 3 as markets are developed and sales stabilize.

Target markets will be the African-American and Hispanic populations that prefer quality, very fresh fish. Church and family reunion fish fry's will be targeted. Sales will be direct to the public with no intermediaries. The live catfish will be positioned as a higher-quality but lower-priced alternative to fillets sold in supermarkets. The strategy will be to advertise to church ministers, invite church groups to visit the farm, organize youth fishing activities, and provide samples, radio advertisements, and flyers.

Planned expenses are as outlined in Engle and Stone (2002). Ponds will need to be built and equipment purchased that will include an ATV, electric paddlewheel aerators, oxygen meter, mowers, a tractor, waders, nets, a feed bin, and a live car for holding fish. Operating costs will include fingerlings, feed, some part-time labor, fuel, electricity and other utilities, and insurance. Total annual costs (including noncash costs such as depreciation) are estimated to be $52,720.50 per year. The break-even price of fish is estimated to be $0.69/lb above operating cost and $0.98/lb above total cost.

The family will serve as the sales force. The part-time labor will also be asked to help spread the word about the farm. Sales the first year are expected to be 12,000 lb, increasing to 54,000 lb by the end of the second year. Anticipated sales price is $1.25/lb, to generate a profit of $0.27/lb, or net returns above all costs of $14,580 per year.

REFERENCES

Asche, F. 2001. Testing the effect of an anti-dumping duty: the US salmon market. *Empirical Economics* 26:343–355.

Carlberg, J. and J. C. Van Olst. 2004. Hybrid striped bass: U.S. production and outlook. *Fish Farming News*, March/April.

Chaston, I. 1983. *Marketing in Fisheries and Aquaculture*. Fishing News Books, London, United Kingdom.

Engle, C.R. 1998. Analysis of regional and national markets for aquacultural food products in the southern region. Southern Regional Aquaculture Center Final Project Report No. 601, Stoneville, Mississippi.

Engle, C.R. and P.J. Kouka. 1995. Potential consumer acceptance of canned bighead carp: a structural model analysis. *Marine Resource Economics* 10: 101–116.

Engle, C.R. and N. M. Stone. 2002. Costs of small-scale catfish production. Southern Regional Aquaculture Center Publication No. 1800, Stoneville, Mississippi.

Engle, C.R., O. Capps, Jr., L. Dellenbarger, J. Dillard, U. Hatch, H. Kinnucan, and R. Pomeroy. 1990. The U.S. market for farm-raised catfish: an overview of consumer, supermarket and restaurant surveys. University of Arkansas Division of Agriculture Bulletin 925, Fayetteville, Arkansas.

FAO. 2004. FISHSTAT+. Food and Agriculture Organization of the United Nations. Accessed at http://www.faoorg

Florida Department of Agriculture and Consumer Services. 2001. National market analysis of Hispanic consumer attitudes towards seafood and aquaculture products. Tallahassee, Florida.

Foltz, J., S. Dasgupta, and S. Devadoss. 1999. Consumer perceptions of trout as a food item. *International Food and Agribusiness Management Review.*

Gempesaw, C.M., J. Richard Bacon, C.R. Wessells, and A. Manalo. 1995. Consumer perceptions of aquaculture products. *American Journal of Agricultural Economics* 77:1306–1312.

Gordon, K.T. 2002. 3 rules for niche marketing. Entrepreneur.com. http://www.entrepreneur.com

Kent, K. 2002. Niche market survival. *Iowa Farmer Today*. http://www.iowafarmer.com

Kinnucan, H.W., R. G. Nelson, and J. Hiariay. 1993. U.S preferences for fish and seafood: an evoked set analysis. *Marine Resource Economics* 8:273–291.

Mintel International Group Ltd. 2002. Fish and seafood market—U.S. report. http://www.marketresearch.com/ research index/802195.html

Morris, J. 1994. Niche marketing your aquaculture products. North Central Regional Aquaculture Center. Iowa State University, Ames, Iowa.

NASS. 2004. Aquaculture Situation and Outlook Report. National Agricultural Statistics Services. United States Department of Agriculture, Washington, D.C. Accessed at www.nass.usda.gov

National Marine Fisheries Service. 2004. Capture species. Accessed at http:www.st.nmfs.gov

Neira, I., C.R. Engle, and K. Quagrainie. 2003. Potential restaurant markets for farm-raised tilapia in Nicaragua. *Aquaculture Economics and Management* 7(3/4):1–17.

Palfreman, A. 1999. *Fish Business Management.* Fishing News Books, Blackwell Science, Oxford, United Kingdom.

Shaw, S.A. 1986. Marketing the products of aquaculture. FAO Fisheries Technical Paper 276, Food and Agricultural Organization of the United Nations, Rome, Italy.

Wessells, C. and J.L. Anderson. 1992. Seafood safety assurances: implications for seafood marketing and international trade. URI/OSU Research paper series RI-103, University of Rhode Island, Kingston, Rhode Island.

Wessells, C.R., S.F. Morse, A. Manalo, and C.M. Gempesaw II. 1994. Consumer preferences for Northeastern aquaculture products; report on the results from a survey of Northeastern and Mid-Atlantic consumers. Rhode Island Experiment Station Publication No. 3100, University of Rhode Island, Kingston, RI 02881.

13
Marketing Research Methodologies

Marketing research is essential to the overall success of any business because the major objectives of any seafood business are to meet consumer demand and operate efficiently at a profit. To stay in business and remain competitive, companies rely on various types of marketing research information to formulate marketing strategies, make marketing decisions, or implement marketing concepts. For example, marketing research will help answer questions such as: "What are the attitudes and desires of consumers?" "Is there a demand for the product?" "What is our volume of sales compared to competitors, that is, what is the market share for the product?" "What products will consumers demand in future?" Answers to such questions are important for business planning because they allow a business to find out more about the current market situation relating to a product of interest as well as to predict future market situations. Market research can also be used to find solutions to specific marketing problems that a company might have.

The American Marketing Association defines marketing research as the function that links the consumer, customer, and the public to the marketer through information—information used to identify and define marketing opportunities and problems; generate, refine, and evaluate marketing performance; and improve understanding of marketing as a process (Bennet 1988). This definition elaborates on the several functions and uses of marketing research. A seafood company that is not doing very well in sales may conduct a marketing research study to obtain information about why the product is not selling and what can be done to improve sales. A new company that wants to introduce a seafood product to the market will first have to find an answer to the question, "Will there be a market for this product or will this product meet a need on the market that has not been satisfied?" Marketing research is therefore conducted for various reasons, and it is essential that the research be conducted appropriately.

An effective market research process can be financially rewarding for a company. If done poorly, however, it could result in failure of the business. Before a company embarks on marketing research, it must clearly define the purpose for doing the research. Any company embarking on market research should know the type of information it needs and the cost of obtaining that information. Sometimes, marketing research may be needed to obtain some general information or market outlook, whereas at other times, it is required to solve specific problems.

TYPES OF RESEARCH AND DESIGN

The process of conducting marketing research consists of gathering, sorting, analyzing, evaluating and disseminating information for timely and accurate market decision making. There is so much information in the marketplace that the focus of the process should be to target information necessary to make informed decisions. Market research can be designed in one of three forms: (1) exploratory research; (2) qualitative research; or (3) quantitative research. The type of research that is most appropriate depends upon the objectives. For example, exploratory research would be most appropriate for a startup business that is taking its first steps in identifying potential markets. Qualitative research would be appropriate for a company attempting to decide whether to change its brand or whether its advertising program should focus more on emphasizing the color, taste, or safety of its fish fillets. Quantitative research could help a company estimate the size of a prospective new market. Each of these types of research is discussed in the next sections.

EXPLORATORY RESEARCH

Through exploratory research, information can be obtained that allows seafood companies to identify and clarify some problems or issues confronting them. It may also provide information that helps a seafood company to identify potential challenges

The Case of Ippolito's Seafood, Philadelphia

Ippolito's is a wholesale seafood company that sells frozen and fresh seafood to hotels and restaurants across the Philadelphia region. The company began as a seafood retailer selling frozen shrimp, lobster tails, and fish fillets. However, the company was struggling to stay in business because of increased competition from grocery outlets and supermarkets. In the early 1980s, through exploratory research, the company realized the need for niche wholesaling in the region that would target the foodservice industry. Large general-line food distributors performed the seafood wholesale functions in the region at the time. Ippolito's also realized that the traditional wholesalers did not deliver seafood on Saturdays. The company therefore launched into seafood wholesaling, offering its traditional products of frozen shrimp, lobster tails, and fish fillets as well as imported Chilean sea bass, New Zealand orange roughy, and fresh octopus and *lupe de mer* from the Mediterranean Sea. In 2001, total sales revenue for the company was $47.3 million and was projected to be more than $50 million in 2002. Ippolito's clients include the Four Seasons Hotel, Rittenhouse Hotel, and the Park Hyatt as well as neighborhood taverns and restaurants in the Philadelphia area (Bennett 2001).

and opportunities and to establish research priorities. Exploratory research raises awareness and provides insights.

In exploratory research, there are no specified objectives; neither is there any structure to the process of gathering market data and information. It is a very informal approach to research that may involve mere observations of things of interest such as observing customers as they shop, consumers' buying patterns, client interactions, sales or revenue figures; reading periodicals; surfing the Internet; visiting the library; and inquiring about certain products, services, prices, market situations, and trends and issues. There is no structure to this form of research and it can therefore be used in a number of situations. Related to exploratory research is what is often referred to as market intelligence, or the art of obtaining updates about relevant developments in the market. The market intelligence system can also be informal or formal, with a focus on searching for information and anything that may be of interest to the company.

Sometimes, the process of gathering information could be purchasing and tasting products of competitors, or scanning periodicals for specific information about the seafood market or seafood products of interest. Many companies subscribe to newspapers, magazines, and industry and trade publications for the purpose of keeping up with industry affairs.

Qualitative Research

Qualitative research also raises awareness and increases insights. However, in qualitative research, theoretical concepts can be tested to provide some definitive explanations. Other textbooks refer to qualitative research by different names, such as descriptive research, subjective research, inductive research, and case study. The name clearly depends on the purpose of the research.

Qualitative research is a more structured and formal type of research that is concerned with obtaining explanations to certain issues or subjects of interest. It deals in words, images, and subjective assessments. For example, qualitative research may be used to describe a purchase behavior or pattern, event, or concept and to understand a market situation from a holistic perspective. This approach is well suited for a store that wants to examine its own brand of products and compare them to national brands of similar products. Qualitative research can also be used where there are concerns about customer opinions, experiences, and feelings. In effect, qualitative research is concerned with finding answers to questions that relate to "Why?" "How?" "What?" Data for this type of research can be collected through direct observations, interviews, or surveys. The data and information are then used to develop concepts that help to understand the marketplace.

Quantitative Research

Quantitative research deals in numbers, logic, or theory and objective measures to provide measurement and statistical predictability of results to the total target population (customers, consumers, and so on). Some level of certainty is required in quantitative research. For example, it may be important to know the size of a target group for a certain product on the market, or the extent of customer satisfaction with a product or service. Quantitative research methods include the use of questionnaire surveys or telephone interviews, and subsequent statistical analyses.

Decisions regarding planning and implementing marketing measures and for making organizational changes can be made with a relatively high level of certainty from quantitative research compared to the other approaches. Good quantitative research requires three elements: a well-designed questionnaire, a randomly selected sample, and a sufficiently large sample. These are discussed in detail next.

Data Collection

Whether the research effort is exploratory, qualitative or quantitative, data need to be collected. Data to be collected should relate directly to the research objectives, research questions, and research hypotheses. There are two basic types of data that can be gathered: primary and secondary data. Primary data collection is very expensive. Companies should carefully weigh the anticipated value of new sales generated as a result of investing in primary data collection with the cost of doing the research. For example, a tilapia company that likely would increase sales by $150,000 would not be wise to invest $500,000 to generate primary data. Even in cases in which the results would be worth the cost of research, spending some time gathering secondary data should be the first step.

Secondary Data

Because primary data can be expensive to collect, it is often worthwhile to access data and information previously gathered by others. This is referred to as secondary data. The benefits of using secondary data are that significant time and financial investment are not required for gathering the data. Moreover, it is always useful to ascertain that the data needed for a research study are not already available.

The major disadvantage of secondary data is that the researcher does not have control over the design of the data-gathering process, the data collection process, or any manipulations of the data. Data may be available only in forms that are not suitable for the specific purpose and therefore require some manipulation in order to be useful. Secondary data are generally published data and can be obtained from a number of sources including: established archives, government and state agencies, private companies, or directly from principal investigators and researchers (see the "Annotated Webliography" section for summaries of various sources of secondary data for marketing aquaculture products).

Primary Data

Primary data are gathered by researchers. Any systematic documentation of personal observations, interviews, surveys, focus groups, or personal experience constitutes primary data. The most common primary data collected in marketing research is the documentation of consumer attitudes and behavior using focus groups, interviews, or surveys. Each of these are discussed shortly. More information on primary data collection methods can be obtained from the American Statistical Association's series on *Survey Research Methods* (American Statistical Association 1997; also see the "Annotated Webliography" section) and from marketing research textbooks such as Blankenship et al. (1998).

Focus Groups

Focus groups are informal techniques to assess consumer preferences and needs, new product concepts, and purchase behavior for a good or service. Focus groups consist of six to 12 carefully selected participants with some common characteristics that relate to the objective of the study. The homogeneity of group participants is vital to generating important data and information from the sessions. With consumer preference studies, the most important characteristics of participants often include income, age, and ethnicity. It is useful to use different groups to obtain a diversity of responses.

Focus group sessions are conducted in the form of a discussion with a moderator who maintains the group's focus. The moderator should promote free-flowing individual participation in the discussion. However, the moderator must follow an agenda on specific issues and goals that relate to the type of information to be gathered and ensure that all group members contribute to the discussion. Discussion questions should be open-ended to allow all possible responses. As much as possible, the moderator should promote give-and-take discussion among participants. These group sessions can last from an hour and a half to two hours.

A well-moderated focus group session can generate new product ideas or concepts, reveal consumer reactions to potential new products, and discover potential market prices for a product. The session can also reveal information about competitive products, product usage, preferred packaging, and effective advertising strategies. Some group dynamics and organizational issues can also be observed during a focus group session.

The major disadvantage of focus group research is that the responses cannot be analyzed statistically or quantitatively. Information obtained from focus group sessions relate more to words and behavior of the participants, who are not representative of a target population. Focus group research is, therefore, a qualitative research.

As an example, in a study in Florida, seafood consumers in three focus groups in June 2000 were asked about their shrimp preferences and purchase behavior (Florida Department of Agriculture and Consumer Services—Bureau of Seafood and Aquaculture Marketing 2001). The focus group studies had five objectives: (1) identify shrimp characteristics important when purchasing shrimp; (2) investigate consumers' shrimp purchasing behavior; (3) explore consumers' attitudes about farm-raised versus wild-harvested shrimp; (4) compare perceptions of Florida farm-raised Pacific white shrimp raised in freshwater with imported shrimp and wild shrimp; and (5) determine consumers' willingness to pay for farm-raised shrimp.

The size of the FDACS-BSAM focus groups ranged from eight to 12 participants, with a total of 30 participating consumers; 67% were female, 77% were white, 60% were between the ages of 35 and 54, 50% had household size of three or more individuals, 57% reported incomes of $20,000–$35,000, and only one member was unemployed. The topics of discussion focused on the objectives of the study. Most participants associated farm-raised shrimp with safety and cleanliness, but others expressed concerns over bacterial contamination. Words and images used by participants to associate farm-raised shrimp included "cleaner," "controlled environment," "safer," "quality control," "uniformity" and "knowing what you are getting." Farm-raised products were considered by participants to be more environmentally friendly. Words and images associated with wild-caught shrimp included "unintended catch such as turtles," "polluted water," and "poor sanitary conditions".

On shrimp prices, 50% of participants felt that farm-raised shrimp should be less expensive than wild-harvested, because the shrimp are all concentrated in one place. Others felt that feeding farm-raised shrimp and cleaning the water could make farm-raised shrimp prices higher than wild-harvested shrimp prices. On willingness to pay, 75% indicated willingness to pay $15.40/kg, and 50% would pay $17.60/kg.

There was a comparison of cooked versus raw Pacific white shrimp from three sources: freshwater farm-raised, Ecuadorian brackish water farm-raised, and East Coast wild-harvested. The participants were shown cooked and raw samples of shrimp from each of those sources. The majority preferred the taste and smell of the Florida freshwater farm-raised Pacific white shrimp and indicated that the flavor was better and the taste was sweeter. The deep, bright orange color of the cooked Florida freshwater farm-raised Pacific white shrimp was striking to the group participants. The color was attractive and had better eye appeal than the other types. The participants indicated that the freshwater farm-raised Pacific white shrimp would look good in a display case because they looked cleaner and fresher. The shrimp from the other two sources looked pale in comparison. This observation suggested that color could be a distinctive product-differentiating feature that could give a competitive advantage to the Florida freshwater farm-raised Pacific white shrimp. The dark color, which was a positive product attribute for the cooked Florida freshwater farm-raised Pacific white shrimp, was a negative attribute for the raw shrimp. The dark color was a purchase deterrent, because the participants equated darker shrimp with being older. The study therefore concluded that marketing and promotion of the Florida freshwater farm-raised Pacific white shrimp should emphasize the bright orange cooked color and the sweet flavor of the shrimp, and that marketing in supermarkets should consist of displays of precooked tails rather than raw product (Florida Department of Agriculture and Consumer Services—Bureau of Seafood and Aquaculture Marketing 2001).

Surveys

Surveys are methods of gathering systematic information from a sample of a target population. In market research, surveys provide a speedy and economical means of determining consumer attitudes, beliefs, expectations, and behaviors about products and services. For example, a seafood product manufacturer will do a survey of the potential market before introducing a new product. Surveys can be conducted in a variety of ways including telephone, mail, or face-to-face. Surveys can also be self-administered. Some surveys may combine several of these methods. For example, a telephone survey can be employed to select eligible respondents and make appointments for in-person interviews.

Personnel involved in market research surveys must have some training in interviewing. Interviewers should possess the ability to approach people in person or on the phone, persuade people to participate in a survey, and collect the needed data. The whole survey process requires skills in survey planning, sample selection, questionnaire development, data processing, data analysis, and reporting. Survey results should always be presented in broad categories such that individual respondents cannot be identified (see Blankenship et al. 1998 for an overview of survey methods).

Mail Surveys Mail surveys have the advantage of being a relatively lower-cost method as compared to the other survey methods. When respondents cooperate, mail surveys can also provide more thoughtful responses to the survey questionnaire. Moreover, there is no potential for interviewer bias with mail surveys. Visualization may be required for respondents to answer survey questions. For example, the use of a color chart or a series of advertisements may make it easier for respondents to understand the questions. Some surveys require respondents to refer to and provide data from records they keep. In these instances, mail surveys can be an effective data collection method.

The major disadvantage of mail surveys is often a low response rate due to lack of cooperation from respondents. Also, if timing is important for the completion of the research problem at hand, mail surveys may not be appropriate because they require more time. Mail survey questions must be clear and simple to understand; otherwise, respondents will give different interpretations and meanings to the same question. This results in unreliable data that are difficult to interpret. Another potential problem with mail surveys are no responses to certain questions and inaccurate responses to particular questions. Respondents may skip questions, some questions may be answered in an incomplete fashion, and responses may even be illegible.

Various techniques have been developed for improving the efficiency and response rate of mail surveys. These include:

1. Notification of recipients well in advance about their participation in the impending survey. This can be done through a letter or post card. This is very common with surveys conducted by the government.
2. Addressing all correspondence using recipient names and not "current occupant," if it is a consumer or household survey.
3. Including a cover letter with the survey questionnaire that outlines the purpose of the survey, the importance of the recipient's response and participation, and the benefits of the study to recipients. An estimated time for completion of the questionnaire should be included in the cover letter. That means that recipients will cooperate and respond to the survey if the time required is short. For household surveys, open-ended and lengthy questionnaires should be avoided. Generally, a mail questionnaire should be short and require straight answers, for example, the questions should have have response categories that can be checked off quickly.
4. Providing incentives for participation. This is a good idea and could be in the form of offering each participant some cash or coupon for participation. This type of incentive should just be a token amount due to the expense, particularly with a large sample size. Alternatively, cash or coupons can be offered as prizes for drawings, where respondents have a chance of winning a prize for participation.
5. Providing postage-paid return envelopes with the survey questionnaires.
6. Sending follow-up postcards or letters to remind recipients about responding to the survey questionnaires. The message should be a shortened version of the cover letter and should include an expression of appreciation to those who have completed and returned the questionnaire. It should also include the willingness to send another questionnaire and a return envelope to the recipient if the first has been misplaced. It is recommended that this be done after the second week of mailing the survey questionnaire, because the majority of responses to the mail surveys are returned within two weeks. This helps with the response rate, especially by getting the attention of recipients who did not respond to the first mailing.

Telephone Surveys Telephone interviews are efficient methods of collecting some types of data and are being used increasingly in marketing research. Compared to mail surveys, telephone surveys are relatively expensive but quicker to administer. Depending on the type of data required, telephone surveys can generate a great deal of quality information. Interviewers exert control over the entire process and can probe for additional information on open-ended questions when a respondent provides an answer that is incomplete or unclear. During the interview process, the respondent does not know what the next question will be, which allows substantially greater flexibility in questionnaire design. Another advantage of telephone surveys is that they lend themselves to proper sampling techniques because almost all households and all businesses have telephones, and when conducted at the appropriate time, the response rate can be very high. Telephone surveys are most suitable when time is of the essence and the length of the survey is limited. However, they cannot be used for elaborate and detailed surveys that require respondents to consult records to provide an accurate response or where visual aids and display materials are associated with survey questions.

Trained interviewers normally conduct telephone surveys using a CATI (computer-aided telephone interviewing) system, with which responses are entered directly into a computer database while the interview is taking place. This approach reduces the setup time and costs. During the telephone interview process, supervisors usually monitor the process and the interviewers to assure the accuracy and integrity of the collected data. The supervisors have facilities that allow them to listen in while the interviewing is proceeding. The telephone interviewing facility usually comprises interviewing stations or booths, high-speed modem autodialing, and, in some cases, a visual and audio monitoring system.

Direct, In-Person Surveys Direct, in-person interviews can be conducted in homes, offices, shopping outlets and shopping malls, or other locations where the interviewer and the respondent can meet face to face. The most common form of this type of survey is that in which interviewers intercept shoppers at shopping outlets or malls (mall-intercept method). In-person surveys are much more expensive than mail or telephone surveys in terms of the cost per interview. The cost can be extremely high if the survey involves travel by interviewers. Another limitation of in-person surveys is bias that may result from the interviewer during the interview process. Interviewers can have their own biases that may affect the responses. This is especially the case with open-ended questions. Interviewers should be as neutral as possible and should not in any way influence the answer provided by the respondent. Bias can also be introduced when selecting the individuals to approach for interviewing. In today's society, where there is always suspicion and mistrust, interviewers may tend to approach neat, safe-looking people to interview from a stream of shoppers at a shopping mall. This is what is referred to as convenience sampling. The primary questions that arise for the researcher are: (1) How do you select people or homes to approach for interview? (2) Will you select every person/home, every other person/home, or some other sampling technique? Sampling techniques are discussed in a later section of this chapter.

Despite the limitations of in-person surveys, the method is convenient when it is necessary to display advertisement, products, packaging, and other materials associated with the survey. In some instances, intercept surveys at shopping outlets and malls can be low cost with no travel cost. Intercept surveys can also provide a good demographic spread and diversity of respondents.

Interviewers require training to be able to effectively solicit and gain cooperation from respondents. The following factors about the interviewer are important for obtaining accurate data and information from face to face interviewing.

1. Appearance: The first impression of the interviewer is very important in determining how cooperative a respondent can be. Although there are no dress codes for the interviewer, the interviewer should not be overdressed or underdressed. Many market research firms provide jackets with signs to indicate what the interviews are about.
2. Good interpersonal-skills: This requires interviewers to have the ability to approach strangers and secure their cooperation for the survey with little opposition. Interviewers should be able to quickly establish a rapport with potential respondents and interest them in the survey.
3. Good judgment: Interviewers should have the skill to make judgments relating to cooperation and noncooperation from respondents. For exam-

The Case of Red Lobster

Market reports in the 1990s indicated a continued rise in chicken consumption while seafood consumption had remained flat. Consequently, the Red Lobster restaurant chain contracted Harris Interactive of Rochester New York in 2003 to conduct a survey to assess the consumption patterns for seafood alongside those of red meats and chicken. The telephone survey polled 1,015 consumers aged 18 and above asking questions that included frequency of seafood consumption, favorite seafood, as well as a series of statements requiring respondents to choose those that best applied to them. The survey results indicated that the favorite seafood was shrimp, and 55% of respondents indicated that they eat it at least once a month. About 70% of respondents also indicated that they consume more seafood because it is a "welcome alternative to beef and chicken" (Seafood Business, 2003). The findings from the study provided enough information for the chain restaurant to continue focusing on its major product, which is seafood.

ple, in an intercept survey, it can be difficult to get cooperation from people who are in a hurry. In the case of a home survey, it will be difficult to get cooperation from households during their meal times or during the Super Bowl.

Self-Administered Surveys Self-administered surveys are administered entirely by the respondent. Potential respondents pick up survey materials, complete them, and return them at their convenience. This kind of survey is common with questionnaires relating to customer satisfaction. Most service firms or even shopping outlets place survey questionnaires at entrances to their facilities for their clients or customers to complete and drop in a box. Others can be taken and return-mailed with prepaid postage. Internet surveys are becoming popular as self-administered surveys, particularly for consumer opinion research. With Internet surveys, potential respondents are directed to a Web site through electronic mail lists or user groups. The Web site includes the questionnaire to be completed.

Self-administered surveys are convenient, less costly, and can provide well-thought-out responses to survey questions. However, respondents might not constitute a representative sample. Some respondents may not be within the intended target population, and responses provided by such respondents are not appropriate. Respondents of self-administered surveys can be people with some strong opinions.

Sampling The basis of quantitative marketing research is to gain information about an entire group of people (that is, population), such as consumers, households, and clients. If we wish to obtain information about all seafood buyers in the nation, that group is our population. Obtaining information about the entire population is ideal. Depending on the size of the target population, it may be possible to survey the entire population. For example, a teacher may want to survey his or her class about its interest in a particular teaching style. In this case the entire class will be the population. In marketing research, however, it frequently is not practicable to survey the entire population of potential consumers, or clients. It is usually necessary to draw a sample (portion of the population) to obtain information about the entire population. The selection of a valid and efficient sample is crucial to the success of applying information obtained about the sample to the entire population. Consequently, we need to find an efficient method to choose the sample from the population. This is referred to as sample design. The accuracy of the survey results depend on the quality of sampling information available at the design stage, and particularly on the implementation of the sampling procedure.

Sample design involves the following steps:

1. **Define the population or group of people to be studied.** This is the intended target group from which you wish to obtain information. For example, in a study of tilapia consumption, the target population could be grocery shoppers, seafood consumers in general, or only those who consume seafood in restaurants. Defining the target population is important, especially when the results of the survey will be used in decisions relating to marketing management and strategy development.
2. **Determine how the potential respondents will be identified.** For in-person surveys, potential respondents are the people who will be contacted in person. These would be shoppers in the case of an intercept survey, heads of households in the case of a home survey, or restaurant managers in a particular geographic area. For telephone and mail surveys, potential names and contact information of respondents are needed. These names and contact information can be obtained from telephone books or can be purchased from market research or communications companies.
3. **Determine sample size.** There is no simple, one-size-fits-all formula for the selection of a sample size to be used in a survey. For large populations, the sample size to use depends on the level of statistical accuracy and reliability that is needed to associate with the survey results. It requires establishing a statistical level of confidence and a margin of error. A high confidence of 95%, and a small margin of error of $\pm$ 1% can be obtained with a large sample. In general, the larger the sample, the better the sample results will reflect the population. The most common approach to determining the sample size for large populations is to assume a normal distribution of the target population and a random sampling procedure. Thus, the sample size can be calculated as:

$$n = \left[\frac{z_\alpha * s}{m} \right]^2$$

where n is the sample size, z_a is the critical value from a standard normal curve based on the desired confidence level α (commonly set 95% or 99% level of confidence for z_α values of 1.96 or 2.575, respectively), s is the sample standard deviation (commonly set at 0.5), and m is the desired margin of error (commonly set not to exceed ± 5%). The preceding formula can be used only under the assumptions that the population to be studied is normally distributed, the sample is generated randomly from the population, and the sample is sufficiently large such that the sample standard deviation is close to the population standard deviation.

For small populations, selecting a sample size can be calculated with the standard error computed with a finite population N correction included.

$$n = \left[\frac{z_\alpha * s}{m} * \sqrt{\frac{N - n}{N - 1}} \right]^2$$

Solving for n becomes:

$$n = \left[\frac{(z_\alpha * s)^2 N}{(z_\alpha * s)^2 + (N - 1)\, m^2} \right]$$

In practice, the use of the preceding formulas can yield a large sample size that will be too expensive to survey. In practice, the choice of sample size is often based on professional experience, available resources, and the purpose of the study when the calculated sample size is high.

4. **Choose a sampling method.** The choice of the sampling method is often determined by the study objectives, population characteristics, time, cost, and sometimes convenience. The various methods available for selecting samples include:

   a. *Simple Random Sampling:* The sample consists of individuals from the population chosen in such a way that each person in the population has an equal chance of selection. This allows the results to be projected reliably from the sample to the larger population.
   b. *Systematic Random Sampling:* The selection procedure consists of selecting every n^{th} individual in the population. If the population is in a random order, systematic random sampling approximates the simple random sampling procedure.
   c. *Stratified Random Sampling:* This procedure is applicable when there is a particular interest in a specific group or subdivision of the population. For example, if individuals of the same age or race are believed to have similar preferences for fish, a stratified random sample allows the researcher to test for this. The population is first divided into subdivisions, called *strata*, and random samples are selected from each stratum. The information collected from each stratum is then combined. This procedure is useful to capture the variability within various strata.
   d. *Multistage or Cluster Sampling:* Cluster sampling is a random selection of individuals, but the sample is chosen in stages. For example, in a survey of households for grocery coupon use, a researcher will first divide the nation into clusters, perhaps counties. A random sample of counties is then selected. From the selected counties, a list of cities and towns is then selected, and from this, a sample of households is selected.
   e. *Ad-hoc Sampling:* The general framework for the preceding four sampling methods is probability, in which each individual of the population has a known chance to be selected. In contrast, ad-hoc sampling is arbitrary and not based on any known probability or chance. Examples include convenience sampling that is often used in intercept surveys at shopping outlets and malls. The sampling is based on those shoppers who pass by. Ad-hoc sampling methods also include sampling based on personal preferences and judgment.

Questionnaire Design

It is important to design questionnaires (survey instruments) carefully. Poorly worded questions, poorly structured questionnaires, and inappropriate questions can result in erroneous and misleading information. There are many good reference sources available that provide detailed instructions on proper questionnaire design. Several marketing research college texts provide sections on questionnaire design and issues.

A questionnaire consists of several components including words, questions, formats, and hypothesis. Word selection can influence the response to a question, therefore the researcher should carefully choose the words for formulating the questions or scales. There should be no ambiguity or abstraction in the wording. There are two question formats: unstructured questions and structured questions. Unstructured questions are open-ended questions that allow respondents to write in their response; structured questions are closed-ended and require the respondent to choose from a predetermined set of responses or scale points.

A questionnaire should begin with easy-to-answer questions. Subsequent questions should flow naturally and in a logical fashion. More sensitive questions that relate to demographic information on age or income, for example, should be at the end. Questionnaires should be checked carefully to eliminate wording that is considered unanswerable, leading (or loaded), double-barreled, or incomprehensible to the respondent (see glossary for details). Validity and reliability tests should also be conducted (see glossary for details). Moreover, pretesting questionnaires is essential and will allow the researcher to correct problems with vague and imprecise wording and misunderstanding.

Different types of questions will generate different levels of information. "Yes/No" questions give some indication of consumer attitudes, but a multiple-choice question will allow for assessment of finer distinctions in attitudes. Other types of questions are the Likert scale, Rank order, Rating Scale, True/False, and Semantic differential scale (see glossary for details).

Response Rate

How many responses are enough? The answer depends on how representative the sample was and the survey method used. An attempt should be made to recontact a small sample of the nonrespondents to be certain that the survey was not biased toward certain groups of people with certain characteristics. Generally, response rates of 35%–50% are considered acceptable as long as no nonrespondent bias was observed.

RESEARCH ON ATTITUDES AND PREFERENCES

In marketing, one of the fundamental axioms that is stressed repeatedly is "know your customers." People are complex biological organisms, and the information needed about consumers depends to a large extent on the intended uses. Those in turn depend on the market conditions and the nature of the products being sold. There are several different forms of collecting information related to consumer attitudes and preferences, and each type requires a different approach.

Suppose that a company is considering adjusting the price of its product. It would be important to obtain information on how the quantity demanded by customers is likely to respond to the price change. If there are other, competing firms selling similar products, it will also be necessary to know how the competitors will respond to its price decisions. More important, the company needs to know the extent to which consumers will substitute one product for another. This is an example of understanding behavioral responses by gauging the effects of a price change.

If customers are not aware of a new product's being sold, or customers are not aware of the new lower price of a familiar product, then advertising could affect the sales of the new product or the old product at the new price. This is commonly practiced by grocery outlets and involves issuing store flyers periodically or doing in-store advertising. Information on how consumers react to advertising strategies will also be needed. For an entirely new product or a change in an existing product, information about consumers will be required.

Gathering information about consumers is not easy. It can be obtained through the marketing research process but is expensive. However, an effective marketing research process can provide good knowledge of consumers. Gathering information that is relevant for some intended uses implies knowing the customer well enough to formulate effective marketing strategies for your product or services.

Theories of Choice Behavior

Understanding behavior and preferences as these relate to choice by individuals or a group of people can be complex. People's choices manifest themselves in many ways, but are particularly expressed through active or passive responses by people, such as through purchasing specific products or services. Individual choices are influenced by factors such as income, habit, experience, advertising, peer pressure, family, and accumulated beliefs. These factors reflect the dynamic nature of human attitudes and preferences. Several theoretical frameworks have been proposed for examining consumer behavior, but there are three basic theories of choice behavior:

Neoclassical Preference Theory

Preferences are expressed as utility, which is a generalized term for the satisfaction obtained by an individual from the choice of a product (good or service). Preference is measured by the price the individual is willing to pay for the product. Total satisfaction obtained from the product is termed *total utility*, and the additional utility obtained from the use of an additional unit of the product is called *marginal utility*. The classical theory assumes that a rational individual purchases a combination of quantities of products that yields the maximum utility subject to constraints of the level of income and prevailing prices (see Pindyck and Rubinfeld 2001).

The behavioral assumptions are that a representative consumer chooses between alternative commodity combinations to maximize utility, has perfect knowledge of all alternative commodity combinations and their prices, and is capable of evaluating the alternatives. The utility is ordinal, that is, a consumer is able to order commodity combinations by level of utility (first, second, third). The utility does not require cardinality (the ability to specify the actual numeric level of utility). A demand schedule for a representative consumer is derived from the behavioral assumption of utility maximization.

Revealed Preference Theory

The neoclassical ordinal utility theory is based upon a set of psychological assumptions, but revealed preference is based on actual behavior. By changing the budget allocation of a consumer and observing the consumer's purchasing pattern, that is, which commodity combinations are actually purchased, we can derive a preference schedule for the consumer through "revealed preference" theory (see Deaton and Muellbauer 1980). Revealed preference utilizes actual behavior of consumers to derive preference curves and consequently a demand schedule.

Hedonic Theory

The classical preference theory assumes that consumer preferences are over quantities of products. With hedonic theory, consumer preferences relate to the bundle of attributes or qualities contained in that product and not the quantities of the product. Hedonic theory assumes that the qualities of the product are the ultimate source of utility for consumers and that a product is described solely by its characteristics (see Lancaster 1971). These characteristics refer to price, flavor, texture, color, and others. This assumption of consumer preference for products provides the ability to derive implicit relative prices of product attributes or qualities and how much consumers are willing to pay for each of the attributes. The relative implicit prices of the attributes, as valued by consumers, differentiate similar products in the market place. Closely related to hedonic theory is the conjoint analysis, which is used in new product research.

PRODUCT RESEARCH

PRODUCT IDEAS

Research in market products begins with a testable product concept. Examples of product concepts that have been translated into successful marketing products over the years may include ready-to-eat products, resealable retail packs, reduced-fat products, or the elimination of preservatives in the products. The concept should try to address some current or emerging consumer need. In particular, product attributes, packaging, positioning, and pricing play a vital role in the development of any product concept.

The development of a new product often depends on the type of product. Whether the product is evolving from a known or an existing product or is an entirely new product will affect its market development path. For example, repackaging a known food product or adding new flavors to an existing product are evolutionary concepts. An existing product can be modified to suit particular needs or improve on particular experiences of customers in the use of the product.

An entirely new product concept can be termed a *revolutionary concept*. Revolutionary product concepts involve discovery of consumer needs that have not been met by existing products. For example, the microwave oven was a revolutionary concept that allowed for the preparation of many ready-to-eat food products to meet the growing busy schedules of the working population.

PRODUCT TESTING

Testing a concept identifies potentially successful new products and determines the probability that consumers will accept the product. Evolutionary product concepts are best tested using qualitative research such as in-depth interviews and focus groups in combination with quantitative, survey-based research methods. The use of the qualitative phase allows for fine-tuning the product concepts and formulating hypotheses for the quantitative phase. The quantitative research provides specific measurements to assist in marketing decisions.

Revolutionary concepts can best be tested with qualitative methods such as focus groups and in-depth interviews. New products are usually developed based on some perception of market need. However, there may be many possible ways to meet that perceived need. The relative strengths and weaknesses of the product concept can be better evaluated through qualitative research. Chances are that there may even be some prototypes available on the market. Qualitative market research will provide information on any product flaws, flavor preferences, size, packaging, and a host of other modifiable attributes before going into mass production.

Product testing could also involve testing the name to avoid confusion with similar products or problems with pronouncing and writing the name. Evaluation of the packaging is also important. A test helps to identify how readily customers would identify the product on the shelf, open the package, and follow the cooking and preparation instructions. A product test involving sensory evaluation will provide understanding of consumer preferences regarding texture, flavor, color, and other basic product attributes.

Generally, product testing allows the prediction of consumer acceptance of new products. With product testing, there is the possibility of achieving product superiority over the competition. Companies that do frequent testing can continuously improve product quality and customer satisfaction, especially as consumer tastes evolve over time, and will also be able to monitor the potential threat levels posed by competitive products. Product testing will provide some understanding of competitive strengths and weaknesses and can also allow the implicit measure of the effects of price, brand name, or packaging on perceived product quality. It is often recommended that tests be conducted in real environment situations. For example, for food products, an in-home usage test is recommended because it provides a more accurate and predictive response.

During the testing process, the critical variables to examine should be the quality attributes of the product from the consumer's perspective. It is important to determine what product attributes are truly important to consumers and what factors determine consumer satisfaction. After consumer acceptance is ascertained for the product, the product can be introduced into a limited geographic area for a period of time (to observe product repeat purchase patterns) before the company ventures into general markets.

MARKET SHARE RESEARCH

Market share is among the important parameters in the marketing research process. It is a critical indicator of relative performance compared to the competition and shows which company's products or services are bought the most and who the competitors in the market are. Therefore, market share research provides measurements of the proportion of the market supplied by the company's specific product. It is the percentage of market unit volume or dollar value held by a company as a proportion of total market size. Market share may be expressed either in unit sales or dollar values, that is,

$$Market\ Share = \frac{Total\ company\ sales\ (units\ or\ dollars)}{Total\ market\ or\ industry\ sales\ (units\ or\ dollars)}$$

In the business world, attaining the highest market share is the objective for most companies. It is believed that regardless of the price of the product or service, a company with a high market share will remain more profitable than the competitors. However, some small companies with small market shares can function profitably in large marketplaces. This is because they develop and service a large share of a small segment of the total market.

Business mergers and acquisitions are common in the marketplace. Therefore, the level of market share also suggests the safety and stability of the position occupied by the company in the market. Large competitors have frequently absorbed smaller competitors to increase market share.

Because of the competitive nature of the business world, it is always important for companies to monitor how their share of the market changes over time. A company's sales may be growing at the same time that market share is decreasing. Monitoring market share over time should be a vital part of a company's overall strategic business, marketing, and sales plan. With good market share information, a company can adjust its marketing strategies and improve its revenue, customer base, and brand value.

To establish the market share of a company's product or service, interviews and surveys can be used to obtain primary sales information from manufacturers, vendors, and customers. In general, market share research has been concerned primarily with the top players in the market for a single product or an entire product line within a single or a segmented market.

ADVERTISING RESEARCH

Advertising is a major part of marketing products. Advertising may be used to convey information about a product or service to a target audience, or it may be used to create awareness or change perceptions about a product or service. Generally speaking, the ultimate purpose of advertising is to influence consumer behavior through either a change in behavior or reinforcement of an impression or perception for the benefit of the advertiser. The change in consumer behavior is a change in purchase of the product or service being advertised that will result in an increase in sales. Thus, in food marketing, changes in sales or similar measures are used as measures of effectiveness of advertising programs.

Companies recognize that consumer preferences are not static but are subject to some degree of randomness and systematic change. Humans, by nature, are dynamic in terms of their preferences. New products continue to be developed and introduced to the marketplace from time to time to meet the changing tastes and needs of consumers. Technological advances also affect new product development and influence food preferences and demand.

Advertising can lead to changes in consumer behavior, but the degree of change will differ by the type of commodity and potentially by the nature and quality of the advertising. There are advertisements that can readily alter the preferences of consumers, and there are others that rarely alter consumer preferences. Such products may be well defined and stable. Food is considered a typical example of a product with a stable preference because of the inelastic nature of demand for food. However, demand for food depends on the prevailing price, price of substitutes, and perhaps the attributes of the product. Thus, consumption depends on consumer knowledge and perceptions about product attributes. Advertising plays the role of influencing knowledge and perceptions of food products.

The marketing literature includes a number of examples of different measures used to determine the effectiveness of advertising. Copy testing is the most common measurement approach and measures the effectiveness of advertising within minutes or hours of exposure to the advertisement. Some of the measurements used in copy testing can be classified into the following (Haley and Baldinger, 1991):

1. *Persuasion Measures:* Using a survey instrument, one can solicit choice of a brand among a product category, overall brand ratings, and purchase interest and intentions of particular products or brands.
2. *Salience Measures:* With this type of measure, respondents are examined for high brand awareness, such as top-of-mind awareness, unaided awareness, and total awareness (unaided and aided) of particular products or brands.
3. *Recall Measures:* These measure the ability of respondents to recall brand from cues of product category and brand category.
4. *Communication Measures:* Sometimes advertising research focuses on the main point of communication (TV, print media, sales point) and nature of advertisement (ad situation, visual characteristics).
5. *Diagnostics Measures:* These relate to reaction to the advertisement. Positive reaction invites responses such as "I learned a lot from this advertisement," "ad tells me a lot about the product," and "I learned something new about the product." A negative response could be "the product does not taste as good as the ad claims."

In practical work, persuasive and recall measures have been found to perform better than others in terms of predicting the effectiveness of advertising.

In farm commodity marketing, advertising has usually been generic. In evaluating generic advertising programs, researchers have typically used (1) advertising expenditures and (2) gross rating point as the primary indicators of effectiveness. Advertising expenditure is used as a proxy for advertising intensity and assumes a positive relationship between the amount spent on advertising and sales. The gross rating product is a product of the reach of the advertisement and the average of its distribution of exposures delivered to a target audience.

SALES CONTROL RESEARCH

Companies are always looking for ways to have a competitive advantage in the marketplace. One of the surest ways of gaining that edge over the competition is to research the market and monitor sales performance of various products. In addition to research on consumer attitudes and preferences, products and services, and market share and advertising, forecasting sales is also an important aspect of market research. Sales forecasting helps to determine trends in the marketplace and how to benefit from such trends.

The process of sales forecasting involves organizing and analyzing information in a way to estimate future sales. Sales are generally affected by several factors including seasons, holidays, special events, direct and indirect competition, labor events, productivity changes, demographic trends, fashions or styles, political events, weather, and other factors. These can be considered as external factors. Within the company, factors that can potentially affect sales include changes in product form, product quality, production capacity, advertising and promotion, sales efforts and strategies, price changes, inventory, distribution methods, and credit policy changes, among others.

A qualitative type of research is required for developing a good sales forecast with internal and external information including information from competitors, neighboring businesses, trade suppliers, business associations, trade associations, and trade publications. The information should be useful in describing a purchase pattern and event and to understand market situations and trends from a holistic perspective.

There are a number of indicators that can be followed to develop sales forecasts:

1. Sales revenues from the same month or quarter in the previous year are good predictors of sales for the same period in succeeding years, but trends and forecasts in the economy and the industry must be accounted for.
2. Actual customer contacts and salespersons closely associated with customers and particular products, services, market or territory can provide some good estimates.

New businesses should begin with the following to develop sales forecasts:

1. Developing customer profiles and determining industry trends.
2. Making some basic assumptions about the customers in the target market by developing a profile of the principal market. For example, assess the business (is it a small- to medium-sized grocery outlet, and what are the sales volumes?). Determine the profile of, say, 20% of the target market (males, ages 20–34, professional, middle income, fitness conscious; or young families, with parents ages 25 to 39, middle income, homeowners).
3. Determining trends by talking to trade suppliers about what is selling well and what is not, and by reading the industry's trade magazines and business periodicals.
4. Establishing the approximate size and location of the business area, using available statistics to determine the general characteristics of the area, unique characteristics, how far the average customer travels to buy from the outlet, or how far to go to distribute or promote the product. Government statistics can be used to estimate the number of individuals, households, or businesses.
5. Listing and profiling competitors in the business area. Study the competitors, visit their stores or locations, analyze the location, customer volumes, traffic patterns, hours of operation, busy periods, prices, quality of goods and services, product lines carried, promotional techniques, positioning, and product catalogues and other handouts, and talk to customers and sales staff.
6. Estimating sales on a periodic basis (monthly or quarterly). The basis for the sales forecast can be the average monthly or quarterly sales of similar-sized competitors operating in a similar market, making adjustments for predicted trends in the industry.
7. Considering how well competitors satisfy the needs of potential customers in that trading area and determining how the new business's products fit in and what niche can be filled (a better location with more convenience, a better price, better quality, or better service).
8. Considering population and economic growth in the trading area and estimating market share.
9. Reviewing forecasts periodically using actual sales figures, and revising the forecast accordingly.

DATA ANALYSIS

From the preceding sections, it is clear that much information and data can be collected in market research. What can be done with all the data and information gathered? Comprehensive analysis of the data is necessary to fully understand the market implications.

The large amount of information and data gathered during the market research process can be useful only if it is presented in a form that makes the information meaningful. There are several software applications that can present the data and information in desired forms. The common statistical software application used in market research is SPSS (Statistical Package for the Social Sciences) by SPSS, Inc. of Chicago, Illinois. There are even some integrated questionnaire design and analysis software programs that allow the researcher to design the interview or survey material and analyze the responses, for example, Survey Pro by Apian Software of Berkeley, California. Whatever data and information are gathered, the analysis needs to relate directly to the nature of the data gathered and the nature of the research objectives, questions, or hypotheses.

To do any analysis with data from a survey instrument, the responses need to be converted into numbers for analysis. This is commonly called *coding*. Code numbers are assigned to particular responses in survey questionnaires. This allows the presenting of market research data in the form of statistical

The Case of the Pacific White Shrimp

In 2000, Florida Department of Agriculture and Consumer Services Bureau of Seafood and Aquaculture Marketing (FDACS-BSAM), the University of Florida Institute of Food and Agricultural Science (UF-IFAS) Aquatic Food Lab, and the University of Florida Indian River Research and Education Center collaborated on a research project to identify and characterize the most attractive direct markets for fresh farm-raised shrimp and to provide shrimp farmers the market information needed to develop successful direct marketing strategies (Florida Department of Agriculture and Consumer Services—Bureau of Seafood and Aquaculture Marketing 2001). The study was motivated by the expanding shrimp farming industry, particularly that of the Pacific white shrimp, *Litopenaeus vannamei*, in Florida and in other southern states. Though demand for shrimp was high, the U.S. farm-raised shrimp could not compete effectively on price with imports in fresh and frozen shrimp markets for the most popular forms and sizes. The study therefore explored the feasibility for U.S. shrimp farmers to market their products directly to restaurants, retailers, and consumers. The potential for direct food markets for live and fresh, head-on, farm-raised shrimp had not been explored.

This study was conducted using a combination of focus groups and interviews to investigate buyer attitudes and preferences toward shrimp. The mail survey included a conjoint analysis experiment to quantify the relative importance of various shrimp attributes for each potential market segment. The survey instrument was administered by mail to 3,038 seafood dealers in the nine southeastern U.S. states. A total of 250 were completed and returned by seafood dealers. Another survey instrument was developed and administered by mail to 2,465 seafood restaurants in the same region. A total of 211 were completed and returned by restaurants. On the consumer side, three consumer focus groups with a total of 30 participants were conducted to explore consumer attitudes and preferences concerning shrimp in general, and, in particular, freshwater farm-raised Pacific white shrimp, *Litopenaeus vannamei*.

Details of the results are provided in Florida Department of Agriculture and Consumer Services—Bureau of Seafood and Aquaculture Marketing (2001). From the results, the study concluded that the shrimp dealer market was not promising for direct sales of whole, farm-raised shrimp, but there was some potential in the seafood restaurant market. According to the study, there was an excellent consumer market potential for cooked, never-frozen Florida freshwater farm-raised Pacific white shrimp. Florida farm-raised Pacific white shrimp processed into tails had the most favorable market acceptance and could be sold at a premium price. The study found that Florida freshwater farm-raised Pacific white shrimp had a distinctive product-differentiating feature that presented a competitive advantage compared to shrimps from other sources. Therefore, the study suggested that in-store displays of Florida freshwater farm-raised Pacific white shrimp should highlight cooked shrimp rather than raw shrimp, at least in the early stages of industry development when consumers were unfamiliar with the product. On willingness to pay for Florida freshwater farm-raised shrimp, 75% would pay $15/kg whereas 50% would pay $17.5/kg.

summaries and inferences, as well as of the relationships among variables.

Statistical Summaries

Statistical summaries can be presented in graphical forms such as charts (for example, line graphs, bar charts, histograms, stem-plots) and in tabular forms. These give a snapshot of all the data gathered. These are some examples of useful statistical measures:

Proportions

Determining the proportion of all respondents that respond in a particular way to specific questions may be useful. For example, after a survey of grocery retailers, one might be interested in what proportion of respondents has fish counters, or what proportion of respondents answered "Yes" to a particular question. This proportion is simply the number of responses of interest over the total number of responses. For example, if 817 out of 950 respondents have fish counters, the proportion is 817/950 = 0.86 or 86%.

Central Measures

The mean or the arithmetic average is commonly used to summarize survey response data. Another measure of the center of the distribution of responses is the median, which is the middle value of all the responses of interest from the lower value to the highest value. A third measure of the center is the mode, which is the value that has the highest number of occurrence. For example, if data were obtained on sales volumes, we might be interested not only in the lowest and highest but also in the mode, median, and average sales volume among the respondents. Suppose that we obtained the total sales value of eight seafood companies in a particular year as $90 million, $37 million, $24 million, $57 million, $68 million, $112 million, $78 million, and $68 million. The calculated mean is $67 million, the mode is 68, and the median is $68.

If the responses are categorical, say, "yes" and "no" answers, the mean and the mode measures are irrelevant. However, the mean of each response is also the proportion of response. For example, 950 people responded to a particular question requiring a "yes" and "no" answer, and the "yes" responses are 456 and the "no" responses are 494. The mean of the "yes" responses is 0.48 (or 48%), whereas the mean of the "no" responses is 0.52 (or 52%). Both results are also the respective proportions of total responses.

Variability Measures

Variability of responses to a particular question is another important measure. It provides information on how much difference and diversity exist among respondents on issues associated with particular questions. One can observe variability and diversity among responses by looking at the distribution of response frequencies and proportion of various responses to a question. If the distribution is over a wider range, it is an indication of differences and diversity. The common measure of variation or diversity in the responses is the standard deviation (commonly represented by σ), which measures how far responses are from the mean value. Figure 13.1 illustrates two normal distribution curves, showing the mean (and the standard deviation σ. The values of the standard deviation are different; the curve with the larger standard deviation is more spread out. The value of standard deviation is always positive. It is zero if there is no difference or diversity in responses. Variance is another measure of differences and diversity.

Statistical Inferences

It was pointed out earlier in the chapter that the basis of marketing research is to gain information about the entire population. However, because this is not often practical, information derived from a sample is extended to the population. Statistical procedures in which measures about the sample are used to make inferences about the population are known as statistical inference.

Statistical procedures involve estimation of parameters using the sample data, testing of hypotheses, and testing of the significance differences between estimated parameters. A parameter is a number that describes a population; a statistic is a number computed from the sample data. Sample statistics are therefore used to make assertions about unknown parameters. For example, if, in a survey of grocery outlets, 817 out of 950 respondents (sample) have fish counters, the sample proportion

$$\hat{p} = \frac{817}{950} = 0.86$$

The sample statistic $\hat{p}$ is then used to estimate the unknown population parameter p. In theory, repeated random sampling or experimentations would result in a sampling distribution of the sample statistic, in this example, $\hat{p}$. With many samples drawn randomly from the population, the mean value of the sample statistic $\hat{p}$ will approach the true value of the population parameter p. In general, if the mean of the sampling distribution of a statistic is equal to the

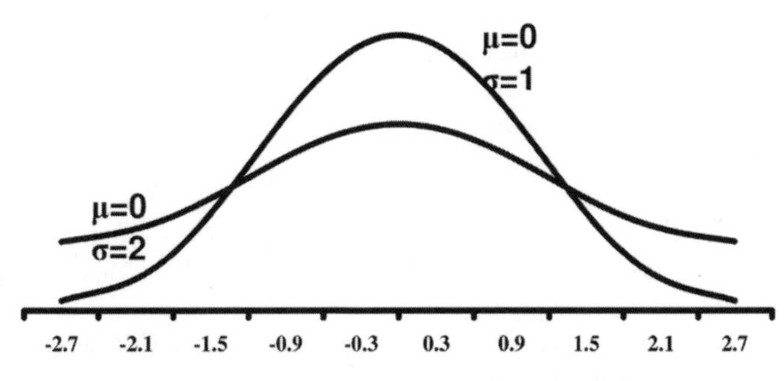

Figure 13.1. Two normal curves, showing the mean μ and the standard deviation σ.

true value of the parameter being estimated, the statistic is said to be unbiased.

Just as variation or diversity measures such as the standard deviation measure how far responses are from their mean, there is also a measure of variability of a statistic describing the spread of its sampling distribution. This measure is called the *standard error*, or s . Researchers often desire to have some level of statistical confidence in the estimated statistic; therefore a confidence interval is stipulated in the form of a percentage. Most researchers stipulate confidence levels of 95%, which corresponds to ± 1.96 standard errors. Others use 90% and 99%, which correspond to ± 1.64 and ± 2.58 standard errors, respectively. The percent levels of confidence are represented as z. Generally, a population estimate has a confidence interval of the form

$$\textbf{estimate} \pm z * s_{\text{estimate}}$$

and $z * s_{\text{estimate}}$ is known as the margin of error.

Relationships between Variables or Responses

Scatter-Plots

The simplest way to examine relationships between two variables is a plot of the data. A scatter-plot reveals the relationship between two variables when one variable is plotted on each axis. Each individual data point appears as a point in the plot fixed by the values of the two variables for that individual. The scatter-plot can be examined for any direction, form, and strength in the relationship between the variables. The relationship may be positive (both variables increase or decrease together in the same direction), or they may be negatively related (as one variable increases, the other decreases or they move in opposite directions). A strong, moderate, or weak association may be observed between the variables or there may be no form of association between them. Where there is a strong association, it may be an indication that one variable depends on the other.

Scatter-plots are graphic depictions of relationships. Researchers may wish to obtain a numerical measure of the relationship rather than mere graphs. A common measure to examine the relationship is to calculate the *correlation*, r. Correlation measures the strength and direction of the linear association between two quantitative variables. A positive value indicates a positive association; a negative value indicates a negative association. The value of r lies between -1 and 1 and indicates the strength of the relationship by how close it is to -1 or 1.

Least-Squares Regression

A least-squares regression is a method of finding a straight line that summarizes the relationship between quantitative variables, where one variable is considered the dependent variable and the other is an explanatory variable. A least-squares regression line tries to fit a straight line as close as possible in a scatter-plot, and the fitted line can be used to predict the value of the dependent variable for a given value of the explanatory variable. Mathematically, this can be expressed as

$$y = \alpha + \beta x + \varepsilon$$

where y is the dependent variable, x is the explanatory variable, ε is an error term, α is the intercept of the fitted line and β is the slope of the fitted line. Not all relationships may be linear or appear to be in a straight line. The relationship may appear in the form of a curve, in which case a nonlinear or curvilinear regression must be applied to fit the relationship. The simplest form of fitting a nonlinear curve is to take the natural logarithm of the variables, that is,

$$\log(y) = \alpha + \beta \log(x) + \varepsilon$$

Cross-Tabulations

A cross-tabulation table is a table showing the relationship between two categorical variables with r rows and c columns. It is sometimes called an $r \times c$ table. Here is an example of responses from grocery supermarkets relating to the statement "customers prefer fresh fish to frozen fish" (see Table 13.1).

Suppose that we want to test whether there are any differences in the responses given by the three groups of grocery outlets. We should first formulate a statistical null hypothesis that the responses from the three types of outlets are the same. Then we compare the observed counts of responses in the table with the *expected counts*. If the observed counts in the table are far from the expected counts, then there is evidence that the responses from the three types of outlets are different, that is, we do not accept the null hypothesis that the responses are the same. The expected count is calculated as:

$$\textbf{Expected Count} = \frac{\textbf{row total} \times \textbf{column total}}{\textbf{table total}}$$

The preceding equation generates the expected count for each cell.

Table 13.1. Example of Cross-Tabulation.

	Strongly Agree	Agree	Disagree	Strongly Disagree	Total
National Chain Store	245	114	126	16	**501**
Regional Chain Store	115	19	135	18	**287**
Independent	66	118	16	12	**212**
Total	**426**	**251**	**277**	**46**	**1000**

The Chi-Square Test

The chi-square test uses the observed counts and expected counts to determine whether any differences are statistically significant. It is a measure of how far the observed counts in a cross-tabulation table are from the expected counts. The formula for chi-square, denoted as χ^2, is

$$\chi^2 = \sum \frac{(\text{observed count} - \text{expected count})}{\text{expected count}}$$

and the summation is over all $r \times c$ cells in the table. The chi-square analysis has been found to be a useful statistic for comparison when at least 20% of all expected counts are 5 or greater and there are no zero values of expected counts. Many statistical computer software applications will give a warning if less than 20% of calculated expected counts are less than 5.

Discrete Choice Analysis

Least-squares regression techniques are appropriate when the dependent variable is quantitative data. However, when the dependent variable is qualitative in nature, as is often obtained in survey data, the analysis requires different techniques. This is what is known variously as qualitative dependent variable analysis, limited dependent variable analysis, or discrete choice analysis. Good references for this type of analysis are Ben-Akiva and Lerman (1987), Maddala (1983), and Greene (1997).

The technique is a linear probability model, in which the dependent variable is interpreted as the probability of occurrence. For example, suppose that we have gathered data from a survey that asked for responses on smoked tilapia. Suppose that our dependent variable is a yes and no response to buying smoked tilapia, and the dependent variables were related to patterns of fish purchase and demographics. Using a linear probability model to fit the data, we can predict the probability of buying smoked tilapia. Probabilistic models are based on the assumption that a choice among alternatives, such as yes and no, is utility driven. In other words, individuals are assumed to choose an alternative that provides more utility than the other alternatives that were not chosen. Following are some examples of probabilistic models.

Logit Model In the example described above, the logit model specifies the probability that an individual will buy smoked tilapia as

$$\textbf{prob(buy)} = \frac{\exp(X\beta)}{1 + \exp(X\beta)} = \frac{1}{1 + \exp(-X\beta)}$$

where exp is exponent, X is a vector of explanatory variables and β is a vector of estimated coefficients. The logit formulation is based on a logistic distribution of the error term. The probability of not buying will then be expressed as

$$\textbf{prob(not buy)} = 1 - \frac{\exp(X\beta)}{1 + \exp(X\beta)} = \frac{1}{1 + \exp(X\beta)}$$

The preceding formulation of the logit model can be expressed in a different way. The ratio of the prob(buy) to prob(not buy) is

$$\frac{\textbf{prob(buy)}}{\textbf{prob(not buy)}} = exp(X\beta) \quad \textbf{or}$$

$$\ln\left[\frac{\textbf{prob(buy)}}{\textbf{prob(not buy)}}\right] = X\beta$$

In terms of predicting the probability of buying, the sign and magnitude of the coefficient indicate the direction and effect of the relevant explanatory variable on the probability of buying. Marginal effects are used to explain changes in probability given a change in the relevant independent variable. The elasticity measure gives the percentage change in the choice probability in response to a percentage change in the explanatory variable.

Probit Model The probit model is similar to the logit model except that it is based on the assumption of a normal distribution of the error term. In this formulation, the probability of buying is specified as

$$\textbf{prob(buy)} = \int_{-\infty}^{X\beta} \frac{\exp(-t^2/2)}{\sqrt{2\pi}}\, dt$$

Interpreting the coefficients and predicting the probability of buying are the same as that of the logit formulation. Marginal effects and elasticity measures can also be used to interpret the effects of explanatory variables on the choice probability.

The logit and probit models are examples of binary choice (two choices) models. Other examples of binary choice models are the Gompertz or log log model, the Burr or Scobit model, and the Complementary log log or Extreme Value model (see Greene, 1997, for differences in the models).

There are instances in which the choices are more than two. For example, suppose that we are interested in knowing which species of fish households will purchase among these alternatives: orange roughy, cod, buffalo, tuna, salmon, and catfish. To analyze multichoice response data, we would use extensions of the logit and probit models that are called Multinomial Logit and Multinomial Probit models.

Discrete choice models are commonly applied in marketing research to problems of how consumers choose among competing products. It helps to determine which attributes matter most to consumers when choosing among alternatives. Whether the project is qualitative research, exploratory research, or examining purchase motivation, product positioning, or market segmentation, discrete choice analysis will help to provide some answers to the question of why consumers buy what they buy.

The major advantage of discrete choice techniques is that they are based on the observation of consumer choices. These can be either real choices or simulated choices. In marketing, consumer choices are ultimately the important factors that companies would want to know. The LIMDEP software by Econometric Software, Inc. of Plainview, NY, is commonly used to perform discrete choice analyses.

Conjoint Analysis

An alternative methodology applied in marketing research to examine consumer choices and preferences is the conjoint technique. *Conjoint* is a generic term that refers to a number of paradigms in psychology, economics, and marketing that are concerned with the quantitative description of consumer preferences or value trade-offs. Conjoint analysis is sometimes referred to as "trade-off" analysis because individuals are forced to make trade-offs among different product attributes when completing conjoint questions. Through the trade-offs, inferences can be made as to how important or valuable different attributes are and how they influence individuals' decision-making processes. Good references for this type of analysis are Green (1974), Green and Srinivasan (1978), and Green and Wind (1975). There are several forms of the conjoint techniques.

The initial step in conjoint analysis is to identify the attributes that are critical to buyers when assessing the product or service. Focus groups and personal interviews of representative buyers can be utilized to determine the attributes. The choice of attributes to include in the experimental design should be distinct among products or services. The next step is to determine the number of attributes and levels to include in the experimental design. For example, three attributes each with four levels would require 64 comparisons ($4 \times 4 \times 4$) so that care must be taken not to include too many attributes and levels. The combinations of the experimental design are included in a stimulus card and the number of stimulus cards are presented to the respondents for evaluation. Each card represents a different combination of levels of attributes selected in the design. A sample of attributes and the levels and a sample of a stimulus card representing one of the profile combinations are shown in Table 13.2 and Table 13.3, respectively. The SPSS software is commonly used to perform conjoint analyses.

Table 13.2. Example of Levels of Attributes for Fish.

Fish Attribute	Levels			
Price /lb	$1.50	$2.00	$2.50	$3.00
Color	off-white	white	pinkish white	pinkish
Flavor	mild	fishy	muddy	musty
Texture	oily	moist	moist & oily	dry

Table 13.3. An Example of a Stimulus Card.

Price:	**$2.00**
Color:	**Pinkish**
Flavor:	**Mild**
Texture:	**Moist & Oily**

Traditional Conjoint In the traditional conjoint technique, respondents are shown different product/service scenarios (stimulus cards) whose attributes vary according to an experimental design. Respondents are typically asked to rate or rank the product scenarios presented to them. Suppose that respondents are presented with 18 such cards. They will be required to rank them from 1 to 18 in terms of the preference. The sequential ranking procedure results in a ranked order for the number of cards from the least preferred combinations of attributes to the most preferred combinations. After ranking data is collected, analysis involves quantifying the values assigned to each level. The method usually employed is the additive model that assumes that the utility of an alternative is formed by a linear combination of the utilities of its parts, that is, individual rankings (part-worths) for each attribute are added together to obtain a total worth value for each combination of the product or service.

Total Worth of a Product/Service	= Part Worth of level *i* for Attribute 1
	+ Part Worth of level *i* for Attribute 2
	+ Part Worth of level *i* for Attribute 3
	+ Part Worth of level *i* for Attribute *n*

The specification of total worth of a product provides a means to estimate the importance and contribution of each attribute to the total utility of an alternative. This approach enables the assessment of the relative importance of various attribute levels in the context of preference and to study the effects of trade-offs among different attributes on consumer evaluations.

Discrete Choice Conjoint In this technique, there is no ranking. Rather, respondents are provided with different pairs of product or service profiles and they are required to select the one they would most likely purchase. For example, respondents might be shown three different profiles of fish and asked to indicate the one they would purchase. Discrete choice conjoint is more commonly applied than the traditional conjoint because it is a more realistic exercise that mimics what actually takes place in the marketplace. The other advantage of discrete choice conjoint studies is that the alternatives often include the option to select "none" of the products, thus indicating that respondents do not like any of the products presented to them. Discrete choice conjoint also allows for much more complex statistical modeling to examine interactions among attribute levels, alternative-specific effects, and cross-effects (Table 13.4).

Best-Worst Conjoint This is the least popular technique among the conjoint techniques. Respondents are typically shown the levels associated with attributes and are asked to select the one that they like best or the one that is most appealing as well as the one they like least. The process is repeated several times with a different set of levels shown each time. After the data are collected, the utilities are calculated that indicate the relative value of the attributes and attribute levels. This conjoint technique is applicable to attributes that are abstract and cannot be easily quantified.

Traditional Demand Estimation

The neoclassical demand theory assumes that an individual consumer possesses a preference ordering for alternative bundles of commodities and that this ordering can be represented by an ordinal utility (U) function, $U = U(X)$, where X is a vector of bundles of commodities. It is required that this preference

Table 13.4. An Example of a Stated Choice Question.

Please Choose ONLY ONE Alternative

FISH ATTRIBUTE	**ALTERNATIVE A**	**ALTERNATIVE B**	**ALTERNATIVE C**
Price /lb	$1.50	$2.50	Neither A nor B is preferred
Color	Off-white	White	
Texture	Oily	Dry	
Flavor	Mild	Fishy	
I would choose	☐	☐	☐

relationship satisfy some six axioms, which indicate rational consumer behavior and facilitate the maximization procedure:

- Reflexivity: Each bundle of commodities is at least as good as itself.
- Completeness: The consumer has ability to rank all the bundles.
- Transitivity: There is consistency in the consumer's ranking.
- Continuity: The utility function is differentiable to the first and second order.
- Nonsatiation: More of the bundle of commodities is always preferred by the consumer.
- Convexity: Ensures diminishing marginal rate of substitution among bundles of commodities.

Details of demand theory and the basis for these assumptions can be found in any standard microeconomics or consumer theory textbook (for example, Pindyck and Rubinfeld 2001). With the preceding assumptions satisfied, the individual consumer is assumed to face the choice of maximizing his or her utility function subject to a budget constraint. The problem of constrained utility maximization can be solved mathematically. The result is the derivation of demand relationships that give quantities as a function of prices and income or total expenditure.

Traditional demand models typically use the quantity demanded as the dependent variable. Multiple regression techniques are used to estimate the quantity demanded as a function of independent variables such as the product's own price, population levels, consumer income, tastes and preferences, and prices of related goods. Because the basic demand model quantifies the relationship between quantity demanded and price, the various elasticity measures explained in Chapter 2 can then be calculated.

An alternative approach to the consumer choice problem is one of selecting commodities to minimize the money outlay necessary to reach a predetermined utility level ($\bar{U}$). The solution to the minimization problem can also be gained mathematically to obtain compensated demand functions. These demand relationships provide a general characterization of the properties of demand functions, which includes adding up, homogeneity, symmetry, and negativity. In empirical analysis, these properties are usually imposed. This kind of analysis requires time series and cross-sectional quantitative data. There are several functional forms for demand model specifications that have been used to examine demand for food. The models include the linear expenditure system, the Rotterdam model, the direct translog, the indirect translog, the almost ideal demand system (AIDS), the quadratic AIDS, the inverse AIDS, the quadratic expenditure system, the general ordinary differential demand system, and Lewbel's demand system (see Pollak and Wales 1992; Lewbel 1990; Deaton and Muellbauer 1980; Theil 1975 & 1976).

EMPIRICAL STUDIES OF INDUSTRIAL ORGANIZATION

Empirical studies of industrial organization in agriculture and agriculture-related industries focus on competitive relationships in agricultural markets. Companies in the industry are assumed to display noncompetitive behavior in the market through the exercise of market power. Noncompetitive markets often exhibit high levels of either buyer or seller concentration, although some industries can be both concentrated buyers and concentrated sellers. Chapters 5, 6, and 8 outlined some of the characteristics of food manufacturing, distribution, and retailing and the attempts by companies to gain economies of size through mergers and acquisitions, and achieve product differentiation.

Empirical work on industrial organization to study industry market power has followed two major approaches: the Structure-Conduct-Performance (SCP) paradigm and the New Empirical Industrial Organization (NEIO) paradigm. Schmalensee (1989) provides a detailed review of SCP studies, whereas Bresnahan (1989) provides detailed discussion of the NEIO approach. Each of these is summarized next.

STRUCTURE-CONDUCT-PERFORMANCE (SCP) PARADIGM

The underlying assumption of the SCP paradigm is the idea that the structure of an industry affects its performance through conduct. The conduct of firms in an industry in terms of market power is measured through the behavior of output prices or price-cost margins. The SCP paradigm assumes that if an industry is concentrated, firms in the industry will exercise market power by depressing prices and making more profit. Consequently, the SCP paradigm uses reduced-form regression models in which the dependent variables could be output price, wholesale-retail margin, farm-wholesale margin, profits, aggregate prices, or transaction price. The major explanatory variable is buyer concentration, but other variables relating to structural and industry characteristics could be incorporated as explanatory variables. An example of the SCP model is as follows:

$$\Pi = \alpha_0 + \alpha_1 H + \sum_{i=2}^{k} \alpha_i z_i + \varepsilon$$

where Π is some measure of profitability (for example, farm-wholesale margin), H is some measure of concentration (for example, concentration ratios), the z_is are some explanatory variables thought to affect the measure of profitability (for example, market share), the α_is are coefficients to be estimated, and ε is an error term. Margins are calculated from published data on prices and cost. With time series data on the variables in the model, the various coefficients can be estimated. The coefficient of interest is α_1, that is, the coefficient of the concentration variable c. A positive value of α_1 implies that there is market power, that is, the higher the concentration in the industry, the higher the profits of firms. The beef packing industry has received greater attention by researchers because of the concentrated nature of the industry. Examples of application of SCP models in the beef packing industry include Menkhaus et al. (1981), Quail et al. (1986), and Marion and Geithman (1995).

The advantages of the SCP paradigm are availability of required data and simplicity of estimation. The major disadvantage however, is the assumption of market power when there is a positive correlation between profitability and concentration. Some researchers have argued that firms become larger because of efficiency, that is, an efficient firm that operates at reduced costs will make higher profits and consequently expand. Therefore, a profitability-concentration correlation is not a good measure of industry conduct or market power.

New Empirical Industrial Organization (NEIO) Paradigm

The NEIO approach measures market power through the gap between output price and marginal cost for oligopoly, and through the gap between input prices and marginal value product for oligopsony. It also uses structural equations involving supply and demand relations and the conduct formulation. To examine oligopoly (market power in the output market), the conduct or market power relation will take the form:

$$p = c - \theta p'(Y)Y + \varepsilon$$

where p is output price, c is the common marginal cost, θ is the market power coefficient, Y is output quantity demanded, and $p(Y)$ is the demand function for the industry's product. Estimation involves specifying: (1) a cost function and deriving an expression for marginal cost c; and (2) a market demand function $p(Y)$. The estimated market power parameter θ measures the competitiveness of a market in a very general way. The parameter is often estimated as a single index of market power, but more generally, it could be a function of exogenous variables to understand the determinants of market power over time. Applications of the NEIO models in the beef packing industry include Schroeter (1988), Azzam and Pagoulatos (1990), Schroeter and Azzam (1990), Azzam (1992), and Muth and Wohlgenant (1999).

Although product prices are readily observable, marginal cost is not observable; therefore, the NEIO approach uses "conjectures" of firms to infer conduct or market power. The main advantage of the NEIO approach is that the price-cost markup and the degree of competition in an industry can be estimated instead of calculating price-cost margins from published data on prices and cost, as is done under the SCP approach. Another advantage is that the NEIO model is applicable to a variety of structure and conduct settings. However, the disadvantages include the extensive data requirements, potential misspecification of the cost function and (or) market demand function, the ad hoc estimated supply relationship, and theoretical criticisms about conjectural variations. NEIO studies typically impose functional-form assumptions, which under static oligopoly models imply strong restrictions on the relationship between price and marginal cost.

AQUACULTURE MARKET SYNOPSIS: ORNAMENTAL FISH

Sri Lanka is credited with starting the collection and export of tropical marine fish in the 1930s (Weibnitz 2004). The trade grew during the 1950s as other countries began to collect fish for export. Lewbert et al. (1999) estimate that in 1998, 1.5–2 million people worldwide kept marine aquaria with a trade in marine ornamentals valued at $200–300 million/yr (Chapman et al. 1997; Larkin and Degner 2001). According to the Global Marine Aquarium Database (GMAD), 1,471 different marine species are traded globally. The Philippines, Indonesia, the Solomon Islands, Sri Lanka, Australia, Fiji, the Maldives, and Palau provided 98% of the marine fish exported from 1997 to 2002 (Weibnitz 2004). The blue-green damselfish (*Chromis viridis*), the clown anemonefish (*Amphiprion ocellaris*), the whitetail dascyllus (*Sascyllus aruanus*), the sapphire devil (*Chrysiptera cyanea*), and the threespot dascyllus (*Dascyllus trimaculatus*) were the most commonly traded marine species from 1997 to 2002.

Ornamental fish are traded in high dollar amounts worldwide, but the United States is one of the largest markets for aquarium fish (Conroy 1975; Hemley 1984; Chapman et al. 1997). Fish are held as pets in aquaria in a majority of households in the United States (American Pet Products Manufacturer's Association 1994). In Europe, Germany is the primary importer (22.5%), followed by the United Kingdom (18%), France (15%), the Netherlands (10%), Italy (8%), Spain (6%), and Belguim (5%) (Ornamental Aquatic Trade Association 1998–2004). Taiwan, Hong Kong, and Japan are also important markets for ornamental fish.

In the United States, ornamental aquaculture production, in terms of dollar value, ranks fourth in importance in U.S. aquaculture (Figure 13.2). There are a wide variety of different species cultured and sold as ornamental fish in the United States. The number of species sold as ornamental fish may be in the thousands of species (Conroy 1975; Hemley 1984; Ramsey 1985).

Additional species are imported and sold through the aquarium and pet store trade. The United States both imports and exports ornamental fish (Figure 13.3). However, the value of imports is more than triple that of exports from the United States. The United States Coral Reef Task Force (2000) estimated that 50% of the marine fish traded as ornamentals are exported to the United States. Although Chapman et al. (1997) demonstrated growing values of imports into the United States from 1982 to 1992, recent data show that the values of both imports and exports appear to have leveled out over the past four years (1999-2003).

The largest volume of imports of ornamental fish come from Southeast Asia and Japan (Chapman et al. 1997). The top five countries supplying the U.S. market are Singapore, Thailand, the Philippines, Hong Kong, and Indonesia. The next largest region supplying the United States was South America, with Colombia, Brazil, and Peru being the leading country suppliers (Fig. 13-4). Chapman et al. (1997) showed that the major port of entry was Los Angeles (39%), followed by Miami (22%) and New York City (16%).

Chapman et al. (1997) identified 1,539 species of ornamental fish that were imported into the United States in 1992. There were 809 saltwater species (809) imported in 1992, as compared to 730 freshwater species. However, in terms of the total volume of imports, 96% were freshwater (80% in terms of value). Of all the species, the guppy (*Poecilia reticulate*) and the neon tetra (*Paracheirodon innesi*) were the most popular species. Guppies and neon tetra, along with the platy (*Xiphophorus maculates*), the betta (*Betta spendens*), Chinese algae-eater (*Gyrinocheilus aymonieri*), and goldfish (*Carassius auratus*), compose half of the total number of ornamental fish imported. Of the ornamental fish exported from the United States, most went to Canada (29%), then to Southeast Asia (25%), Europe (20%), and Japan (18%).

Most of the freshwater fish imported from Southeast Asia are cultured, whereas the freshwater fish from South America are caught from the wild. However, most of the saltwater fish imported are wild caught. Philippines and Indonesia supply the greatest volume of saltwater ornamentals into the United States.

Distribution channels for ornamental fish are complex. Hong Kong and Singapore are world pur-

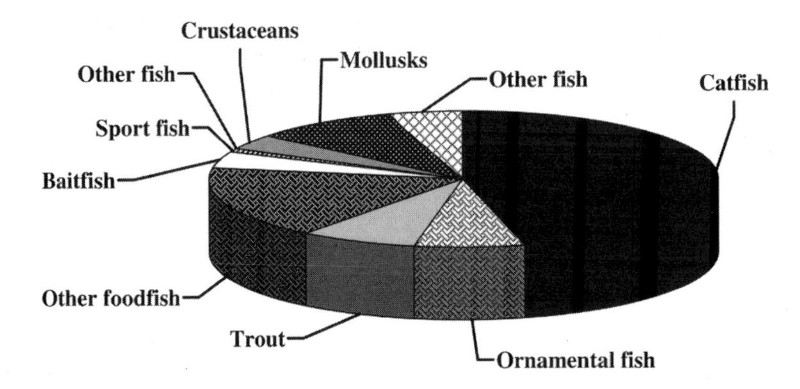

Figure 13.2. Major aquaculture species in the United States, in terms of value. (Source: USDA 1998.)

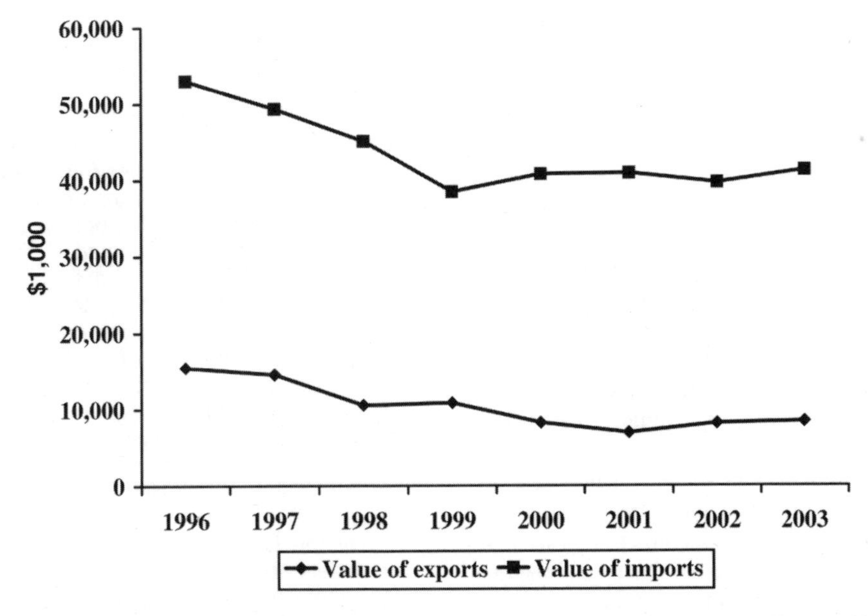

Figure 13.3. United States trade in exports and imports of ornamental fish, 1996–2003. (Source: NASS 2004.)

Figure 13.4. *Microgeophagus ramerizii*. Blue Ram Cichlid. Photo by Joe Richards, University of Florida.

chase and transhipment centers for ornamental fish. Ornamental fish arriving in the United States through the ports of Los Angeles, Miami, New York, and Tampa arrive at broker-wholesale warehouses for subsequent delivery to single retail pet stores and warehouses (Chapman et al. 1997). The smaller ports for ornamental fish, such as Chicago, New Orleans, and San Francisco, are thought to service primarily pet dealers and wholesalers in the immediate localities.

Some of the most popular types of ornamental fish kept as pets in the United States are koi carp (*Cyprinus carpio*) and the goldfish (*Carassius auratus*). Breeding of unusual color varieties of these species dates back to China more than 1,000 years ago (Watson et al. 2004). Today, there are more than 100 different varieties of these fish. Varieties may include various color and scale patterns or unusually shaped body parts. "Orandas" have caps on their heads, and "Bubble Eyes" have sacs under their eyes.

Koi carp and fancy, or ornamental, goldfish are raised commercially on farms in the United States. U.S. production is geared primarily toward the low-medium–priced products such as "Black Moors," "Shibunkins," "Fantails," "Calicos," and "Comets" (Watson et al. 2004). There is increasing interest on the part of U.S. growers in production of the higher-valued varieties such as the "Orandas," "Bubble Eyes," and "Ryunkins." However, most of the higher-priced goldfish are currently imported into the United States from China.

Koi carp have become more popular in the United States as water gardening has grown. The U.S. koi industry has traditionally focused on lower-priced koi that can be sold in higher volumes. However, as with the ornamental goldfish industry, interest in producing the high-value koi is growing. Grading of koi carp must be done by hand because fish are selected individually based on color patterns. Recognizing the most highly valued color patterns requires training and expertise. To produce a top-of-the-line koi carp, 90% of the fish may have to be culled (Watson et al. 2004).

Challenges for the ornamental fish industry include concerns over the ecological sustainability of wild capture of species for export. Those who oppose the international trade in ornamental fish cite damages caused by collecting techniques, overharvesting of target species, and high levels of mortality along the supply chain (Weibnitz 2004). Supporters maintain that proper conservation and management of coral reefs and other aquatic resources, and well-managed shipping and husbandry practices, can alleviate these problems. Moreover, the capture of ornamental fish for export creates employment in rural areas.

Other concerns include those over the use of genetically modified organisms in the aquatic trade (Fossa 2004). Genetically modified fluorescent zebra fish and medakas were introduced into the market in 2003, and more will likely come. The concern is the consequences to wild populations if genetically modified ornamental fish would escape into the wild. To breeders, these techniques allow them to create new products to supply to the aquarium trade.

Interest in the culture of marine ornamental fish has grown. Aquaculture might be the solution to supplying the demand for these fish without undue pressure on natural populations and resources.

Sources of Market Information in the U.S. and Japan

For many types of aquaculture products, there is quite a lot of secondary information available that can be used to understand overall market conditions and trends. Much of the information is available online, and many information sources are listed in the Annotated Webliography in this book.

For global market trends, the Fishstat Plus database that is available on the United Nations Food and Agriculture Organization's Web site is the best source. A number of databases can be downloaded to be searched through FISHSTAT+, including databases on both quantities and value of aquaculture and capture fisheries species worldwide, by region, and on international trade.

For the U.S. market, the USDA, through the National Agricultural Statistics Service and the Economics Research Service, publishes an Aquaculture Situation and Outlook Report as well as reports on specific aquaculture commodities. Urner Barry Publications publishes very detailed price reports of a wide variety of seafood items in specific markets. The Institutional Distribution Magazine publishes the top 50 broadline distributors, and the National Restaurant Association publishes foodservice sales forecasts. The National Fisheries Institute publishes the top ten seafoods in terms of per-capita consumption in the United States, and the National Marine Fisheries Service publishes U.S. seafood trade data in its annual Fisheries of the United States.

In Japan, the Japan Fisheries Agency publishes data on imports of seafood aquaculture production and wild-caught supply of seafood (Japan Fisheries Agency). The Japan National Marine Fisheries publishes Japanese import data by species.

SUMMARY

Marketing research helps to find answers to questions such as: "What are the attitudes and desires of consumers?" "Is there a demand for the product?" "What is the volume of sales compared to our competitors, that is, what is the market share for the product?" "What products will consumers demand in future?" Answers to such questions allow a business to find out more about the current market situation relating to a product of interest as well as predict future market situations. The research can be designed in one of three forms: (1) exploratory research, (2) qualitative research, or (3) quantitative research. The type of research that is most appropriate depends upon the objectives. Whether the research effort is exploratory, qualitative, or quantitative, data need to be collected. Data to be collected should relate directly to the research objectives, research questions, and research hypotheses.

There are two basic types of data that can be gathered, primary and secondary data. Secondary data are information previously gathered by others, whereas primary data are any systematic documentation of observations. The most common primary data collected in marketing research is the documentation of consumer attitudes and behavior using focus groups, interviews, or surveys. Surveys can be conducted in a variety of ways including telephone, mail, or face-to-face. Surveys can also be self-administered and may also combine several of these methods.

The basis of quantitative marketing research is to gain information about an entire group of people (that is, population), for example, consumers, households, and clients. It is usually necessary to draw a sample (portion of the population) to obtain information about the entire population using a sample design. The sampling methods can be a simple random sampling, systematic random sampling, a stratified random sampling, a multistage or cluster sampling, or ad-hoc sampling.

Designing questionnaires should be done with care. Poorly worded questions, poorly structured questionnaires, and inappropriate questions can result in erroneous and misleading information. The choice of words, questions, formats, and hypothesis should be done with care. A questionnaire should begin with easy-to-answer questions. Subsequent questions should flow naturally and in a logical fashion. More sensitive questions on demographic information on age or income should be placed at the end. Questions can take the form of "Yes/No" questions, multiple-choice questions, Likert scale, Rank order, Rating Scale, True/False, and/or Semantic differential scale.

There are several different forms of collecting information related to consumer attitudes and preferences, and each type requires a different approach. People's choices manifest themselves in many ways but are particularly expressed through active or passive responses of purchasing specific products or services. People's behavior can be examined using the neoclassical reference theory, revealed preference theory, and hedonic theory.

The development of a new product often depends on the type of product. Whether the product is evolving from a known or an existing product or is an entirely new product will affect its market development path. Repackaging a known food product or adding new flavors to an existing product are evolutionary concepts, whereas an entirely new product concept is a revolutionary concept. With a new product, testing is essential to identify its potential successful and determine the probability that consumers will accept the product. It is always important for companies to monitor how their share of the market changes over time.

Data analysis can take the form of graphics that give a snapshot of all the data gathered; statistical summaries such as proportions; central measures such as mean, mode, and median; and variability measures such as variance and standard deviation. Data can also be subjected to scatter plots, least square regression, and cross tabulations. Chi-square tests can be used to determine whether any differences between the observed counts and expected counts are statistically significant in a cross tabulation. Other quantitative methods for data analysis include discrete choice analysis, such as logit and probit, and conjoint analysis.

STUDY AND DISCUSSION QUESTIONS

1. Define marketing research. Outline three functions that marketing research can help to accomplish in developing a market for a new fish product.

2. A new seafood distribution company is about to be set. What form of market research would you recommend to the owners, and why?

3. What are the advantages of secondary data over primary data? Suggest some ways that you can use to address some of the potential problems associated with using secondary data.

4. A focus group consists of six to 12 participants, a sample too small to be representative of the population. What is the usefulness of the information gathered from such a small sample of the population?

5. Response rate is a problem with mail surveys. What techniques can be applied to improve on the response rate for mail surveys?

6. Mall-intercept interviews solicit the cooperation of shoppers to complete a survey. In today's society in which there is always suspicion and mistrust, how can an interviewer gain the cooperation of busy shoppers?

7. Suppose that the number of adult Hispanics living in Arkansas is 10,000 and we want to survey them to determine their attitudes toward catfish. Calculate the sample size to use for the survey assuming a 95% confidence level and a margin of error of ± 0.5.

8. What are the five types of sampling methods? Under what circumstances would each type be applicable?

9. What are the theoretical frameworks for examining consumer behavior? Which framework is more applicable for investigating the market for a new seafood product?

10. Why is it necessary for a company to monitor its market share regularly?

11. Describe the various measures of advertising effects. Which of these performs better in predicting the effectiveness of advertising?

12. Outline the steps a new business can follow to develop sales forecasts.

A sample of 15 clam consumers indicated the number of times that they have purchased clams from the grocery store within the past one year as:

Customer no.	Times purchased
1	27
2	50
3	33
4	25
5	86
6	25
7	85
8	31
9	37
10	44
11	20
12	36
13	59
14	34
15	28

Calculate the mean and mode of these observations. Calculate the variance and standard deviation (that is, find the deviation of each observation from the mean, square the deviations, and then obtain the variance and standard deviation).

Here are data from a survey of consumers of Asian origin in three cities in California concerning whether they buy fish from fish shops:

	Doesn't buy fish from fish shop	Buys fish from fish shop
City 1	400	1,380
City 2	416	1,823
City 3	188	1,168

(a) Make a two-way table of cities according to whether or not they purchase from a fish shop.
(b) Calculate the proportion of Asians who buy fish from fish shops in each city.
(c) Find the expected counts, and check whether the chi-square can be used. What null and alternative hypothesis does the chi-square test?
(d) What can you conclude from the data?

REFERENCES

American Pet Products Manufacturers Association (APPMA). 1994. National pet owners survey. American Pet Products Manufacturers Association, Scarsdale, New York.

American Statistical Association. 1997. ASA Series: what is a survey? American Statistical Association, Alexandria, Virginia.

Azzam, A. 1992. Testing the competitiveness of food price spreads. *Journal of Agricultural Economics* 43:248–256.

Azzam, A. and E. Pagoulatos. 1990. Testing oligopolistic and oligopsonistic behavior: an application to the U.S. meat packing industry. *Journal of Agricultural Economics* 41:362–370.

Ben-Akiva, M. and S. R. Lerman. 1987. *Discrete Choice Analysis: Theory and Application to Travel Demand*. The MIT Press, Cambridge, Massachusetts.

Bennett, E. 2001. Hooked on seafood. *Philadelphia Business Journal* 8.

Bennet, Peter D. 1988. Glossary of marketing terms. American Marketing Association, Chicago, Illinois.

Blankenship, A.B., G.E. Breen, and A. Dutka. 1998. *State of the Art Marketing Research*, Second Edition, American Marketing Association, NTC Business Books, Chicago, Illinois.

Bresnahan, T. 1989. Empirical studies of industries with market power. In: Schmalensee, R. and R. Willig, (eds.), *Handbook of Industrial Organization*, North-Holland, Amsterdam.

Chapman, F.A., S. A. Fitz-Coy, E.M. Thunberg, and C. M. Adams. 1997. United States of America trade in ornamental fish. *Journal of the World Aquaculture Society* 28(1):1–10.

Conroy, D.A. 1975. An evaluation of the present state of world trade in ornamental fish. FAO Fisheries Technical Paper No. 146, Food and Agriculture Organization of the United Nations, Rome, Italy.

Deaton, A. and J. Muellbauer, 1980. *Economics and Consumer Behavior*. Cambridge University Press, Cambridge, Massachusetts.

Florida Department of Agriculture and Consumer Services. 2001. Identifying and assessing potential direct markets for farm-raised shrimp grown on small farms. Bureau of Seafood and Aquaculture Marketing, Florida Department of Agriculture and Consumer Services. http://www.fl-seafood.com/industry/reports/shrimp/index.htm

Fossa, S. 2004. Genetically modified organisms in the aquatic trade? Ornamental Fish International. Accessed at www.ornamental-fish-int.org

Green, P.E. 1974. On the design of choice experiments involving multifactor alternatives. *Journal of Consumer Research* 1:61–68.

Green, P.E. and V. Srinivasan. 1978. Conjoint analysis in consumer research: issues and outlook. *Journal of Consumer Research* 5:103–123.

Green, P.E. and Y. Wind. 1975. New ways to measure consumers' judgments. *Harvard Business Review*, p. 89–108.

Greene, W. H. 1997. *Econometric Analysis*, Third Edition. Macmillan, New York City, New York.

Haley, R.I. and A.L. Baldinger. 1991. The ARF copy research validity project. *Journal of Advertising Research* 31(2).

Hemley, G. 1984. U.S. imports millions of ornamental fish annually. *Traffic USA* 5(4):1.

Lancaster, K.J. 1971. *Consumer Demand.* Columbia University Press, New York, New York.

Larkin, S. and R. Degner. 2001. The U.S. wholesale market for marine ornamentals. *Aquarium Sciences and Conservation* 3:13–24.

Lewbel, A. 1990. Full rank demand systems. *International Economic Review* 31: 289–300.

Lewbert, G., M. Stoskopf, T. Losordo, J. Geyer, J. Owen, D. White Smith, M. Law, and C. Altier. 1999. Safety and efficacy of the Environmental Products Group Masterflow aquarium management system with Aegis Microbe Shield™.

Maddala, G. S. 1983. *Limited Dependent and Qualitative Variables in Econometrics*. Cambridge University Press, United Kingdom.

Marion, B.W. and F.E. Geithman. 1995. Concentration-price relations in regional fed cattle markets. *Review of Industrial Organization* 10:1–19.

Menkhaus, D.J., J.S. St. Clair, and A.Z. Ahmaddaud. 1981. The effects of industry structure on price: a case in the beef industry. *Western Journal of Agricultural Economics* 6:147–153.

Muth, K.M. and M. K. Wohlgenant. 1999. Measuring the degree of oligopsony power in the beef packing industry in the absence of marketing input quantity data. *Journal of Agricultural and Resource Economics* 24:299–312.

Ornamental Aquatic Trade Association (OATA). 1998–2004. Ornamental Aquatic Trade Association Worldwide. Accessed at www.ornamentalfish.org

Pindyck, R.S. and D.L. Rubinfeld. 2001. *Microeconomics*, Fifth Edition. Pearson Prentice Hall, Upper Saddle River, New Jersey.

Pollak, R.A. and Wales, T.J.. 1992. *Demand System Specification and Estimation*. Oxford University Press, Oxford, United Kingdom.

Quail, G., B. Marion, F. Geithman, and J. Marquardt. 1986. The impact of packer buyer concentration on live cattle prices. Working Paper, North Central Project 117, North Central Agricultural Experimental Stations, Madison, Wisconsin.

Ramsey, J.S. 1985. Sampling aquarium fishes imported by the United States. *Journal of the Alabama Academy of Science* 56(4):220–245.

Seafood Business (2003). Red Lobster research corroborates rise in U.S. seafood consumption. 22(11):6.

Schmalensee, R. 1989. Inter-industry studies of structure and performance. In: Schmalensee, R. and R. Willig, (eds.), *Handbook of Industrial Organization*, North-Holland, Amsterdam.

Schroeter, J.R. 1988, Estimating the degree of market power in the beef packing industry. *Review of Economics and Statistics* 70:158–162.

Schroeter, J.R. and A. Azzam. 1990. Measuring market power in multi-product oligopolies: the U.S. meat industry. *Applied Economics* 22:1365–1376.

Theil, H. 1975. *Theory and Measurement of Consumer Demand*, Vol. I. North-Holland Publishing Company, Amsterdam, The Netherlands.

Theil, H. 1976. *Theory and Measurement of Consumer Demand*, Vol. II. North-Holland Publishing Company, Amsterdam, The Netherlands.

United States Coral Reef Task Force. 2000. International trade in coral and coral reef species: the role of the United States. Report of the Trade Subgroup of the International Working Group to the USCRTF, Washington, D.C.

USDA. 1998. Census of aquaculture (1998). 1997 Census of Agriculture Volume 3, Special Studies, Part 3, United States Department of Agriculture, Washington, D.C.

Watson, C.A., J.E. Hill, and D.B. Pouder. 2004. Species profile: koi and goldfish. Southern Regional Aquaculture Center Publication No. 7201, Mississippi State University, Mississippi.

Weibnitz, C. 2004. From ocean to aquaria: the trade in marine ornamentals worldwide. Cambridge, United Kingdom. Accessed at www.ornamental-fish-int.org

Annotated Bibliography of Aquaculture Marketing Information Sources

This bibliography is organized alphabetically into broad categories.

AGRICULTURAL MARKETING

Books

Kohls, R.L. and J.N. Uhl. 1985. *Marketing of Agricultural Products.* Macmillan Publishing Company, New York.

This book has been a classic agricultural marketing textbook for a number of years with various editions. It is a good source for studying the fundamental principles of agricultural marketing.

Rhodes, V.J. *The Agricultural Marketing System*, Fourth Edition. Gorsuch Scarisbrick, Publishers, Scottsdale, Arizona.

This book presents a good overview of the agricultural marketing system in the United States. It is written in a concise and reduced form that allows the reader to concentrate on the most critical information. The book does expect the reader to have an understanding of fundamental economic principles.

Periodicals

Meat and Seafood Merchandising, P.O. Box 2074, Skokie, Illinois 60076.

This is a magazine put out by Vance Publishing. It covers the latest trends in the retail grocery sector related to displaying and advertising all types of meat and seafood items.

AQUACULTURE MARKETING

Books

Shaw, S.A. 1986. *Marketing the Products of Aquaculture*. FAO Fisheries Technical Paper 276, Food and Agriculture Organization of the United Nations, Rome, Italy.

This manual provides practical advice on choosing products and markets, product forms, and retail issues such as displaying fish for customers.

Journals that Publish Scientific Articles Related to Marketing Aquaculture Products

Aquaculture Economics and Management

This journal is the only one devoted exclusively to issues related to the economics of aquaculture. This includes marketing issues. This journal has devoted several special issues to aquaculture marketing.

The literature on aquaculture marketing is widely dispersed among the various aquaculture and agricultural economics journals. To conduct a thorough literature analysis, the following journals should be searched:

Journal of the World Aquaculture Society
Journal of Applied Aquaculture
Aquaculture
North American Journal of Aquaculture
American Journal of Agricultural Economics
Journal of Applied and Resource Economics
Agribusiness

CONJOINT ANALYSIS

Green, P.E. 1974. On the design of choice experiments involving multifactor alternatives. *Journal of Consumer Research* 1:61–68.

Green, P.E. and V. Srinivasan. 1978. Conjoint analysis in consumer research: issues and outlook. *Journal of Consumer Research* 5:103–123.

Green, P.E. and Y. Wind. 1975. New ways to measure consumers' judgments. *Harvard Business Review* 89–108.

Louviere, J.J., D.A. Henser, and J.D. Swait. 2000. *Stated Choice Models: Analysis and Application.* Cambridge University Press, Cambridge, UK.

DATA SOURCES FOR AQUACULTURE PRODUCTS AND MARKETS

The Food and Agriculture Organization of the United Nations puts out the most comprehensive statistical reports on world aquaculture and world capture fisheries. These are revised and reprinted every six years with interim updates. The last printed version was in 2002. The FAO reports fisheries and aquaculture production by species, by country, by region, and by type of environment (freshwater, marine, and so on). Although available in printed form, complete information is also available on the Internet. For details, see the Annotated Webliography in this book.

The United States Department of Agriculture publishes data on several of the leading aquaculture industry segments on a regular basis. For example, catfish data are provided on supply, sales, prices, inventory (broodfish, fingerlings/fry, stockers, small food-size, medium food-size, and large food-size), processor sales (by product form), acres, numbers of farms, and inventory estimates. Similarly, data are compiled on trout sales, weight, and the value of foodfish, stockers, fingerlings, and eggs. Quantities and value of ornamental fish, trout, salmon, shrimp, oysters, mussels, clams, and tilapia imported into the United States for the past several years are reported. Imports of tilapia, salmon, and shrimp by country are also reported. U.S. export quantities and value are reported for the past several years on oysters, mussels, clams, ornamental fish, trout, salmon, and shrimp.

USDA. 1998. Census of aquaculture (1998). 1997 Census of Agriculture Volume 3, Special Studies, Part 3, United States Department of Agriculture, Washington, D.C.

The USDA conducted its first ever census of aquaculture in 1997. The aquaculture census data were published in 1998 and are available in hard copy. Preparations are under way to conduct a follow-up survey. Data collected include acreage, total production, and value of production of many aquaculture species by state and by region.

DEMAND ANALYSIS

Deaton, A. and J. Muellbauer. 1980. *Economics and Consumer Behavior.* Cambridge University Press, Cambridge, UK.

Lewbel, A. 1990. Full rank demand systems. *International Economic Review* 31:289–300.

Pollak, R.A., Wales, T.J. 1992. *Demand System Specification and Estimation.* Oxford University Press, Oxford, UK.

Theil, H. 1975. *Theory and Measurement of Consumer Demand*, Vol. I. North-Holland Publishing Company, Amsterdam, The Netherlands.

Theil, H. 1976. *Theory and Measurement of Consumer Demand*, Vol. II. North-Holland Publishing Company, Amsterdam, The Netherlands.

DISCRETE CHOICE ANALYSIS

Ben-Akiva, M. and S. R. Lerman. 1987. *Discrete Choice Analysis: Theory and Application to Travel Demand.* The MIT Press, Cambridge, Massachusetts.

Greene, W. H. 1997. *Econometric Analysis*, Third Edition. Macmillan Publishing Company, New York.

Maddala, G. S. 1983. *Limited Dependent and Qualitative Variables in Econometrics.* Cambridge University Press, Cambridge, UK.

EXTENSION MATERIALS ON HOLDING, TRANSPORTATION OF FISH FOR MARKET

There are many extension materials on marketing alternatives for fish farmers available both in written form and downloadable from the Internet. One of the most accessible sets is through the Regional Aquaculture Center (RAC) networks. The RAC networks of scientists and extension specialists have developed a series of Fact Sheets that are available free of charge. The following is a sampling of Fact Sheets that provide detailed information and recommendations on specific components of marketing channels for aquaculture. These are available through extension aquaculture specialists in each state, through the Cooperative Extension Service, and can be downloaded from the Internet at the www.aquanic.edu site.

Cichra, C.E., M. P. Masser, and R.J. Gilbert. 1994. Fee-fishing: an introduction. Southern Regional Aquaculture Center Publication No. 479, Stoneville, Mississippi.

Cichra, C.E., M. P. Masser, and R.J. Gilbert. 1994. Fee fishing: location, site development and other considerations. Southern Regional Aquaculture Center Publication No. 482, Stoneville, Mississippi.

Cole, B., C.S. Tamaru, R. Bailey, C. Brown, and H. Ako. 1999. Shipping practices in the ornamental fish industry. Center for Tropical and Subtropical Aquaculture Publication Number 131. University of Hawaii, Hilo, Hawaii.

Engle, C.R. and N. M Stone. 1997. Developing business proposals for aquaculture loans. Southern Regional Aquaculture Center Publication No. 381, Stoneville, Mississippi.

Gilbert, R.J. 1989. Small-scale marketing of aquaculture products. Southern Regional Aquaculture Center Publication No. 350, Stoneville, Mississippi.

Higginbotham, B.J. and G.M. Clary. 1992. Development and management of fishing leases. Southern Regional Aquaculture Center Publication No. 481, Stoneville, Mississippi.

Jensen, G.L. 1990. Sorting and grading warmwater fish. Southern Regional Aquaculture Center Publication No. 391, Stoneville, Mississippi.

Jensen, G.L. 1990. Transportation of warmwater fish. Southern Regional Aquaculture Center Publication No. 390, Stoneville, Mississippi.

Jensen, G.L. 1990. Transportation of warmwater fish: procedures and loading rates. Southern Regional Aquaculture Center Publication No. 392, Stoneville, Mississippi.

Masser, M.P., C.E. Cichra, and R.J. Gilbert. Fee-fishing ponds: management of food fish and water quality. Southern Regional Aquaculture Center Publication No. 480, Stoneville, Mississippi.

Regenstein, J.M. 1992. Processing and marketing aquacultured fish. Northeastern Regional Aquaculture Center Fact Sheet No. 140-1992. University of Massachusetts, Amherst, Massachusetts.

Riepe, J.R. 1999. Marketing seafood to restaurants in the North Central region. North Central Regional Aquaculture Center Fact Sheet Series #110, Iowa State University, Ames, Iowa.

Riepe, J.R. 1999. Supermarkets and seafood in the North Central Region. North Central Regional Aquaculture Center Fact Sheet Series #112. Iowa State University, Ames, Iowa.

Strombom, D.B. 1992. Business planning for aquaculture—is it feasible? Northeastern Regional Aquaculture Center Fact Sheet No. 150-1992. University of Massachusetts, Amherst, Massachusetts.

Swann, L. 1993. Transportation of fish in bags. North Central Regional Aquaculture Center Fact Sheet Series #104. Iowa State University, Ames, Iowa.

GENERAL MARKETING SOURCES

Chisnall, P.M. 1986. *Marketing Research*, Third Edition. McGraw Hill, London.

As its name implies, this book presents summaries of marketing research techniques.

Chisnall, P.M. 1991. *The Essence of Marketing Research*. Prentice-Hall, Hemel Hempstead.

Crimp, M. 1990. *The Marketing Research Process,* Third Edition. Prentice-Hall, Englewood Cliff, New Jersey

This book contains a clear summary of marketing research techniques.

Curtis, T. 1994. *Business and Marketing for Engineers and Scientists*. McGraw-Hill, Maidenhead.

This book provides a scientist's perspective on marketing.

Kotler. 1997. *Marketing Management, Analysis, Planning, and Control*. Prentice Hall, Paramus, New Jersey.

This book presents a comprehensive, clear, and informative survey of general marketing.

Rosthorn, J., Al. Haldane, E. Blackwell, and J. Wholey. 1994. *The Small Business Action Kit,* Fourth Edition. Kogan Page, London.

This book is a very good guide to marketing and other aspects of small business management.

INDUSTRIAL ORGANIZATION

Books

Carlton, D.W. and J.M. Perloff. 2000. *Modern Industrial Organization*. Addison-Wesley Longman, Inc., Reading, Massachusetts.

This book presents the latest theory on the organization of firms and industries and combines it with practical evidence. Although it discusses the traditional approach of focusing on structure, conduct, and performance of markets, it also addresses the modern approaches such as transaction-cost analysis, game theory, contestability, and information theory.

Bresnahan, T. 1989. Empirical studies of industries with market power. In: Schmalensee, R. and R. Willig (eds.), *Handbook of Industrial Organization*, North-Holland, Amsterdam.

Schmalensee, R. 1989. Inter-industry studies of structure and performance. In: Schmalensee, R. and R. Willig (eds.), *Handbook of Industrial Organization*, North-Holland, Amsterdam.

Articles

Azzam, A. 1992. Testing the competitiveness of food price spreads. *Journal of Agricultural Economics* 43:248–256.

Azzam, A. and E. Pagoulatos. 1990. Testing oligopolistic and oligopsonistic behavior: an application to the U.S. meat packing industry. *Journal of Agricultural Economics* 41:362–370.

Marion, B.W. and F.E. Geithman. 1995. Concentration-price relations in regional fed cattle markets. *Review of Industrial Organization* 10:1–19.

Menkhaus, D.J., J.S. St. Clair, and A.Z. Ahmaddaud. 1981. The effects of industry structure on price: a case in the beef industry. *Western Journal of Agricultural Economics* 6:147–153.

Muth K. M., and M. K. Wohlgenant. 1999. Measuring the degree of oligopsony power in the beef packing industry in the absence of marketing input quantity data. *Journal of Agricultural and Resource Economics* 24:299–312.

Quail, G., B. Marion, F. Geithman, and J. Marquardt. 1986. The impact of packer buyer concentration on live cattle prices. Working Paper, North Central Project 117, North Central Agricultural Experimental Stations, Madison, Wisconsin.

Schroeter, J.R. 1988. Estimating the degree of market power in the beef packing industry. *Review of Economics and Statistics* 70:158–162.

Schroeter, J.R. and A. Azzam. 1990. Measuring market power in multi-product oligopolies: the U.S. meat industry. *Applied Economics* 22: 1365–1376.

PRINCIPLES OF ECONOMICS

There are a large number of good books that cover the principles of economics. Titles may include *Principles of Economics*, *Microeconomics*, or *Macroeconomics*. A few examples of currently available books include:

Baumol, W.J. 2002. *Macroeconomics: Principles and Policy with Xtra!* Student CD-ROM and InfoTrac College Edition. South-Western College Publishers.

Baumol, W.J. 2002. *Microeconomics: Principles and Policy with Xtra!* Student CD-ROM and InfoTrac College Edition. South-Western College Publishers.

Hall, R.E. and M. Lieberman. 2002. *Microeconomics: Principles and Applications*. South-Western College Publishers.

Mankiw, N.G. 1997. *Principles of Economics*. Harcourt Brace and Company Publishers.

Pindyck, R.S. and D.L. Rubinfeld. 2001. *Microeconomics*, Fifth Edition. Prentice Hall.

Rubenfeld, D.L. 2004. *Microeconomics*. Prentice Hall Publishers.

SEAFOOD MARKETING TRADE INFORMATION

Books

Johnson, H.M. 2004. *2004 Annual Report on the United States Seafood Industry*. H.M. Johnson & Associates. Jacksonville, Oregon.

This report is put out on annually and provides a good overview of trends in the U.S. seafood sector. The report compiles price and quantity data from a variety of sources for both wild-caught and aquaculture species and highlights various trends in the retail and wholesale sectors that handle seafood.

National Marine Fisheries Service. 2003. *Fisheries of the United States 2003*. National Marine Fisheries Service, National Oceanic and Atmospheric Administration, Washington, D.C.

This book summarizes trade statistics for seafood products imported and exported into and from the

United States. It includes per capita consumption statistics for countries around the world.

Anderson, J.L. 2003. *The International Seafood Trade*. Woodhead Publishing Limited, Cambridge, England.

This book is an excellent summary of international trade theory, statistics, and issues related to the international market for seafood. It begins with an overview of the worldwide market for seafood with discussions of major importing and exporting nations and net trade flows. It summarizes trade in the major species groups of seafood traded around the world. A chapter on institutions involved in international trade summarizes regulations, and the framework within which trade disputes are resolved among countries.

Periodicals

There are several good seafood trade periodical publications.

Seafood Business

Seafood Business is published by Diversified Business Communications in Portland, Maine. It covers the entire seafood market and is increasing print space to include information on aquaculture products. It periodically includes updates by species. A retailer survey and restaurant survey are conducted each year to provide updates on trends in these two market segments. Feature articles highlight recent newsworthy events. A typical issue will include sections such as: News, Market, Product Spotlight, Species Focus, Top Story, Seafood Star, Trend Watch, On The Menu, Seafood University, Equipment, and Highlights of the Boston Seafood Show.

Seafood International

Seafood International is published by Quantum Publishing Ltd. from Surrey, United Kingdom. Its focus is more on European markets for seafood, but it provides a good perspective on trends in seafood markets from a different point of view. This company also publishes *Fishing News International* and *Fish Farming International*. A typical issue will include: News, Markets, New Products, Publications, Events, Last Bites, and feature articles.

The Catfish Journal

The Catfish Journal is published monthly by the Catfish Farmers of America. Although it is the voice of the U.S. catfish industry, it also includes information on processing companies, some market developments, and occasional commentary on price trends, particularly as these relate to catfish prices.

SURVEYS

Books on Surveys

American Statistical Association. 1997. *ASA Series: What Is a Survey?* Alexandria, VA.

Blankenship, A.B., G.E. Breen, and A. Dutka. 1998. *State of the Art Marketing Research*, Second Edition. American Marketing Association, NTC Business Books, Chicago, Illinois.

Chisnall, P.M. 1986. *Marketing Research*, Third Edition. McGraw Hill, London, UK.

Crimp, M. 1990. *The Marketing Research Process*, Third Edition. Prentice-Hall, Englewood Cliff, New Jersey.

THEORIES OF CHOICE BEHAVIOR

Rubenfield, D.L. 2004. *Microeconomics*. Prentice Hall Publishers.

Deaton, A. and J. Muellbauer. 1980. *Economics and Consumer Behavior*. Cambridge University Press, Cambridge.

Hall, R.E. and M. Lieberman. 2002. *Microeconomics: Principles and Applications*. South-Western College Publishers.

Lancaster, K.J. 1971. *Consumer Demand*. Columbia University Press, New York, N.Y.

Mankiw, N.G. 1997. *Principles of Economics*. Harcourt Brace and Company Publishers.

Pindyck, R.S. and D.L. Rubinfeld. 2001. *Microeconomics*, Fifth Edition. Pearson Prentice Hall.

Annotated Webliography of Sources of Data and Information for Aquaculture Marketing

E-COMMERCE

GlobalNet Xchange (GNX)

https://www.gnx.com/reg/index.jsp

GlobalNet Xchange (GNX) is an e-business solution and service provider for the global retail industry. GNX solutions was set up to help retailers, manufacturers, and their trading partners reduce costs and improve efficiency by streamlining and automating sourcing and supply chain processes. GNX hosts a platform that enables members to buy, sell, trade, or auction goods and services over the Internet using standardized Web browsers.

http://www.seafood.com

This site is a seafood trading site that is free to buyers. The site offers the opportunity to purchase seafood for overnight delivery. Sellers must pay a fee to list products and manage the product inventory online, and the site provides an opportunity to advertise. The site provides contact information for seafood and related services companies.

EUROPEAN UNION

Common Organization of the Market in Fishery and Aquaculture Products

http://europa.eu.int/scadplus/leg/en/lub/166002.htm

The European Union established a common fisheries policy with foundations laid in 1970. The Web site details provisions under the following five categories:

- Common marketing standards
- Consumer information
- Producer organizations
- Price support system based on intervention
- Arrangements for trade with third countries

www.europa.eu.int/eur-lex

This site prints all the legislative actions and full texts of regulations coded by number as well as information and notices. This is the site where the *Official Journal of the European Union* is published. This site provides full text of decisions of the Commission as published in the *Official Journal of the European Union*. Actions taken on antidumping orders in the European Union are published on this site.

www.europa.eu.int

Gateway to the European Union. This site provides overviews of European Community agencies including the European Food Safety Authority. This site contains the standards, logo, and certification program for organic products in the European Union.

European Commissions' Health and Consumer Protection Directorate General

www.eurunion.org/legislat/home.htm

This is the site of the European Commission's Health and Consumer Protection Directorate General. It contains the full text of the White Paper on Food Safety that contains the major policy provisions for food safety in the European Union.

European Food Safety Authority

www,efsa.eu.int

This is the site of the European Food Safety Authority (EFSA). The latest opinions and reports of the various scientific panels can be found on this site.

FRANCE

IFREMER

http://www.ifremer.fr/cofepeche/referencetexten/text/marche.htm

Ifremer (French Research Institute for Exploitation of the Sea) publishes market studies on pilot projects, pricing, sector studies, socioeconomic studies, and market appraisals in France, Europe, Africa, Asia, and Latin America.

INDUSTRY ASSOCIATIONS

AT-SEA PROCESSORS ASSOCIATION (APA)

http://www.atsea.org/

This is a Web site for the At-sea Processors Association (APA). The APA represents U.S.-flag catcher/processor vessels that participate in the groundfish fisheries of the Bering Sea. Their principal fishery is the mid-water pollock fishery—the largest fishery in the United States. Members both harvest and process fish at sea.

EFFICIENT FOODSERVICE RESPONSE PROJECT

http://www.efr-central.com/

The Web site for the Efficient Foodservice Response (EFR) project. EFR is an industry-wide effort to improve efficiencies in the foodservice supply chain linking manufacturing plants to distribution warehouses to the retail end of the foodservice industry. It simplifies the flow of products, information, and funds within the foodservice supply chain. The EFR project is sponsored by five foodservice industry associations: Canadian Council of Grocery Distributors, International Foodservice Distributors Association, International Foodservice Manufacturers Association, National Restaurant Association (NRA), and Uniform Code Council (UCC).

ELECTRONIC FOOD SERVICE NETWORK

http://www.efsnetwork.com/

The EFS Network, Inc. provides supply-chain solutions for the foodservice industry by combining collaborative workflow technology, hosted application modules, and robust data management services. EFS Network serves suppliers, distributors, operators, and other supply-chain participants such as sales agencies and carriers. EFS Network's customer base includes Ben E. Keith Company, BiRite Foodservice, Bunn Capitol, Cargill, Inc., Dot Foods, eMac Digital, L.L.C., FoodHandler, Inc., Harker's Distribution, Inc., HJ Heinz, Heritage Bag, Kraft Foods, Martin Brothers Distribution Co., Inc., Mattingly Foods, Inc., McCain Foods Limited, Nestlé FoodServices, Performance Food Group, Quality Foods, Inc., Rich Products, Ritz Foodservice, The Schwan Food Company, Sysco Corporation, Thoms Proestler Company, Tyson Foods, Inc., and Ventura Foods LLC.

EURO-RETAILER PRODUCE WORKING GROUP (EUREP)

www.eurep.org

The Euro-Retailer Produce Working Group (EUREP) is made up of leading European food retailers. It is an established a mechanism for drawing up production standards for commodities entering the retail trade through their outlets. Extension to the products of aquaculture started in 2001. Products will not enter the retail trade unless they meet the retailers' standard. The EUREP-GAP programme focuses on production process quality, labeling, traceability, and food safety. Third-party verification by an accredited certification body is required.

FOODCONNEX

http://www.foodconnex.com/

This e-commerce platform is hosted by Integrated Management Solutions, a leading provider of technology solutions to the Food Distribution and Processing Industries. Foodconnex comes with a software.

THE GLOBAL AQUACULTURE ALLIANCE (GAA)

www.gaalliance.org

The GAA is an international nonprofit trade association dedicated to advancing environmentally responsible aquaculture. The GAA program focuses mainly on the management of shrimp farming and processing operations. Third-party verification is required, and certified operations can label their products with the GAA logo. The GAA Individual Codes of Practice Food Safety can be found on this site.

INTERNATIONAL FOODSERVICE DISTRIBUTORS ASSOCIATION (IFDA)

http://www.ifdaonline.org/index.html

This is the Web site for the International Foodservice Distributors Association (IFDA). It is a trade organization representing foodservice distributors throughout the United States, Canada, and internationally. In 2004, IFDA had 135 members that include broad-line and specialty foodservice distributors that supply food and related products to restaurants, institutions, and other food-away-from-home foodservice operations. IFDA advocates the interests of the foodservice distribution community in government and industry affairs through research, education, and communication.

ISO Programs

www.iso.org

This site summarizes the ISO programs and members and offers copies of a variety of technical summaries and brochures of the more than 14,000 International Standards for business, government, and society.

Ornamental Aquatic Trade Association Worldwide

www.ornamentalfish.org

This site includes marketing and trade statistics for the ornamental fish trade. It also includes a Code of Conduct for businesses, and includes water quality criteria and a customer charter.

www.members.tripon.com/Tanganyika

This site includes contact information for ornamental fish trade companies around Lake Tanganyika, online magazines, books, and photos.

Ornamental Fish International (OFI)

www.ornamental-fish-int.org

The Ornamental Fish International (OFI) was founded in 1980 and currently has 38 members that represent wholesalers, collectors, breeders, retailers, importers, exporters, plant specialists, airlines, consultants, and manufacturers. It has a Code of Ethics on the site, as well as the OFI *Journal*, which is published three times a year.

SECODIP

www.tns-sofres.com

TNS SECODIP specializes in market research related to consumer spending, including consumer panels conducted repeatedly over time to measure changes in consumer spending. SECODIP is considered the primary source of consumer panel survey data for France. This is the main source of quantitative data on French seafood consumption in France.

Uniform Code Council Net

http://www.uccnet.org/

A Web site for the Uniform Code Council's (UCC) subsidiary UCCnet™. UCCnet™ makes use of industry standards in the development of powerful tools to synchronize item information and the transfer of information in a business-to-business environment. UCCnet uses standards-based e-commerce to provide nonproprietary collaborative capabilities among trading partners.

The U.S. Trout Farmers Association

www.ustfa.org

The Trout Producer Quality Assurance Program can be found on this page.

INTERNATIONAL AGENCIES AND ASSOCIATIONS

The Food and Agriculture Organization of the United Nations

http://www.fao.org

The Food and Agriculture Organization of the United Nations maintains a Web site with the most current global statistics available on aquaculture and fisheries. The site lists data sets from 1991 to 2000 on quantities and values of aquaculture products by groups of species, by categories of production areas (inland, marine, and so on), by principal species, by country, total international trade, international trade by principal importers, and exporters.

The FAO Web site also includes articles that summarize trends as well as summary statistics. Examples include the following:

Rana, K. and A. Immink. 2003. Trends in global aquaculture production: 1984–1996. Available at www.fao.org/fi/trends/aqtrends/aqtrend.asp

Grainger, R. 2003. Recent trends in global fishery production. Available at www.fao.org/fi/trends/catch/catch.asp

Yearbook of Fishery Statistics. Available at www.fao.org/fi/statist/summtab/default.asp

The FAO Web site further offers two databases:

FISHSTAT+: A set of fishery statistical databases downloadable to personal computers together with a data retrieval, graphical, and analytical software.

Available databases for use with FISHSTAT+ are the following:

- Aquaculture production: quantities
- Aquaculture production: values
- Capture production
- Total production
- Fishery commodities production and trade
- Eastern Central Atlantic capture production
- Mediterranean and Black Sea capture production

Fishery Data Collection in FAOSTAT of WAICEN (World Agricultural Information Center): The FAO Web site includes information on fish processing on a variety of different levels. Specifics on fish freezing are included, as are planning and engineering data for fish processing businesses.

- Fish Production: This domain presents the volume of fish production (catches and aquaculture) by country, by 50 groups and species of the FAO International Standard Statistical Classification of Aquatic Animals and Plants (ISSCAAP) and 29 FAO major fishing areas.
- Fishery Data: This domain contains time series data by country on volume of annual production (catches and aquaculture) from all waters, production of processed and preserved products, and external trade of these groups of products in volume and value. The data are provided for seven aggregates of species and eight main types of product preservation, divided into fishery primary products and fishery processed product.

On the basis of production utilization and trade data, balance sheets by individual countries are prepared which also provide indications on the role of fish in consumption.

Commodities and Trade Division

The Commodities and Trade Division of the Food and Agriculture Organization of the United Nations includes articles and statistics on trade in aquaculture and fisheries at:

www.fao.org/waicent/faoinfo/economic/ESC/esce/cmr/cmrnot

These include commodity notes, tables of apparent consumption, estimated value of fishery production by groups of species, trade flow by region, international exports by species and year (1996–2000), and the relative importance of trade in fishery products in 2000.

FAO-Uruguay Round Agreement on Agriculture

www.fao.org/docrep/003/x7353e/x7353e03.htm

This is a page on the FAO site that includes information on the Uruguay Round Agreement on Agriculture.

Code of Conduct for Responsible Fisheries

www.fao.org

FAO Code of Conduct for Responsible Fisheries available on this site. This document lays the foundation for responsible management of aquaculture and fisheries stocks.

FAO-Codex Alimentarius

www.fao.org/DOCREP

This page presents an international regulatory framework for fish safety and quality. It discusses the World Trade Organization (WTO) agreements on the Application of Sanitary and Phytosanitary Measures (SPS) and the FAO Codex Alimentarius.

FAO-HACCP

www.fao.org/figis

Fact sheet on HACCP from FAO.

Globefish

http://www.globefish.org/presentations/presentations.htm

Globefish is a publications unit within FAO that publishes a wide variety of reports and analyses related to fish and seafood markets around the world. It includes global overviews, world market reports by species, specific market situation analyses, international trade, fishmeal, and trade barriers.

Infofish

www.infofish.org

Infofish publishes articles on capture fisheries and aquaculture, processing, packaging, storage, transport, and marketing, and includes announcements of upcoming meetings and seafood shows.

International Institute of Fisheries Economics and Trade (IIFET)

http://oregonstate.edu/dept/IIFET

IIFET is the International Institute of Fisheries Economics and Trade. This organization is an international group of economists, government managers, private industry members, and others interested in the exchange of research and information on marine resource issues. IIFET holds biannual meetings and publishes a newsletter and proceedings of its various meetings. The newsletters and proceedings can be ordered on the Web site.

Convention on International Trade in Endangered Species of Wild Fauna and Flora

www.cites.org

Official site of the Convention on International Trade in Endangered Species of Wild Fauna and Flora. Includes species and trade databases, registers, export quotas, reports, contacts, resolutions, and reports of the standing, animals, plants, and nomenclature committees.

World Trade Organization

www.wto.org

Official site of the World Trade Organization, the only global international organization dealing with the roles of trade among nations. Includes a training package, videos, list of members, publications, calendar of events, news releases, committee reports, and international trade statistics.

The World Aquaculture Society (WAS)

http://www.was.org

The World Aquaculture Society is an international nonprofit society founded in 1970 with the object of improving communication and information on aquaculture worldwide. WAS sponsors numerous professional meetings, including the international triennial meetings as well as annual chapter meetings in the United States, Latin America, Asia, and Europe. The WAS has an extensive publications unit with books, the *Journal of the World Aquaculture Society*, and *World Aquaculture Magazine*.

INTERNATIONAL TRADE: THEORY AND BACKGROUND

Suranovic, S.M. 1997-2004. International theory and policy analysis. The International Economics Study Center.

http://internationalecon.com/v1.0

This site is introductory course/text on international trade theory and policy. It can be purchased at moderate rates ($25 for the entire file in 2004) for an electronic version. The file is very large and may cause technical difficulties in access, but it is well written. It is easily understood by those with no economics background and is illustrated with a number of clear examples of trade issues, policies, and tools. It lays out clearly the advantages and disadvantages of both free trade and protectionist policy and positions. It is a good starting point for understanding international trade issues.

Deardorff's Glossary of International Economics. Alan Deardorff (UMichigan) collection of citations and definitions regarding international economies.

This Web site includes basic information on international economics. It includes a glossary of terms and concepts from international economics, trade, and finance. A bibliography is available as well as a picture gallery of diagrams of international economics.

UNITED STATES

Aquanic

http:/aquanic.org

Aquanic is the U.S.-based gateway to the world's electronic aquaculture resources. It includes pages of:

- Discussion groups
- Species
- Systems
- Job services
- Contacts
- Sites
- Publications
- Newsletters
- Media
- Educators
- News
- Calendars
- Classified ads
- Online courses
- Feedback

Aquanic includes a page by the Joint Subcommittee on Aquaculture that lists Federal Marketing Services available through the USDA. Not all of these programs directly apply to aquaculture. Aquanic also lists programs of the Agricultural Marketing Service and the Foreign Agricultural Service.

The Economic Research Service (ERS), USDA

http://www.ers.usda.gov/Data/

ERS produces data products in a range of formats, including online databases, spreadsheets, and Web files. Data and reports include: farm income, trade, food prices, food markets, diet and health, natural resources, and food consumption trends. The food consumption database includes historical data on the U.S. population and the daily per-capita amounts of food energy, nutrients, and food components in the U.S. food supply. The trade data includes types of export subsidies, expenditures on export subsidies, and the quantity of subsidized exports during a given year by World Trade Organization (WTO) members. Domestic support data detail the type and amount of support that WTO members have provided annually. Market access data contain information on tariff commitments and their implementation by presenting bound tariff levels and tariff-rate quotas agreed to in the Uruguay Round, as well as applied annual tariff rates (that is, the tariff rates published by national customs authorities for duty administration purposes).

Federal Statistics

http://www.fedstats.gov/

This site provides statistical profiles of states, counties, cities, congressional districts, and federal judicial districts; comparison of international, national, state, county, and local statistics; descriptions of the statistics on agriculture, demographics, economics, environment, health, natural resources and others, and links to relevant Web sites, contact information, and key statistics.

Fishery Market News

www.st.nmfs.gov/st1/market_news/index.html

The National Marine Fisheries Service (NMFS) in the United States has maintained its "Fishery Market News" since 1938 with the objective of providing accurate reports on trade in fish products.

This Web site includes the following under Headquarters Reports:

Monthly Imports of Shrimp
Monthly Imports of Frozen Fish Blocks
Monthly Imports of Selected Fishery Products
Monthly Exports of Selected Fishery Products
Quarterly Fish Meal and Oil Production
Market News Abbreviations

The NMFS Northeast Region Reports include:

New York: Fulton Fish Market Fresh Prices (Daily)
Weekly New York Frozen Prices (Friday)
Boston: New England Auction Prices (Daily)
Boston Lobster Prices (Daily-except Wednesday)
Weekly Boston Frozen Market Prices (Wednesday)
Weekly New England Auction Summary (Friday)

The NMFS Southeast Region Reports include:

Weekly Gulf Shrimp Landings by Area and Species (Monday)
Weekly Ex-vessel Gulf Fresh Shrimp Prices and Landings (Monday)
Weekly Gulf Finfish and Shellfish Landings (Monday)
Weekly Fish Meal and Oil Prices (Thursday)
Monthly Gulf Coast Shrimp Statistics
Monthly Menhaden Purse Seine Landings

The NMFS Southwest Region Report includes:

Canned Tuna Import Update
San Pedro Market Fish Receipts
Japanese Shrimp Imports
Japanese Fishery Exports
Japanese Fishery Imports
Japanese Cold Storage Holdings
Tokyo Wholesale Prices
Fish Landings and Average Ex-Vessel Prices
Sales Volume and Average Wholesale Prices

The NMFS Northwest Region Report includes:

Oregon Weekly Prices with Comparison Report
Seattle Wholesale Producer Prices

This site includes graphs of nominal and real wholesale prices from 1991–2001 for clam, cod, crab, croaker, flounder, lobster, oyster, pollock, squid, swordfish, and whiting; annual cold storage reports for 1990–2002; annual foreign trade reports for 1996–2002; and an annual summary of Fulton Fish Market fresh prices 1987–2002 and New York frozen wholesale prices, annual summary, 1990–1997.

International Trade Commission

www.usitc.gov

This is the official site of the International Trade Commission. It includes information on antidumping and countervailing duty orders for product group, country, and data. The site lists events from the daily and weekly reports and tariff schedules.

National Agricultural Library, ARS, USDA

www.nal.usda/atmic/pubs/srb9303.htm

Seafood Marketing Resources includes postings by the USDA National Agricultural Library related to aquaculture, trade, databases, hearings, legislation, journals, U.S. government contacts, trade associations and organizations, seafood shows and expositions, and lists of distributors/exporters/importers, both foreign and U.S.

The National Aquaculture Development Act

www.aquanic.org/publicat/state/md/perm1.htm

The National Aquaculture Development Act became law in 1980. The Act states that is "in the national interest, and it is the national policy, to encourage the development of aquaculture in the United States." This act indicates that the principal responsibility for the development of U.S. aquaculture is on the private sector, but it also assigned USDA, USDOC, and USDI responsibility. In a later interagency agreement, USDA was given responsibility for research and support activities for private freshwater aquaculture. The NAA has been reauthorized twice, in 1985, establishing USDA as "the lead federal agency with respect to the coordination and dissemination of national aquaculture information" and designating the Secretary of Agriculture as permanent chair of the JSA.

www.cllie.plus.com/byrd.pdf

Collie, D.R., H. Vandenbussche. 2004. Anti-dumping duties and the Byrd amendment.

www.adcvd.com

A complete guide to U.S. antidumping and countervailing duty law.

National Marine Fisheries Service, U.S. Department of Commerce

http://www.nmfs.noaa.gov/trade/DOCAQpolicy.htm

This Web page outlines the mission statement and the vision of the U.S. Department of Commerce for U.S. aquaculture. The statement outlines the specific objectives by the year 2025.

Rural Development Agency

http://www.rurdev.usda.gov/rbs/coops/csdir.htm

This is a federal government Web site that provides information on cooperative programs administered by USDA. They provide information on cooperative spotlights, cooperative data, charts on cooperatives, publications on cooperatives, and funding opportunities for research in cooperatives. You can also get an electronic copy of *Rural Cooperative Magazine*, a magazine published every other month that focuses on cooperatives and issues facing cooperatives.

Southern Regional Aquaculture Center

Located on the Aquanic site, this page includes a large number of extension fact sheets on a wide variety of topics related to aquaculture and includes marketing and economics fact sheets.

U.S. Census Bureau

http://www.census.gov/epcd/ec97/industry/E311712.htm

The Web site provides detailed national statistics for the fresh and frozen seafood processing industry from the 1997 Economic Census. Data provided include number of firms, employees, payroll, and revenue by employment-size of the enterprise.

http://www.census.gov/epcd/susb/1999/us/US311712.htm

The site provides statistics of U.S. Fresh and frozen seafood processing, including Employment size of enterprise, number of firms, number of plant establishments, number of paid employees, and annual payroll ($1,000). The statistics are from 1998 through 2001.

http://www.census.gov/cir/www/mqc1pag2.html

The Web site provides results from the Survey of Plant Capacity Utilization conducted jointly by the U.S. Census Bureau, the Federal Reserve Board (FRB), and the Defense Logistics Agency (DLA). The survey collects data for the fourth quarter and includes number of days and hours worked, estimated value of production at full production capability, and estimated value of production achievable under national emergency conditions. Data is from 1994 through 2002.

U.S. Department of Agriculture - Agricultural Marketing Service

http://www.ams.usda.gov/cool/

This is a USDA Web site that provides information about the country of origin labeling relating to the 2002 Farm Bill provision, federal register rulemaking, and news releases.

This is the site of the Agricultural Marketing Service, an agency within the United States Department of Agriculture that is handling the Country of Origin Labeling rule. The interim rule posted in October 2004 can be found at the site. Definitions of terms used in the rule, including specific definitions of "retailer," "food service establishment," "covered commodities," "processed food item," and others are listed on the site. Copies of related rulemaking efforts, resources related to the COOL rule, talking points, overviews, examples of records that may be useful for the COOL verification process, and copies of news releases can be found on the site.

U.S. Department of Agriculture - National Agricultural Statistics Service (NASS)

http://www.usda.gov/nass/

NASS provides statistical information on agriculture that includes publications, charts and maps, historical data, statistical research, and census of agriculture.

U.S. Department of Commerce

www.commerce.gov

This is the official site of the U.S. Department of Commerce. It provides information on the state of the U.S. economy. This site provides export-related assistance and market information, lists export regulations, and includes summaries of trade statistics.

http://ia.ita.doc.gov

Federal register notice that includes the regulations on antidumping and countervailing duty proceedings to conform to the Department of Commerce's regulations to the Uruguay Round Agreements Act.

The USDA Economics and Statistics System

http://usda.mannlib.cornell.edu/usda/

The site contains nearly 300 reports and datasets from the economics agencies of the U.S. Department of Agriculture. These materials cover U.S. and international agriculture and related topics. Aquaculture falls under Specialty Agriculture. Data and reports on aquaculture that include Aquaculture Outlook (by ERS), Catfish Processing: Dataset (by NASS), Catfish Processing: Report (by NASS), Catfish Production (by NASS), and Trout Production (by NASS).

World Outlook Board, USDA

http://www.usda.gov/agency/oce/waob/index.htm

This board serves as the focal point for economic intelligence on the outlook for U.S. and world agriculture. It forecasts supply and demand for major commodities at the world level, and for livestock products and refined sugar at the U.S. level. The forecasts are in the form of a balance sheet that matches supply (beginning stocks added to the anticipated crop) with demand (how much will be consumed at home, exported, or remain as ending stocks).

National Oceanic & Atmospheric Administration (NOAA), Fisheries

http://www.nmfs.noaa.gov/aquaculture.htm

This site provides information on U.S. Aquaculture, Bycatch, Grants, International Interests, Legislation, Permits, and Recreational Fisheries. The site also provides information on the Department of Commerce's Aquaculture Policy, National Aquaculture Act of 1980, NOAA Aquaculture Policy, Policy Paper on the Rationale For a New Initiative in Marine Aquaculture, Department of Agriculture's National Aquatic Animal Health Plan, the Environmental Protection Agency's final aquaculture effluents rule, and a draft Code of Conduct for Responsible Aquaculture Development in the U.S. Exclusive Economic Zone. The reports on Fishery Market News and Fisheries Statistics including domestic and international trade.

American Statistical Association (ASA)

http://www.amstat.org/sections/srms/whatsurvey.html

The site provides brochures about Survey Research. The ASA Series includes: "What is a Survey?"; "How to Plan a Survey"; "How to Collect Survey Data"; "Judging the Quality of a Survey"; "How to Conduct Pretesting"; "What Are Focus Groups?"; "More About Mail Surveys"; "What Is a Margin Of Error?"; "Designing a Questionnaire"; and "More About Telephone Surveys."

U.S. Food and Drug Administration

www.fda.gov

This is the site of the U.S. Food and Drug Administration. It includes the mission statement, summaries of what FDA regulates, and its history.

www.cfsan.fda.gov

This is a page on the U.S. FDA Web site that deals with seafood HACCP. This site provides an overview of HACCP as it relates specifically to seafood. It includes a summary of the provisions in the rule as well as the final rule, full text, for the seafood HACCP rule.

MARKETING PLANS AND STRATEGIES

There are quite a few Web sites that offer assistance in development of marketing plans and strategies. A simple Web search will turn up quite a few sites. Some offer free services, sample market plans, and templates for developing market plans and strategies, whereas others offer services for fees, workshops, books, and software. These are dynamic sites, but a few examples are listed here.

www.morebusiness.com

This site includes templates for developing marketing plans and sample market plans, and includes software for business planning.

www.entrepreneur.com

This site contains a market planning checklist, tools and services to enhance marketing success, marketing tips, business coaches, and business services.

www.money.howstuffworks.com

This site discusses how marketing plans work.

www.paloalto.com

This site contains sample market plans and includes tutorials on how to write a marketing plan.

NONGOVERNMENTAL ORGANIZATIONS

Marine Aquarium Council

www.aquariumcouncil.org

The Marine Aquarium Council (MAC) is an international, not-for-profit organization that brings marine aquarium animal collectors, exporters, importers, and retailers together with aquarium keepers, public aquariums, conservation organizations, and government agencies. The mission is to conserve coral reefs and other marine ecosystems by creating standards and certification for those engaged in the collection and care of ornamental marine life from reef to aquarium.

Global Marine Aquarium Database (GMAD)

www.unep-wcmc.org/marine/GMAD

The United Nations Environment Programme-World Conservation Monitoring Centre along with the Marine Aquarium Council have compiled a database of data on 2,399 species from 45 representative wholesale exporters and importers. The database can be queried by genus and then species, year, and either imports or exports.

Monterey Bay Aquarium

www.montereybayaquarium.org

This is the site of the Monterey Bay Aquarium. The aquarium issues a pocket guide for fish consumers that judges how sustainable each type of fish is. There is a report on each seafood species available on this site.

Seafood Business

www.seafoodbusiness.com/archives/02feb/news_trade

This site provides a summary of the out-of-court settlement between Great Eastern Mussel Farms of Maine and mussel producers from Prince Edward Island, Canada. The settlement followed antidumping petition filed by Great Eastern Mussel Farms.

UNIVERSITIES

University of Wisconsin

http://www.wisc.edu/uwcc/

This is the University of Wisconsin Center for Cooperatives (UWCC) Web site that provides information on all aspects of cooperatives including business principles, organizing cooperatives, cooperative financing, cooperative structure, cooperative management, leadership and governance, and related topics for both agricultural and consumer cooperatives.

Glossary

Absolute quotas: regulations that limit the quantity of an imported good to a certain time period and volume.

Administered pricing: system in which prices are announced as nonnegotiable selling or buying prices.

Ad valorem tariff: tax levied on value of commodity, expressed as a percentage.

Advertising: organized programs and presentations designed to communicate product attributes to consumers to encourage sales.

Agent: individual or firm that represents either buyers or sellers in the marketplace; agents do not take title to goods.

Agricultural cooperative: an agricultural cooperative is a user-owned and user-controlled business from which benefits are derived and distributed equitably on the basis of use by the owners.

Antidumping duties: levies on products that are deemed to be imported at less than fair market value.

Asymmetric information: condition in which one participant in the market has greater knowledge of prices and quantities than do other market participants.

ATA: American Tilapia Association.

Autarky: condition of such restrictive trade policies and restrictions that no trade occurs; the country's economy exists in isolation from the rest of the world.

Away-from-home consumption: food dispensed for immediate consumption outside the consumer's home. Includes all food consumed in foodservice facilities, such as restaurants, hotels, cruise ships, and schools.

Bouchot mussels: mussels cultured in France on fixed wooden poles. Seed mussels that have been collected on ropes are wrapped around wooden poles that have been driven into the ocean bottom. They are transferred to plastic net tubes that are wrapped again around the poles.

Brackish water: water with salinity between 0.5 and 35 ppt (parts per thousand).

Bretton Woods Agreement: agreement made in Bretton Woods, U.S.A., in 1944 that established a post-war fixed currency rate between countries and the International Monetary Fund.

Broad-line distributors: merchant wholesale operators that handle a broad line of groceries, health and beauty aids, and household products. Also referred to as general-line and full-line distributors.

Broker: middlemen who arrange transactions for a commission but do not assume ownership; are paid by the party that hired them, and do not carry inventory.

Brokers and agents: wholesale operators who buy or sell as representatives of others for a commission and typically do not physically handle the products or take title to the goods.

Business-to-business (B2B): refers to direct market transactions between two independent businesses.

Captive supplies: livestock acquired by meat packers through forward basis contracts.

Cardinal utility: enables a consumer to specify the actual numeric level of utility or "satisfaction" obtainable.

Carryover stocks: stocks left from one marketing year and held for sale in the next.

Ceteris paribus: Latin expression meaning "holding all other factors constant," or "all else being equal."

CFA: Catfish Farmers of America.

Chain stores: a company with more than 11 stores under one ownership and name.

Checkoff programs: programs that add a fee to either feed sales or product sales for use in advertising or research related to that particular commodity.

Collaborative planning, forecasting, and replenishment (CPFR): supply-chain technology that involves sharing sales forecasts of the manufacturer with the retailer, and tailoring orders and deliveries accordingly.

Commission merchant: middleman who takes a load of a commodity to market, sells it for the best price, deducts a commission, and sends the balance back to the growers.

Commodity: economic good that can be legally produced and sold by a large number of individuals as opposed to differentiated products that belong to a specific seller.

Comparative advantage: an economic principle that states that a country should specialize in producing and exporting those goods that it can produce at relatively lower cost and that it should import those goods for which it has a relatively high cost of production.

Competition-oriented pricing: prices are set based on prices for similar and competing goods.

Competitive market: market in which numerous firms supply a product that is homogenous or standardized.

Complementary product: products that consumers tend to consume at the same time.

Computer-aided telephone interviewing (CATI): interviewing system in which responses are entered directly into a computer database while the interview is taking place.

Concentration: the degree to which decreases in the number of firms in the industry control a high portion of the sales.

Conjoint analysis: sometimes referred to as "trade-off" analysis because respondents are forced to make trade-offs among different product attributes; inferences are made from the quantified trade-offs as to how important or valuable different attributes are and how they influence respondents' decision-making processes.

Consolidation: reduction in the number of firms in an industry as a result of mergers.

Convenience stores: small, self-service stores located near residential areas that offer a limited line of goods.

Conventional distribution channel: a channel consisting of one or more independent producers, wholesalers, and retailers, each a separate business seeking to maximize its own profits even at the expense of profits for the system as a whole.

Cooperative: business that is owned and controlled by those working in it and whose benefits are allocated equally among the owners/members; an organization that is owned and controlled by the people who use its products, supplies, or services.

Cost-plus pricing: pricing system in which a set margin is added to costs of production to determine selling price.

Countervailing: an action designed to offset (countervail) the effect of another action.

Countervailing duties: levied on imported products that receive an unfair subsidy from a foreign government.

Cross price elasticity: responsiveness of quantity demanded in one good to changes in price of a related good.

Delivery rights: a tradable share, which requires delivery of a certain quantity and quality of a product for a specified period at some negotiated price. Some contracts for delivery rights specify production standards.

Demand: various quantities of a good or service that consumers are willing and able to take off the market (purchase) at varying prices.

Demand-oriented pricing: accompanies market segmentation in which higher prices are charged for those products considered to be of higher quality and lower-cost products are sold to market segments that seek out lower prices.

Department stores: large retail outlets with entire departments of different categories of consumer goods.

Determinants of demand: factors that determine the specific relationship between price and quantity demanded.

Differentiated product: economic good that belongs to a single seller and that has unique characteristics.

Discount pricing: price reductions offered from advertised prices.

Discount stores: stores that offer lower-priced merchandise.

Disintermediation: bypassing intermediaries to sell directly to final buyers.

Double-barreled question: a question that asks the respondent to address more than one issue at a time.

Dressed fish: fish that has been deheaded, eviscerated, and skinned.

Dressout: refers to the fish product after cleaning. *Dressout weight* refers to the weight of the fish after cleaning, whereas *dressout yield* refers to the weight of the fish after cleaning as a percentage of the fish prior to cleaning.

Dumping: selling products at prices below the cost of production and below normal domestic prices.

Economics: allocation of scarce resources to meet the unlimited needs and wants of human beings.

Economies of scale: condition in which average per-unit costs decrease as the size of businesses increases; decreasing average costs with increasing output levels.

Economy of size: larger companies can operate at relatively lower costs by having cost advantages.

Efficient consumer response (ECR): a collaborative relationship in which any combination of retailer, wholesaler, broker, and manufacturer works together to seek out more ways manner to distribute manufactured food products. The purpose of ECR is to drive the order cycle and all the other business processes with point-of-sale data and other consumer-oriented data, giving an accurate read on consumer demand.

Efficient foodservice response (EFR): technology system in the foodservice supply chain that links food manufacturers to distribution warehouses and to restaurant outlets.

Elasticity of demand: degree of responsiveness of quantity demanded to a given change in price.

Elasticity of supply: degree of responsiveness of the quantity supplied to changes in the price of the good.

Electronic data interchange (EDI): a technological system that allows businesses to order merchandise, streamline delivery, and reduce overall costs. The system requires that suppliers and retailers use compatible computer systems.

Entrepreneurship: assuming control over the decision-making, organization, and operation of a business including the associated risks and benefits.

Equilibrium price: price at which buyers and sellers agree on the quantity to be offered and that desired; all product clears the market at the equilibrium price.

Eviscerate: removal of internal organs and other internal body contents.

Evolutionary concept: product concept involving an existing product that can be modified to suit particular needs or improve on particular experiences of customers in the use of the product.

Exclusive dealing: when a processor or supplier forbids an intermediary to carry products of competing suppliers or processors.

Exclusive economic zone (EEZ): imposition by countries of a 200-mile fishing zone along their coast lines that is reserved for fishermen from their own country; fishing exploitation rights reside exclusively with that country.

Existing demand: quantities that would be purchased of a particular product for a range of specific prices.

Exploratory research: informal research that has no structure to the process of gathering data and information, for example, observing, reading periodicals, and surfing the Internet.

Export subsidies: payments by a government to a business that exports certain products.

Farm-retail price spread: difference between the retail price of food products and the farm value of an equivalent quantity of food sold by farmers.

Focus groups: informal techniques to assess consumer preferences and needs, new product concepts, and purchase behavior for a good or service.

Food marketing bill: difference between total consumer expenditures for all domestically produced food products and what farmers receive for equivalent farm products.

Form utility: value added to products as they are transformed into products for final sale.

Free trade: voluntary exchange of goods between and among different countries that occurs in the absence of regulations that either promote or constrain the exchange of goods.

Freshwater: water with salinity less than 0.5 ppt (parts per thousand).

Futures contracts: standardized, legally binding agreements to either deliver or receive a certain quantity and grade of a specific commodity during a designated delivery period.

Futures market: a contractual agreement made between two parties through a regulated futures exchange where the parties agree to buy or sell an asset at a certain time in the future at a mutually agreed upon price.

General Agreement on Tariffs and Trade (GATT): international agreement negotiated originally in Geneva, Switzerland, with the intent to increase international trade by reducing barriers to trade.

General equilibrium analysis: analysis in which a number of variables are allowed to vary, and changes in price may affect other prices.

Generalized system of preferences: framework under which developed countries give preferential treatment to manufactured goods imported from certain developing countries.

General-line foodservice wholesaler: businesses that provide products to restaurants, hospitals, schools, hotels, and other foodservice establishments.

General-line grocery wholesaler: business that purchases both food and nonfood products for sale to retailers that do not have warehouses.

Generally Recognized as Safe (GRAS): FDA category for food additives that have been shown through scientific studies or experience in common use to be safe for human use in food.

General tariff: duty levied on imported products that applies to countries not eligible for preferential or most-favored-nation status.

Generic advertising: promotion of a general type of commodity without use of particular brands or processors.

Genetically Modified Organisms (GMO): signifies a product that has been subjected to genetic engineering methods.

Giffen good: product for which the quantity demanded goes down as prices go down , or for which quantity demanded goes up as prices go up.

Grid pricing: price discovered for each animal where higher-quality cattle receive higher prices and lower-quality cattle receive lower prices.

Hedonic theory: consumer theory that assumes that the qualities of a product are the ultimate source of utility for consumers and that a product is described solely by its characteristics.

Growout: production of food-sized aquatic animals. In the case of finfish, fingerlings are stocked in growout facilities to produce food-sized products. Synonymous with "on-growing" in Europe.

H & G: headed and gutted.

HGS: headed, gutted, and skinned.

Homogeneous product: products with nearly identical characteristics.

Horizontal marketing system: a distribution channel in which two or more companies at the same level of the marketing chain (with similar marketing functions) join together to pursue a new marketing opportunity.

Hypermarkets: largest of supermarket-type grocery stores with up to 200,000 sq. ft. of selling space in groceries, sporting goods, auto supplies, and so on, selling up to 40% of sales of general merchandise.

Income elasticity: measure of the response of the quantity demanded to changes in income.

Incomprehensible question: a question that respondents cannot understand, probably because of the concept or wording.

Industry concentration: percentage of business (share of total value of shipments) accounted for by a number of businesses in the industry.

Inferior good: product for which demand decreases as incomes increase, or for which demand increases as incomes decrease.

Intermediary: middlemen in the marketing chain who add value to the product by either assembling units into large volumes, processing or transporting products, or identifying and servicing customers at the next level of the marketing chain.

Laissez-faire policy: no regulations that would either restrict or encourage exchange of goods between and among different countries.

Law of One Price: a commodity has a single market price, after commodity arbitrage (active trading), when expressed in common currency units and transportation and marketing costs are accounted for.

Leading (or Loaded) question: a question that forces or directs a respondent to a response that he or she may not normally give.

Likert Scale questions: a technique that presents a set of statements to respondents to which a respondent expresses agreement or disagreement using a scale, usually a 5-point scale. The most common scale is where 1=strongly disagree, 2=disagree, 3=not sure, 4=agree, and 5=strongly agree.

Livehaulers: buy live fish from producers, transport live fish, and sell to fee-fishing businesses, grocery stores, or other outlets.

Loss-leaders: product priced below cost to draw customers into the store to have opportunities to sell other, more profitable goods.

Magnuson Fishery Conservation and Management Act: Public Law 94-265, which defines United States rights and authority regarding fish and fishery resources, including agreements regarding foreign fishing and international fisheries, and the national fishery management program.

Marginal utility: the "satisfaction" gained from the consumption of one extra unit of a good.

Market: Location where goods are exchanged; where goods and services are bought and sold.

Market equilibrium: the price and quantity at which all product is removed from the market.

Market failure: market failure occurs when the market is characterized by destructive competition;

structural imperfections such as monopoly and monopsony; externalities relating to commodity promotion, grades, and standards; and uncertainty relating to information needs, for example, asymmetric information.

Market intelligence: the art of obtaining updates about relevant developments in the market.

Market power: ability of a particular firm to raise price above competitive levels in output markets; in an input market to the extent that it can profitably reduce price below competitive levels.

Marketing: performing all functions related to assembling, processing, transporting, and advertising of goods from the point of production through consumption by the end user.

Marketing bill: A USDA measure of the amount of total consumer dollar expenditures incurred by marketing functions as compared to that received by farmers.

Marketing channels: routes of product flows and customer value delivery systems in which each channel member adds value for the customer; a combination of interrelated intermediaries (individuals and organizations) who direct the physical flow of products from producers to the ultimate consumers.

Marketing margin: costs (including profit) incurred from services and value added as products move through the marketing chain.

Marketing order: marketing orders and agreements are legal instruments issued by the United States Department of Agriculture (USDA) Secretary that are designed to stabilize market conditions for certain agricultural commodities by regulating the handling of those commodities in interstate or foreign commerce. Marketing orders and agreements are administered by the Agricultural Marketing Service (AMS), an agency within the USDA, and are authorized by the Agricultural Marketing Agreement Act of 1937, as amended, 7 U.S.C. §§ 601-14; 671-74.

Merchant wholesalers: operators of firms primarily engaged in buying groceries and grocery products, and reselling to retailers, institutions, and other businesses.

Miscellaneous wholesaler: establishments specializing in the wholesale distribution of a narrow range of dry groceries such as canned foods, coffee, tea, or spices. Also referred to as systems distributors.

Monopolistic competitive market: a relatively large number of firms operate competitively in the market by supplying differentiated products.

Monopolistic market: market with only one firm as supplier of a unique product for which there is no close substitute.

Monopsony: one-buyer control.

Multistage or cluster sampling: a random selection process in which the sample is chosen in stages.

Neoclassical preference theory: the theory assumes that the decision-making process involves a comparison of two alternatives, *a* and *b*, in a choice set *C* using a preference ordering.

NFI: The National Fisheries Institute.

Normal good: product for which the quantity demanded goes up as the price goes down.

Oligopolistic market: few firms operate in this market, and products may be differentiated.

On-growing: production of food-sized aquatic animals. In the case of finfish, fingerlings are stocked in growout facilities to produce food-sized products. Synonymous with "growout" in the United States.

Opportunity cost: the value foregone from spending one's resources on a particular project.

Ordinal utility: ordinal utility enables a consumer to order commodity combinations by level of utility or "satisfaction" obtainable (1st, 2nd, 3rd).

Partial equilibrium analysis: analysis in which most of the key parameters are held constant in order to understand the relationships of other variables one at a time.

PBO: pinbone-out products.

Perceived-value pricing: pricing the product based on nonprice factors such as quality, healthfulness, environmental sustainability, or prestige.

Pinbone: set of small bones found behind the ribs in a strip of flesh extending one-third of the fillet's length along its lateral line.

Poikilothermic: organisms without the ability to control body temperature.

Population: entire group of individuals from whom information is required.

Post-larvae (PL): term used to describe the size and stage of shrimp stocked into growout ponds. This stage in the shrimp's life cycle is the first one in which the shrimp has transformed from a floating, planktonic stage to a bottom dweller with walking legs.

Potential demand: quantities that consumers might purchase at specific prices if the product were available.

Price checkoff: mandatory or voluntary program that requires the affected individuals or businesses to pay a flat fee per unit of sale or some specified percentage of sale value of the products sold by the individuals or businesses.

Price elasticity of demand: same as elasticity of demand.

Price penetration: pricing strategy in which a low price is charged to gain increased market share.

Product differentiation: where products are distinguishable through physical attributes, functional features, material make-up, packaging, advertising, and branding.

Production capacity utilization rate: the ratio of total capacity utilized relative to the total processing capacity available.

Psychological pricing: establishing prices that either look better or convey a certain message to the buyer.

Qualitative research: formal and structured research that deals with words, images, and subjective assessment. It is concerned with obtaining explanations to certain issues of subjects of interest.

Quantitative research: formal and structured research that deals in numbers, theory, and objective measures to provide statistical predictability of results to the target population.

Questionnaires (survey instruments): formalized framework consisting of a set of questions and scales designed to generate primary raw data.

Quick response (QR) system: a supply-chain system for the grocery retail industry used to shorten the retail order cycle; that is, the total time from the point merchandise is recognized as being needed to the time it arrives at the store.

Quotas: limit to the total quantity that can be imported of a particular good for a given period of time.

Rank order questions: such questions require the respondent to rank a set of factors in a certain order, for example, low to high, usually using numbers. These types of questions allow certain product attributes or brands to be ranked based upon specific characteristics. Example: "Rank the following shrimp attributes in terms of importance to your purchase decisions, where 1 is the most important and 4 is the least important: quality ___, freshness ___, price ___, and size ___."

Rating scale question: a question that requires a respondent to rate a product or brand along a well-defined and evenly spaced continuum. Rating scales are often used to measure the direction and intensity of attitudes. Example: "Which of the following categories best describes the taste of lime-flavored marinaded tilapia fillet? very tasty __; somewhat tasty __; neither tasty or sour __; somewhat sour __; very sour __."

Rational individual: individuals are assumed to have preference orderings that satisfy six axioms: reflexivity, completeness, transitivity, continuity, nonsatiation, and convexity.

Retailing: selling product to the end consumers.

Revealed preference theory: preference theory that utilizes actual behavior of consumers to "reveal" the preference of consumers.

Revolutionary concept: product concept that involves discovery of consumer needs that have not been met by existing products.

Safeguard remedies: actions taken when increasing volumes of imports threaten to injure a U.S. industry or the creation of a U.S. industry.

Saltwater: water with salinity levels of 35 ppt.

Sample: the part of the population that is studied in order to gather information about the entire population.

Sample design: the method used to choose the sample from the population.

Sample size: the number of samples to use that is assumed to be representative of the population.

Scan-based trading (SBT): an electronic-based sales-sharing system that tailors orders and deliveries using retailer checkout counter scan systems. It is a technological system that provides food manufacturers instant information on their inventory in retailer outlets when the goods are scanned and sold. Inventory is therefore on a consignment basis from vendors. The system allows food manufacturers to monitor inventory levels for replenishment and bill retailers for their inventory only after the goods are scanned. SBT is also known as pay-on-scan (POS).

Self-distribution retailers: large independent retailer or small independent retailers that band together in the form of a cooperative that provides its own wholesaling. Self-distributing retailers own distribution centers and buy directly from food manufacturers and producers.

Semantic differential scale question: semantic differential scales ask respondents to rate a product, brand, or an attribute based upon some point-scale that has two extreme adjectives at each end. Example:

Which of the following categories best describes the taste of lime-flavored marinaded tilapia fillet? (Check only one)				
Very tasty				Very sour
_____	_____	_____	_____	_____
(_5_)	(_4_)	(_3_)	(_2_)	(_1_)

Simple random sampling: the sample of individuals is chosen from the population in such a way that each person in the population has an equal chance of selection.

Skimming: introducing the product at a relatively higher price for more affluent, quality-conscious consumers, and then lowering the price as the market becomes saturated.

Slotting fees/allowances: slotting allowances and slotting fees describe a family of marketing practices that involve payments by manufacturers to persuade downstream channel members to stock, display, and support new products.

SMI: Salmon Marketing Institute.

Specialty wholesalers: establishments primarily engaged in the wholesale distribution of items such as frozen foods, bakery, dairy products, poultry products, fish, meat and meat products, or fresh fruits and vegetables.

Specific tariff: fixed charge per unit of imported good, regardless of its value.

Speculative stocks: inventory held in anticipation of higher prices.

Spot market: a market in which a commodity is bought or sold for immediate delivery or delivery in the very near future.

Stock-keeping unit (SKU): an identification system, usually alphanumeric, of a particular product that allows it to be tracked for inventory purposes.

Strata: subdivisions of similar individuals of a population. Each subdivision is known as a stratum.

Structured question: closed-ended question that requires the respondent to choose from a predetermined set of responses or scale points.

Subsidies: payments by a government to a business that produces a particular good.

Substitute product: competing product.

Superior good: product for which demand increases (decreases) as income levels increase (decrease).

Superstores: large supermarkets that seek to supply all the products, food and nonfood, that consumers want.

Supply: quantity of goods and services that producers are willing and able to offer in the marketplace at specific prices.

Supply-chain management: managing the flow of resources, final products, and information among input suppliers, producers, resellers, and final consumers.

Surimi: minced, washed fish product formed into various seafood analog products with flavorings.

Surrogate: market economy country that is at a level of economic development comparable to that of the nonmarket economy and is a significant producer of comparable merchandise.

Surveys: methods used to gather systematic information from a sample of a target population.

Tariffs: taxes levied on imports, often passed to reduce quantities of imports.

Tariff-rate quotas: regulations that allow a certain volume to be imported at a reduced tariff rate.

TCI: The Catfish Marketing Institute.

Terms of trade: terms of trade measure the rate of exchange of one good or service for another when two countries trade with each other.

TMI: The Tilapia Marketing Institute.

Trade barriers: Policies, regulations, programs, or laws that make it more difficult for imports to enter a country.

Trade liberalization: reducing or eliminating policies that restrict, encourage, or otherwise change what the trade would be without government intervention.

Tying agreement: an agreement in which a supplier supplies a product to a channel member with the stipulation that the channel member must purchase other products as well.

UCCnets: registry and synchronization service of UCC that helps to improve the accuracy of members' supply-chain product and location information. Suppliers provide product, location, and trading partner information to the UCCnet Registry service and the system then validates the data with demand side partners, ensuring that all trading partners are using identical UCC standards.

Unanswerable question: a question that requires some specific information to respond but the respondent does not have access to the information.

Uniform Code Council (UCC): the Uniform Code Council, Inc. is a not-for-profit standards organization that administers the Universal Product Code (U.P.C.) and provides a full range of integrated standards and business solutions for more than 250,000 member companies doing business in 25 major industries.

Unstructured question: open-ended question formatted to allow respondents to respond in their own words. There is no predetermined list of responses available to aid or restrict the respondents' answers.

Uruguay Round: agreement that created the World Trade Organization (WTO) after negotiations among 100 nations from 1986 to 1993.

USTFA: United States Trout Farmers Association.

Utility: refers to the level of "satisfaction" obtainable from "consuming" a bundle of goods or products.

Vertical coordination: methods by which goods and services may be exchanged between different stages of production. Units at different stages of production owned by the same firm and product flows coordinated through administrative means.

Vertical distribution channel: a distribution channel structure in which producers, wholesalers, and retailers act as a unified system. One channel member owns the others, has contracts with them, or has so much power that they all cooperate.

Vertical integration: when a firm operates at more than one level of a series of levels leading from raw materials to the final consumer in the business chain.

Volume discounts: a reduction in price based on the purchase of a large quantity.

Voluntary export restraints (VERs): regulation established by a government to limit the volume of a good that can be exported.

Warehouse club: a hybrid wholesaler and retailer that sells food, appliances, hardware, office supplies, and similar products to members (both individuals and small businesses) at prices slightly above wholesale.

Warehouse food store: discount supermarket that sells at lower prices than traditional supermarkets but with fewer services offered to customers.

Wholesale club: sells annual membership fees and a variety of grocery and nonfood items at deep discounts.

Wholesaling: assembling smaller units of product into larger volumes to facilitate larger sales to larger companies.

World Trade Organization (WTO): replaced the GATT institutions in 1995; was created by the Uruguay Round Agreement; administers the provisions of the GATT.

Name Index

Subject Index